T0135134

Studies in Computational Intelligence

Volume 499

Series Editor

J. Kacprzyk, Warsaw, Poland

For further volumes:
http://www.springer.com/series/7092

Stanisław Ambroszkiewicz
Jerzy Brzeziński · Wojciech Cellary
Adam Grzech · Krzysztof Zieliński
Editors

Advanced SOA Tools and Applications

Editors
Stanisław Ambroszkiewicz
Institute of Computer Science
Polish Academy of Sciences
Warsaw
Poland

Jerzy Brzeziński
Institute of Computing Science
Poznań University of Technology
Poznań
Poland

Wojciech Cellary
Department of Information Technology
Poznań University of Economics
Poznań
Poland

Adam Grzech
Division of Teleinformatics
Institute of Informatics
Wrocław University of Technology
Wrocław
Poland

Krzysztof Zieliński
Department of Computer Science
AGH University of Science
and Technology
Cracow
Poland

ISSN 1860-949X ISSN 1860-9503 (electronic)
ISBN 978-3-662-52235-6 ISBN 978-3-642-38957-3 (eBook)
DOI 10.1007/978-3-642-38957-3
Springer Heidelberg New York Dordrecht London

Softcover reprint of the hardcover 1st edition 2014

Printed on acid-free paper

Springer is part of Springer Science+Business Media (www.springer.com)

Preface

The main purpose of this book is to present advanced software development tools for construction, deployment, and governance of SOA applications. Novel technical concepts and paradigms, formulated during the research stage and during development of such tools, are presented and illustrated by practical usage examples. Hence, the book should be of interest not only to theoreticians but also to engineers who cope with real-life problems. Additionally, each chapter contains an overview of related work, enabling comparison of the proposed concepts with exiting solutions in various areas of the SOA development process. This should make the book interesting for students and scientists who investigate similar issues.

The book consists of an introduction and five chapters addressing the following topics:

- **Specification and Deployment of Business SOA Applications within a Configurable Framework Provided as a Service**. This chapter introduces the main components and features of an integrated framework supporting business processes in a distributed ICT environment based on the Service-Oriented Architecture (SOA) paradigm. The framework's features are divided into areas that span the whole life cycle of business-oriented ICT applications. The framework provides applications and tools for business requirement analysis and comparison, selection and composition of services, efficient communication and provisioning of computational resources, allocation of distributed, virtualized environments, as well as monitoring and management of resource utilization and quality of service indicators.
- **Platform for Development of Electronic Markets of Sophisticated Business Services**. A general architecture of a platform which enables development of electronic markets of sophisticated business services is proposed. The platform consists of the following elements: (i) Service Broker, (ii) Service Registry, (iii) Quote Agents, (iv) Interface Repository, (v) Service servers, (vi) Planner and Task Manager, (vii) User interface, and (vii) Transaction Coordinator.
- **Applying the Service-Oriented Architecture at the Inter-Organizational Level and for Automation of Administrative Procedures**. This chapter introduces a multilevel SOA architecture, describing its core principles, technical layers, and individual levels (known as the Team Level and the Interorganizational

Level). An approach to automation of administrative procedures is also presented along with relevant models. The traditional approach to modeling and execution of administrative procedures is contrasted with dynamic composition. The applicability of the proposed concepts is studied in the context of a real-world SOA application from the construction domain.

- **Dependability Infrastructure for SOA Applications**. The basic services which comprise a dependability infrastructure are presented (this includes group communication, replication, and recovery). Management aspects of SOA dependability infrastructures, including self-configuration and knowledge repository, are investigated. Security problems are also addressed and support for security policy deployment in the form of the ORCA framework is proposed. The applicability of all these concepts is analyzed in the context of the Healthcare Integration Platform and the Medical Event and Data Registration Platform.
- **Implementation, Deployment, and Governance of Adaptive SOA Systems**. The goal of this chapter is to propose a pragmatic methodology for adding and managing adaptability in multiple layers of the S3 stack, and to present the AS3 Studio package, which is a complete suite of tools supporting extensions of SOA systems with adaptability features. The adaptability aspects presented in this chapter are placed in a broader context in order to address a wide range of SOA applications composed of software services (also known as Virtual Services) and hardware components (called Real-World Services).

Contents

Chapter 1
Introduction

Stanisław Ambroszkiewicz, Jerzy Brzeziński, Wojciech Cellary, Adam Grzech and Krzysztof Zieliński

Abstract The main purpose of this book is to present advanced software tools for development, deployment and governance of Service Oriented Architecture (SOA, for short) applications. Novel technical concepts and paradigms, formulated during the research stage and during the development of such tools, are presented and illustrated by examples of practical use. Hence, the book should be of interest not only to theoreticians but also to practitioners and engineers who cope with real-life problems. Additionally, each chapter contains an overview of related work, enabling a comparison of the proposed concepts with exiting solutions in various areas of the SOA development process. This should make the book interesting for students and scientists who investigate similar issues.

S. Ambroszkiewicz(✉)
Institute of Computer Science, Polish Academy of Sciences, ul. Jana Kazimierza 5, 01-248 Warsaw, Poland
e-mail: Stanislaw.Ambroszkiewicz@ipipan.waw.pl

J. Brzeziński
Institute of Computing Science, Poznan University of Technology, pl. M. Skłodowskiej-Curie, 60-965 Poznan, Poland
e-mail: Jerzy.Brzezinski@put.poznan.pl

W. Cellary
Department of Information Technology, Faculty of Informatics and e-Economy Poznan University of Economics, Ul. Mansfelda 4, 60-854 Poznan, Poland
e-mail: cellary@kti.ue.poznan.pl

A. Grzech
Division of Teleinformatics, Faculty of Computer Science and Management, Institute of Informatics, Wrocław University of Technology, ul. Łukasiewicza 5, 50-371 Wrocław, Poland
e-mail: adam.grzech@pwr.wroc.pl

K. Zieliński
Department of Computer Science, AGH University of Science and Technology, Al. Mickiewicza 30, 30-059 Cracow, Poland
e-mail: kz@agh.edu.pl

S. Ambroszkiewicz et al. (eds.), *Advanced SOA Tools and Applications*,
Studies in Computational Intelligence 499, DOI: 10.1007/978-3-642-38957-3_1,

The main purpose of this book is to present advanced software tools for development, deployment and governance of Service Oriented Architecture (SOA, for short) applications. Novel technical concepts and paradigms, formulated during the research stage and during the development of such tools, are presented and illustrated by examples of practical use. Hence, the book should be of interest not only to theoreticians but also to practitioners and engineers who cope with real-life problems. Additionally, each chapter contains an overview of related work, enabling a comparison of the proposed concepts with exiting solutions in various areas of the SOA development process. This should make the book interesting for students and scientists who investigate similar issues.

The book consists of five chapters. Each chapter is self-contained and can be read separately but the chosen order guides the reader from abstract, high-level concepts to technical and implementation-oriented discussions.

In Chap. 1 the concept of development and management of SOA-based applications within a configurable service platform is presented. All components of the modular and reconfigurable platform are implemented as services, enabling exploitation of the SOA paradigm for configuration and management purposes. The framework is divided into features which cover the entire lifecycle of business-oriented ICT (Information and Communication Technology) application development. Applications are responsible for business requirement analysis and comparison, selection and composition of services, efficient communication and computational resource provisioning as well as allocation in a distributed, virtualized environment. Topics such as optimal resource utilization, monitoring and management are also addressed. The unique features of the platform include: business process compatibility, easy reconfiguration of composition schemes, visual support for requirements and service definition, QoS (Quality of Service) assessment (including communication services) and service execution control. Moreover, this chapter shows how to develop effective tools for SOA management and how the SOA paradigm may be applied to achieve better interoperability and flexibility.

The presented framework covers service request processing, from arrival of a request all the way to completion and evaluation of the requested task. The platform's building blocks (modules) assure service request processing, composition of services (taking into account their functional and non-functional requirements), efficient provisioning of communication and computational resources, execution of services in a virtual distributed environment and the evaluation of the quality of composite services. All of the aforementioned modules can be applied as independent data processing units or as aggregations of such units, supporting specific areas of the requirements and service-matching processes for any given business scenario supported by SOA-based or legacy information systems. All applications within the platform are designed and implemented as open, modular, scalable, and heterogeneous components (services), integrated in accordance with SOA principles. Such an approach allows for integration of selected applications to support various decision-making processes.

The main contribution of this chapter is a novel approach to service selection, composition and execution. First, the basics of the business process description in the context of services are explained and an original notation is proposed to support service-oriented process description and SLA (Service Level Agreement) generation. Subsequently, flexible and adaptive composition and provisioning of composite services within the original architecture is presented, allowing the use of various communication and computational services by domain-oriented services. A scalable service execution environment, including an original service-based execution engine, is discussed. Service validation within the layered model of SOA governance is addressed, along with a method for composite service security assessment. Finally, an approach to automatic generation of user interfaces for semantically-described composite services is considered.

A different approach to automation of sophisticated business processes is presented in Chap. 2. It is based on the idea that automation can be performed if the process of using a single business service is automated first. Four basic phases of using a business service are identified in this context: (1) the client sends a query to the service provider and obtains a quote; (2) an order is placed and a contract made; (3) the service executes in accordance with the contract; (4) payment is effected (this may occasionally precede step 3). These phases are interrelated, so that if the first phase (query-quote) can be realized in an universal way, then the whole process can be automated. For this purpose a representation of the service environment (XSD schemata of documents processed by services) is proposed as a semantic model (grounding) for a description language which can be applied in the first phase, matching service performance with client requirements.

The authors also propose a set of extensions for the business service architecture related to Web services and the Semantic Web. Within this architecture there are two methods of communication. The first method, used in the query-quote phase, consists of exchanging messages containing formulas in a description language. The second mode of communication involves document processing using WSDL and is applied during the order-contract, execution, and payment phases. The concept of a business service and the automation of the query-quote phase enable automation of the whole sophisticated business processes, involving many services.

The proposed technology is specified in the form of protocols. The corresponding prototype (a communication platform for developing electronic markets, called *SOA-enT*) is generic and may be applied to any application domain if appropriate XSD schemata (related to the problem domain) are introduced into the system.

The presented technology is of particular use in electronic markets, due to the fact that such markets extensively rely on automation of sophisticated business processes comprising many services. Automation should involve not only service publication and discovery but also automatic planning, arrangement and execution of complex and time-consuming business processes. Although much work has been done on this front since 2000, the problem of automatic service composition remains a challenge.

It should be noted that while electronic markets for consumer products are relatively well developed, the corresponding service markets are often restricted and focus on simple services provided on a mass scale. The main problem here is that offers and commissions are to be specified in a plain language, implying that the descriptions of services related to these offers are imprecise and not formal enough to allow automatic processing. As a result, service markets are usually used only to submit offers and agree upon commissions, while transactions take place using more traditional methods.

Electronic service markets are also a meeting venue for business partners. As such, they should support cooperation between these partners, discussions, joint planning and realization of business ventures in the form of electronic business processes. It is therefore reasonable and necessary to introduce social media to modern electronic markets.

These aspects are further explored in Chap. 3 in which the flexibility and adaptation of collaborative processes occurring among organizations in a Service Oriented Architecture are considered at an inter-organizational level. First, an in-depth discussion of applying the service-oriented approach to inter-organizational collaboration is presented. The need for SOA at the inter-organizational level is explained by the ubiquity of services and the economically motivated need of organizations to collaborate to gain and retain competitive advantage. However, the heterogeneity and dynamism of SOA ecosystems at the inter-organizational level has to be taken into account by information systems supporting service-oriented collaboration. Two main problems have to be addressed to appropriately support inter-organizational collaboration in heterogeneous and dynamic SOA ecosystems: flexibility and adaptation.

Flexibility means the lack of a full specification of a process model and the corresponding need to construct such a model uniquely and separately for each active instance at runtime.

Adaptation means adjusting a process model for an active instance to exceptional circumstances which may or may not be predicted *a priori*. Two SOA-based methods are proposed in this context: the CMEAP method (Composable Modeling and Execution of Administrative Procedures) supporting flexibility, and a method for adaptation of service protocols supporting process adaptation. The CMEAP method allows administrative procedures to be automatically composed in a flexible manner, based on modeling legislative provisions in the form of elementary processes, decision rules, and domain ontology. While CMEAP is proposed as an approach to support flexibility in administrative procedures, service protocols are proposed as an approach to support adaptation of collaborative processes in a dynamic environment. The proposed methods for adaptation of service protocols allow collaborators to modify the process model governing their collaboration at runtime, taking into account their social relations.

Following this discussion, two prototype systems—the PEOPA and the *ErGo*, both implementing the proposed methods for collaborators in the construction sector—are detailed. Finally, a case study illustrates how the prototype may support construction processes. The CMEAP approach is used to develop a detailed model of the administrative procedure for issuing a building permit and obtaining approval for the design of a residential building. The investment is characterized by various circumstances influencing the approval process, such as the complete lack of local zoning guidelines, construction in a zone listed in the communal record of historical monuments but not listed in the corresponding national registry, placement in the vicinity of a Nature 2000 site, and exemptions from technical and building regulations in the building design. The proposed method for adaptation of service protocols is first applied to identify the collaborators required to finance and perform groundwork at the construction site, and then to react to an unforeseen discovery of hazardous chemicals at the site.

Chapter 4 addresses system-related aspects of SOA applications, including tools and methods for improving the dependability of such applications. Dependability is a general concept connected with fault tolerance, which—in the face of inevitable failures of system components—arises as one of the essential properties of distributed systems, especially SOA-based ones. It comprises availability (readiness to perform a service), reliability (continuity of correct operation), safety (absence of catastrophic consequences), integrity (absence of improper system alterations), maintainability (ability to undergo modifications and repairs) and—to some extent—security. The SOA paradigm itself addresses some of these attributes through the concept of decoupling system components, leading to their independence and thereby limiting the consequences of failures. Dependability is further enhanced by redundancy.

A straightforward way to incorporate redundancy into SOA-based systems is through replication of services (service instances), which directly enhances availability. The approach to replication considered here is two-fold. First, a group communication mechanism is provided as a tool for building internally replicated services, i.e. services organized in the form of replicated components or lower-layer items. The primary aim of this replication technique is to improve service reliability, which entails preservation of strict consistency at the expense of partition tolerance. The second technique addresses availability via partitioning. It combines optimistic and pessimistic replication through appropriate specification of operations at their submission.

Reliability and availability of servers are crucial aspects of dependability, although within the business process layer an important contributing feature is the reliability of interactions between service providers and their clients. The problem of reliable interaction results from possible inconsistencies between the server state and its client-side representation, which can appear following server or client crash and subsequent recovery. This creates the need for a rollback-recovery mechanism that restores a consistent state of distributed processing whenever any component fails. The proposed recovery service is tailored for SOA-based systems and remains transparent both for clients and service providers.

Security, being a critical property in many IT systems and related to dependability, encompasses availability (for authorized actions), and integrity, i.e. absence of unauthorized system alterations. This set of features is often extended to include additional properties such as confidentiality (prevention of unauthorized disclosure of information). Typical security management mechanisms used to provide these attributes include authentication and authorization, access control and information flow control, as well as communication protection. A supplementary issue, tackled here, is the concept of security policies, which are indispensable for efficient management of security requirements in large-scale distributed computing systems. A security policy consists of rules controlling interactions between system components. However, due to the complex structure of modern service-oriented systems, policy mechanisms often suffers from problems such as inconsistencies, which gravely degrade the efficiency of enforcement.

The presented concept and methods for the development of dependable services are embodied by two toolkits: *ReSP (Reliable SOA Platform)* and *DyMST (Dynamic Management SOA Toolkit)*. Their usage is illustrated by enhancing the dependability of applications from the healthcare domain: *Healthcare Integration Platform* for the exchange of patients' medical data among various healthcare units, and *Medical Event, and Data Registering Platform* facilitating day-to-day operations in a healthcare unit.

Chapter 5 explores additional aspects of the execution infrastructure in SOA-based applications. It introduces a pragmatic methodology for adding and managing adaptability in multiple layers of the SOA application execution infrastructure. Adaptability mechanisms and techniques are investigated by referring to the MAPE-K pattern, which is viewed as the most representative solution for adaptive and autonomous systems. Implementing an adaptive SOA system remains a challenging issue: the adaptation process requires suitable mechanisms to be built into applications or the execution environment itself. Satisfying service orientation principles means that self-adaption of SOA systems can be considered a self-contained aspect and introduced during development or at runtime. An adaptive application can be developed by transforming a preexisting application and hardware components (which does not involve changes in the application's business logic). This is consistent with the fact that the investigated adaptive systems remain business-agnostic and instead focus on ensuring the required nonfunctional parameters.

The SOA solution stack (S3) developed by IBM is selected as the basis for the application execution infrastructure model. This makes the proposed concepts easier to understand, while not detracting from their general nature. The adaptability aspect is considered in a broad context, with attempts to address, in a uniform way, all SOA applications composed of software services (Virtual Services) and hardware components (Real World Services). Such an approach is justified by the increasing importance of pervasive systems, bringing interaction from enterprise systems back to the real world. In this context the adaptive behavior of Real Word Services is viewed as a critical element, combining adaptive interaction, adaptive composition and task automation by involving knowledge of user profiles, intentions and previous usage patterns. The proposed methodology is supported by the *AS3 Studio* package which is a complete suite of tools providing extensions of SOA systems with adaptability features. This methodology is presented as a crucial part of the governance process of SOA applications. Finally, a case study which illustrates the proposed approach is described.

Chapter 2
Specifications and Deployment of SOA Business Applications Within a Configurable Framework Provided as a Service

Adam Grzech, Krzysztof Juszczyszyn, Grzegorz Kołaczek, Jan Kwiatkowski, Janusz Sobecki, Paweł Świątek and Adam Wasilewski

Abstract This chapter presents the concept of development and management of SOA applications within the configurable service platform which supports all phases starting from business process definition. The unique features of the platform include: business process compatibility, easy reconfiguration of composition schemes, visual support for requirements and service definition, QoS assessment (including communication services) and service execution control. Moreover, it illustrates how effective tools for SOA management may be developed within the SOA paradigm itself, and how this paradigm may be used to achieve their interoperability and flexibility.

1 Introduction

This chapter is devoted to presenting the main components and functionalities of PlaTel (platform for ICT planning and monitoring solutions), an integrated framework supporting business processes in distributed ICT (Information and Communication Technologies) environment based on the Service Oriented Architecture (SOA) paradigm. The key architectural assumptions of the PlaTel are: modular organization, reconfiguration capability, full compliance with the SOA paradigm. In result, the users are offered extended capabilities of building SOA applications starting from the definition of business processes which are translated to composite services' requirements. All components of PlaTel are implemented as services which allows to utilize the features of SOA paradigm in platform configuration and management. The framework scope of functionalities is divided into applications that cover the entire life cycle of business-oriented ICT applications. The framework's applications

A. Grzech (✉) · K. Juszczyszyn · G. Kołaczek · J. Kwiatkowski · J. Sobecki · P. Świątek · A. Wasilewski
Faculty of Computer Science and Management, Institute of Computer Science, Wrocław University of Technology, Wroclaw, Poland
e-mail: adam.grzech@pwr.wroc.pl

S. Ambroszkiewicz et al. (eds.), *Advanced SOA Tools and Applications*, Studies in Computational Intelligence 499, DOI: 10.1007/978-3-642-38957-3_2,

and tools are responsible for business requirement analysis and comparison, service choices or the composition of services, efficient communication and computational resource provisioning and allocation in a distributed, virtualized environment as well as for the utilization of resources and the quality of monitoring and management services.

The presented framework covers service request processing, beginning with a service request's arrival until the completion of the request and evaluation of the delivered service. The platform's building blocks (modules) assure service request processing, composition of composite services—taking into account the functional and non-functional requirements, the provision for efficient communication and computational resources, the execution of services in a virtual distribution environment, and the evaluation of the quality of composite services. All of the aforementioned modules can be applied as independent data processing units or as sequences of such units that support selected sections of the requirements and service-matching processes for any given business process supported by SOA-based or legacy information systems. All the applications within the platform are designed and implemented as open, modular, scalable, and heterogeneous components (services) that are integrated according to the SOA paradigm. Such an approach allows for the integration of all or selected applications to support various decision-making processes.

So far, we can find many approaches to the problem of business process execution by means of available software services [25]. In work [26], it is mentioned that, rather than starting with a complete business process definition, the composition system could begin with a basic set of requirements and in the first step to build the entire process, whereas many approaches [26] require a well-defined business process to generate such a complex service. Current work often raises the topic of the semantic analysis of user requirements, service discovery (meeting the functional requirements), and the selection of specific services against non-functional requirements (i.e. execution time, processing and communication costs, security, etc.). However, the presented solutions have some disadvantages, i.e. these methods have not yet been successfully combined to jointly and comprehensively solve the problem of the composition of complex services that satisfy both functional and non-functional requirements. In many cases, only one aspect is considered. For example, the work [57] focuses on the selection of services based only on one functional requirement at a time. Other works show that non-functional requirements are considered to be of key importance; however, many approaches ignore the aspect of building a proper complex service structure that is key to the optimization of execution time, for example.

To date, the service composition problem has been approached from different perspectives. Some have presented specialized methods for selection services or composite service QoS-based optimization [26]. However, despite the importance of their contribution, those solutions are not widely used. Some propose complete end-to-end composition tools that introduce the concept of a two-stage composition: the logical composition stage to pare back the set of candidate services and then the generation of an abstract workflow [2]. METEOR-S presents a similar concept of binding web services to an abstract process and selecting services that fulfill the

QoS requirements. Notions of building complete composition frameworks are also clear in SWORD, which was one of the initial attempts to use planning to create web services [47]. However, the proposed approaches are closed and do not support the implementation of other methods and, because of this, it is difficult to call them frameworks. Also, a framework-based approach is what is currently needed in the SOA field in order to create composition approaches that are suited to different domains and problems that are characteristic for them.

However, the service composition task is only a partial solution of the deployment of the SOA business applications problem. Numerous works point out those SOA business applications responsible for the realization of business processes should be constructed in accordance with business process description techniques [20]. Moreover, they should maintain the quality requirements imposed on the business process realization by means of composite services. It is especially important in the event that hardware devices and telecommunication services are involved.

Taking the conditions above into account, the PlaTel framework presented in this chapter offers a general solution for addressing all of the key issues involved in the deployment of SOA business applications: a process-level description of services, service composition, provisioning, dimensioning and execution, service validation and security assessment as well as the generation of user interfaces for semantically described services. All of these features constitute the generic PlaTel process for the deployment of SOA applications (Fig. 1) and are discussed in detail in the following sections.

It is started with the business process query, complemented by its associated non-functional requirements (cost, security, time constraints, etc.). The query is processed by means of comparing it with existing processes stored in the reposito-ry which contain associated Service Level Agreements (SLA), defining the requirements for composite services. In the event that there are no adequate processes, a new SLA is created. On the basis of SLA, an SOA application (composite service) is composed and dimensioned according to its functional and non-functional requirements. Composed and dimensioned services are executed by the native PlaTel service execution engine environment. The entire process undergoes validation and quality assessment in which the results are stored in a repository to be used to process future queries.

All stages of the SOA business application deployment are performed by dedicated PlaTel services that use common repositories. The PlaTel framework provides a multi-criteria (functional and non-functional) composition of Web services and QoS assurance for utilized ICT services (bez zmian) that allows for the construction of various service selection and composition scenarios. It also utilizes configurable composite services to provide for its functionality. In order to ensure the interoperability with external service and ontology repository specialized services, Mediators, responsible for database access and providing standardized interfaces for internal PlaTel services— had been created. As a result, external resources can be integrated within the PlaTel framework (which was equipped with access control functions). All of these combined features offer a flexible and extensible environment for communication and composition between the components of SOA applications. To the best

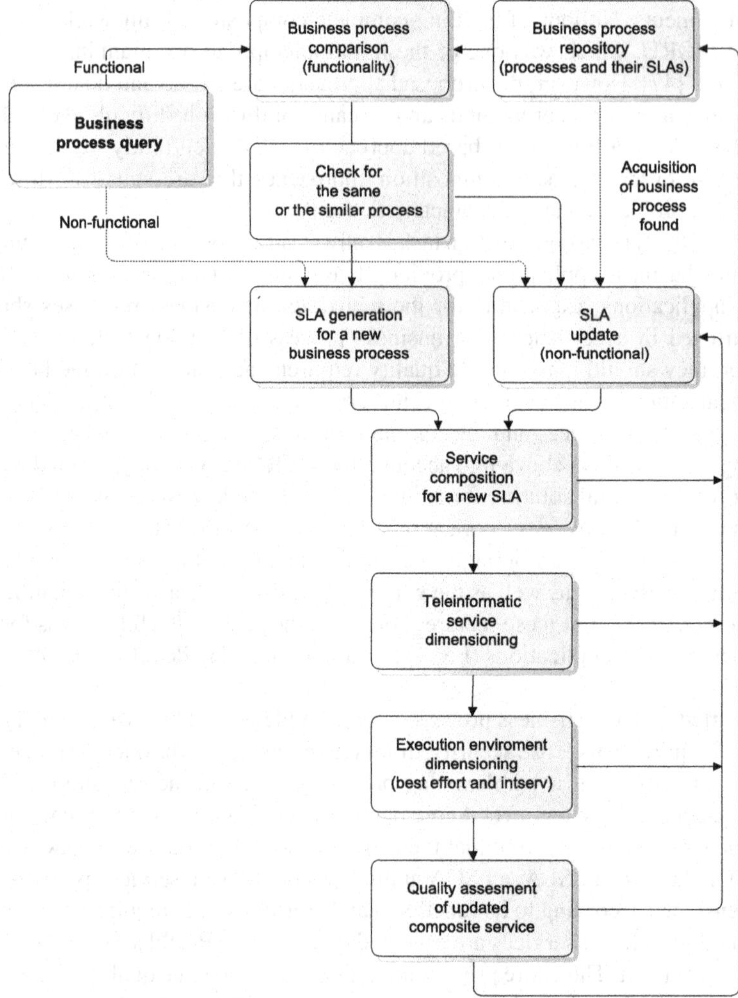

Fig. 1 Deployment of SOA business applications within the PlaTel framework

of our knowledge, it is the first framework for service composition that implements this kind of holistic and flexible approach.

There are several unique features of the PlaTel framework, associated with the various stages of SOA business application deployment:

- All the key components of the PlaTel (namely: service composer, communication mapper, and execution engine) are composite services themselves, which implies that their execution schemes may be easily reconfigured or extended—and a specialized graphic user interface is provided to support such actions. This approach allows for the creation of repositories of dedicated composite services for the com-

position, mapping, and retrieval of the Web services, which may be used according to current needs. They may be stored and conditionally chosen from the repository.
- To address the non-functional requirements of Web services, an extensible description language (SSDL—Smart Service Description Language), which allows for the expression of arbitrary non-functional parameters of a composite service, is proposed.
- The execution engine interprets the SSDL definitions of composite services and maintains the non-functional parameters. It has two methods of execution—interpretation (the components of a composite service are explicitly executed) and composite service emulation (in this case, the execution engine reads the SSDL and communicates like a service described in the SSDL definition, thus supporting the automatic deployment of composite services). The framework also allows the use of external execution engines (in this scenario, the PlaTel engine serves as an SSDL-driven interface for them).
- The issue of communication between Web services is being addressed by providing a communication mapper—a dedicated service which supports a service composer and substitutes each node of the composite service scenario graph with a sub-graph representing the order of the execution of atomic ICT services (communication and computational) necessary to link domain atomic services (communication services provide an appropriate medium for the transmission of data between computational services, while computational services are responsible for data processing tasks such as: encryption, encoding, signal merging/splitting, etc.).

The main contribution of this work is at present a novel approach to service selection, composition, and execution , in which the abovementioned functionalities are provided as services. The PlaTel framework allows also the composition of communication services and supports flexibility via configuration mechanisms for basic processes (composition, mapping, and the execution of services).

In the following sections the details of the main activities required to complete a business process query request depicted in Fig. 1 are provided. The next section presents the basics of the business process description in the context of services and proposes an original PlaTel Event-driven Process Control notation that supports the service-oriented process description and SLA generation. Section 3 covers the composition and provisioning of composite services within the original architecture, allowing for a flexible approach to the composition task and the use of various communication and computational services which are used by the domain-oriented services. Section 4 is devoted to the PlaTel service execution environment and presents its original service-based execution engine and scalable execution environment. The next section discusses service validation within the layered model of SOA governance along with a method for composite service security assessment. Finally, Sect. 6 proposes an approach to the automatic generation of user interfaces for semantically-described composite services.

2 Service-Based Business Processes Description

This section is devoted to the presentation of the advantages and disadvantages of well-known and successfully applied business process modeling attempts and notations in terms of their suitability for the simultaneous description of the functional and non-functional requirements that are important for composite service composition. Results of a critical and comprehensive analysis of contemporary modeling approaches leads to the specification of invented modeling notation PEPC (PlaTel EPC), thus expanding the functionality of the well-known notations by adding new opportunities required in composite service composition processes.

Models of business processes are used as an input for service composition. A PlaTel customer expects a composite service that meets his/her functional and non-functional requirements. Functional requirements may be stated as a business process flow or a functional part of the SLA (Service Level Agreement). Non-functional requirements in typical approaches to BPM (ARIS, BPMN) have to be included in the SLA. In PEPC, notation for non-functional requirements can be set for each object in the business process model.

In the first step of the PlaTel workflow (Fig. 1), the customer's functional requirements are compared with the functional requirements defined for the business process models available within the business process repository. Next, the set of known business process models containing the expected functionality is compared due to the structural similarity with the customer's business process. If an acceptable model is found in the repository, then the non-functional requirements are updated. If there is not a suitable business process model in the business process repository, then required business processes have to be modeled using templates (patterns) from the business process repository or beginning with the blank model (PEPC modeler supports both options).

A new business process model provides information regarding functional requirements. A customer's non-functional requirements are added to the SLA (or to the PEPC model) and PlaTel is ready for service composition.

2.1 Business Process Modeling (BPM)

Each organization has its own formal or informal rules and patterns of conduct that result from the experience, knowledge, and competence of employees. Usually taken actions (activities) are arranged in a sequence with clearly-defined start and end triggers, targets, and human or system owners. Such sequences—business processes—require the involvement of many people from different organizational units. This means that the process actors do not have complete vision of the process and knowledge of the actions taken at all process stages. The solution to this problem is business analysis and then the mapping and modeling of business processes called Business Process Modeling (BPM). BPM supports standardization of the management of business processes that pass through multiple data repositories, applications, departments, or even companies [41]. Business analysis lets for the understanding of how organizations work to accomplish their purposes and defining the capabilities the organization requires to provide products and services to external stakeholders [1]. On that basis, BPM allows the representation of enterprise processes and the capture of an ordered sequence of business activities and supporting information.

Key benefits of BPM are the following [21]:

- formalization of existing processes and spotting needed improvements,
- facilitation of an automated, efficient process flow,
- increase of productivity and decrease in the number of employees,
- opportunity for people to solve the difficult problems.

2.1.1 Basic Rules of BPM

Models of business processes may be designed to answer the question "Where are we now ("as-is" perspective) or "Where do we want to be?" ("to-be" perspective). The first option gives a common, baseline model which represents an accurate depiction of the actual situation within the organization. When the model is developed, it may be used to analyze and improve the business process. The second option lends itself to providing a vision of an organization's development. Such diagrams present how the future processes might appear after incorporating the proposed improvements. Moreover, such models may be used to demonstrate and simulate the new processes before their implementation.

Before starting any business process modeling, it is important to be clear about several basic questions that can be extended by detailed questions (Table 1).

Comparing the ARIS methodology (especially EPC diagrams) and BPMN can draw the following conclusions:

- ARIS methodology enables a more accurate modeling of an organization's information system, but modeling of the information system requires diagrams from every ARIS perspective,
- BPMN enables more a accurate modeling of the process flow,
- BPMN does not support modeling of the context of business processes, including the linkages between the processes.

In addition:

- The ARIS methodology is widely-used for business process modeling for the implementation of integrated information systems (in particular, SAP products for which ARIS offers dedicated diagrams)—this means that many organizations (especially large and medium size) prepare maps of processes using the ARIS methodology during the implementation of the ERP (Enterprise Resource Planning) system,

Table 1 Questions to answer before business process modeling

	ARIS (EPC)	BPMN
Process flow	Yes	Yes
Decision points (connectors, logical operators)	Yes	Yes
Splitting/merging process flow	Yes	Yes
Sub-processes	Yes	Yes
Process flow including activities (functions) and events	Yes	Yes
Different types of events (start, mid, end)	No	Yes
Requirement of mid-events	No	No
Different sub-types of events (message, exception, timer, etc.)	No	Yes
Different sub-types of activities/functions (human, systems, etc.)	No	Yes
Participants and roles in the process	Yes (bound with Organizational Structure diagram)	Yes (as swim lanes)
Difference between process flow and message flow	No	Yes
Milestones	No	No
Exceptions' modeling	No	Yes
Definitions of inputs and outputs	No	No
Relationship between processes	Yes (Value-added diagram)	No
Relationship between participants (organizational units, roles, persons, etc.)	Yes (Organizational Structure diagram)	No
Data structure	Yes (ERM—Entity Relationship Model diagram)	No
Relationship with the environment	Yes (Organizational Structure diagram)	Yes (as pools)

- BPMN seems to be very useful for documentation of the organization, redesigning of the organization, using knowledge management, and supporting continuous process management,
- EPC diagrams are easier to understand by those in the organization [34],
- EPC and BPMN strongly support patterns to reduce the perceived model's complexity,
- transforming EPC diagram into the BPMN model is possible but with some limitations [22].

2.2 *BPM for Service Composition*

Service composition uses available services that meet the requirements of a specific user in order to deliver composite services. According to the SOA paradigm, services should fulfill business needs (in terms of functional and non-functional properties). If the business process model contains the necessary information, it may be the basis for service composition.

2.2.1 Requirements of Service Composition

The successful composition of composite services requires information about the business process, including:

- sequence of activities to be performed (process flow),
- types of activities, including human-centered activities that need human-computer interaction,
- business logic (including the distribution of the process flow on different paths)
- functional and non-functional requirements,
- input and output data,
- the set of services that provide required functionality (by process, sub-process or activity)—this knowledge is available from the service repository.

Table 2 shows how the ARIS methodology (EPC diagrams) and BPMN meet the needs arising from the composition of composite services.

Analysis of the ARIS methodology and BPMN notation shows that their application for composition of composite services is limited because most of the service composition requirements are not met. Sometimes IT tools or extensions of notations (e.g. for performance measurement [39]) enables one to solve some problems (e.g. allows the indirect indication of functional and nonfunctional requirements

Table 2 Fulfillment of service composition requirements

	ARIS (EPC)	BPMN
Sequence of activities (functions)	Yes	Yes
Activity types	No	Yes (partly, as swim lanes)
Business logic	Yes	Yes
Definition of functional requirements	No	No
Definition of non-functional requirements	No	No
Binding with ontology	No	No
Input and output data	No	No
Relationship between activities (functions) and services	No	No

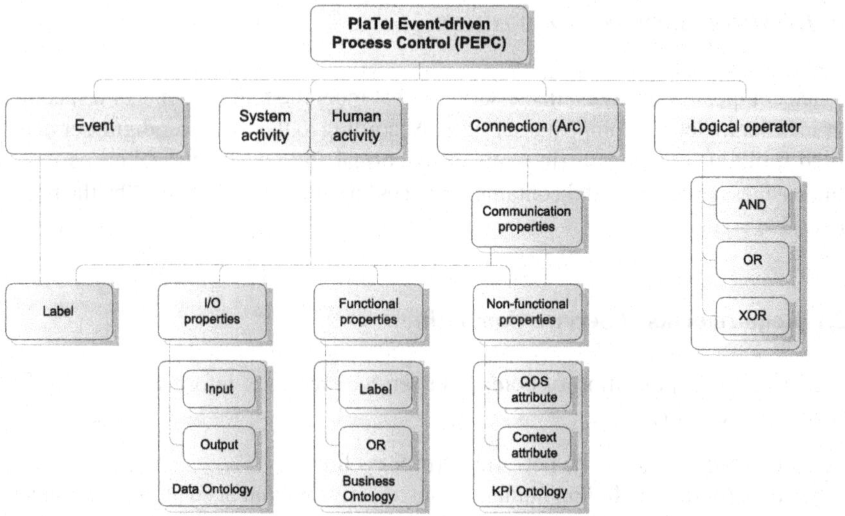

Fig. 2 Meta-model of PlaTel event-driven process control (PEPC) notation

and input/output data), but to get closer to the service composition requirements, expensive commercial tools are needed.

To fulfill the requirements of business process modeling for composite service composition, the concept of dedicated notation PEPC (PlaTel Event-driven Process Control) was developed. PEPC is based on the EPC diagram but has several extensions that lead to the modeling of business processes as a convenient addition for automated service composition.

2.2.2 PEPC Notation

PEPC notation supports business processes modeling that provides models for the automatic composition and execution; these are used as composite services.

The key attributes of the proposed modeling PEPC concept are as follows: (Fig. 2):

- process flow modeling typical for EPC diagrams (activities are followed by events, logical operators—AND, OR, XOR—and are used similarly to the EPC notation),
- different graphic objects for system activities (without human-computer interaction) and human activities (with human-computer interaction),
- connections between activities, events, and logical operators,
- names of activities (labels) and key words based on business ontology—Ş-the functional properties of activities,
- activities have defined inputs and outputs based on given data ontology,
- activities and connections have non-functional properties based on known KPI (Key Performance Indicator) ontology,

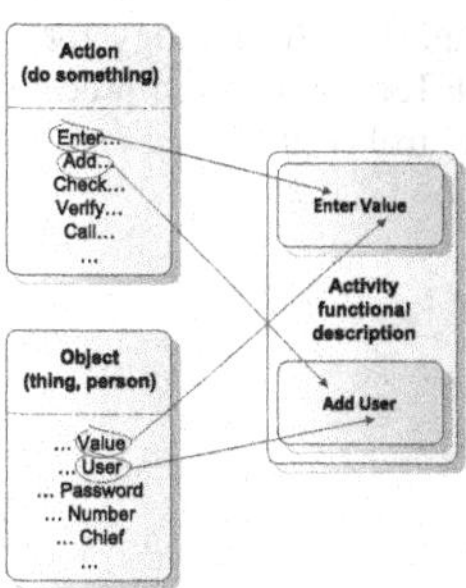

Fig. 3 Using business ontology in PlaTel event-driven process control (PEPC)

- activities may be added optionally for information (WSDL file) with the definition of corresponding ICT service or services that provide functionality required in the activity.

PEPC notation uses the concept of Semantic Business Process Modeling (SBPM), a modern Semantic Web approach that adopts well-defined ontologies in the models of business processes [11].

PEPC includes data, business and KPI ontologies.

Data ontology stores information about a set of inputs and a set of outputs. Each input and output has a name and type that is necessary in order to link process activities with services. According to WSDL specifications, each service operation should be described by inputs and outputs so it is possible to find service or services and business process activities that correspond due to inputs and outputs automatically.

Business ontology includes set of terms that describe actions and objects. A combination of those terms is used to set the functional properties of the activity (Fig. 3). Functional properties are stored as activity labels or activity key words.

KPI ontology includes non-functional properties of the activities and connections in a hierarchical structure (Fig. 4).

Non-functional attributes of activities and connections allow for the mapping of SLA (Service Level Agreement) requirements for the business process or any of its components (sub-process, activity, connection).

PEPC optionally assigns an atomic or composite service or services to business process activities. It offers the possibility for one to indicate which services should be used in the composition of composite service (e.g. for reasons of safety or signed contracts with the provider of specific services) or precisely define the requirements for such services. Such services should be previously registered in the service repository to ensure the functional consistency of the composite services.

PEPC incorporates the functional description of the composite services with various QoS indicators thus defining a complete SLA and allowing for service composition and provisioning.

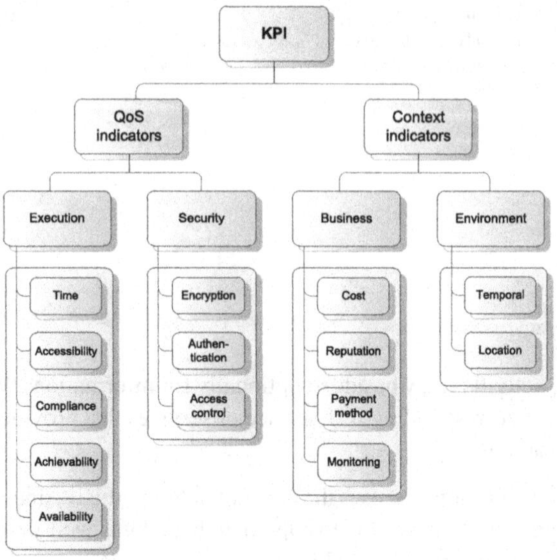

Fig. 4 KPI ontology for PlaTel event-driven process control (PEPC)

3 Service Composition as a Service

On the basis of the SLA that results from the business process query, a service composition process begins. To generate a composite service means to find a set of atomic services and bind them together so that they, as a new service, fulfill all user functional and non-functional requirements. The services stored in repositories are functionally and non-functionally described with terms derived from the ontologies mentioned in the preceding section. A typically automated composition process requires a semantic query (description of a required composite service; often this description is referred to as a Service Level Agreement—SLA) and in PlaTel consists of two parts: service composition and service mapping. The first is responsible for the functional composition of the composite service, while the second adds the necessary communication services that connect the functionalities while preserving the QoS defined in the SLA.

Recently, developing information and communication technologies (ICT) have enabled entrepreneurs to develop monolithic structures into distributed ones. The entrepreneurs could focus on translating business processes into composite services; thus the Service Oriented Architecture (SOA) becomes a crucial paradigm in designing service-oriented systems (SOS) [65].

In SOS, the key element is a service that provides certain and well-defined functionalities and is characterized by parameters that describe the quality of a required and delivered service. Furthermore, services may be instantiated and assembled dynamically, leading to the changing structure, behavior and location of a software application in runtime. However, to ensure high satisfaction of clients and service providers' profits, the services should be provided in order to guarantee a high quality of service (QoS) [60].

3.1 Service Description

The PlaTel approach to the service description problem assumes the use of the native service description language, SSDL (Smart Service Description Language), proposed as a solution allowing a simple description of composite service execution schemes and the support for the functional and non-functional description of services—there exist similar solutions, like ROsWeL language, which utilizes a declarative, resource oriented approach [23]. Its functionality includes the Web Service Description Language (WSDL) but also offers important extensions. SSDL is dedicated to service execution support, including the service guarantees of QoS parameters and dynamic service composition at runtime.

3.1.1 Languages

WSDL applicability to service composition

Web Service Description Language (WSDL) is a format that aims to describe the functionality offered by Web services, mainly by describing their interfaces. It provides a machine-readable description of how to call Web Services, what parameters to forward, and what to expect from it in return. The WSDL file describes only a single Web Service and its functionality, whereas SSDL utilizes the WSDL as a base to describe composite services and requirements (or functionalities) of those services. Each SSDL file that describes a composite service consists of a set of nodes which, connected together in a graph-like structure, form a composite service execution plan determining a proper way to execute the services and to pass on the data between them. Each SSDL node contains a pointer to a WSDL file that describes how to access a particular Web Service in a composite service.

The main difference between WSDL and SSDL is not only the scope of description (atomic service vs. composite service) but also the fact that WSDL is a complete instruction on how to call a Web Service via HTTP (namely using SOAP protocol), while SSDL is an execution plan needing an engine to execute (interpret) it.

SSDL

When working with the service composition problem, it was noticed that there is a need for a language that allows for the definition of requirements and a service execution plan in a unified manner. Composition methods transform user requirements to the form of a composite service that fulfills those requirements. The literature has shown no description language designed to utilize this particular feature.

The most widely-known candidates considered were BPEL and OWL-S—both designed with different applications in mind. BPEL, a business standard adopted on many occasions, focused on execution of services and has no semantic description of what the services actually do or should do. The latter is especially essential for the dynamic composition of semantically-described services. On the other hand, OWL-S, which offers a semantic description of web services, was designed for the description of atomic services; its offer for composite scenarios while preserving the non-functional QoS requirements is limited.

A decision was made to design a description language that would have the following design notions, have low complexity, allow for the definition of both functional and non-functional requirements of services for composition purposes, and at the same time the designer could combine the service requirements with specific services. In the process of development of this so-defined language, its other aspects became apparent. As a language defined for services requirement specifications, it had to allow for the distinction of various well-defined kinds of requirements and services. As a consequence, a specialized execution engine was designed whose primary purpose was to execute composite services described in such a manner. To fulfill the premise of being fully executable, the description language had to be fully interpretable. This interpretation is done in real-time by composition algorithms invoked by the engine.

The designed language was named SSDL after its elementary concept of a Smart Service. The idea of Smart Services extends the concept of a composite service. A Smart Service is, in fact, a type of a composite service; however, its elements do not necessarily have to be atomic services. In general, a Smart Service is an interpretable and simultaneous definition of atomic services and requirements of different types in one execution plan.

At the bottom of a Smart Service definition process stands the following usage case. Typically, users will want to define and then execute new composite services. In this case, with the use of the provided specialized graphic user interface (Sect. 6), they will input their requirements and appropriate composition mechanisms will be executed to produce an optimal composite service. All this can be saved in user profile and executed in well-defined situations when a composed service is requested.

However, sometimes a user will want his orher service to be more flexible than just a standard composite service. Users can design their composite services to change with time or be composed in real-time. Saving parts of the Smart Service as requirements and not services will lead to dynamic interpretation at the point of execution later. The execution engine, inferring by the type of user requirements, will be able to perform different composition scenarios or discover and execute appropriate services for each requirement.

Technically, a Smart Service is represented by a combination of interconnected nodes (Smart Service Graph), some of which can represent concrete services: some sets of services and some various types of requirements. Nodes of a Smart Service are connected by identifying data sources—then they define data flow—and order of nodes defining control flow. Various types of nodes can be defined (ultimately in a specialized ontology) as long as the execution engine is configured to interpret and execute them. Note that it is the user who can define node types and their interpretation of what makes the SSDL language dynamic and highly susceptible to personalization.

A definition of SSDL node types contains all basic data types that allow for the functional and non-functional description of a service, its execution requirements, and the description of complex services with the conditional execution of their atomic components. The SSDL language was designed to enable the description of complex services which is directly interpreted and executed by the service engine as described in Sect. 4.1. SSDL allows a user to define alternatives from which dynamic composition can choose candidate services (in case the base one returns an exception) or introduce more levels of complexity of a service by introducing sub graphs. Furthermore, some control instructions could be added to further instruct the execution engine to interpret the Smart Service description.

3.1.2 Functional and Non-Functional Service Requirements in SSDL

A single SSDL node, which is used to describe a basic functionality requirement for a service has several important sections, which are extensively used during service selection, composition, and the final execution plan optimization:

- Physical description—used by every type of data that links to a specific Web Service, thus representing the optional description of service execution conditions (requirements leave this section empty),
- Functional description—used to semantically describe the capabilities of a service or required capabilities when defining a functionality-type node
- Non-functional description—used to describe the non-functional parameters of a service or requested services such as: time, cost, availability, etc.; non-functional parameters that can be requested for composition purposes are not limited in any way—external validation can be performed using user-defined ontology and rules, for example.

Each of them is associated with the number of sub-nodes allowing for precise description of a service. Below, we show an example of the information stored within the *nonFunctionalDescription* node:

```
<nonFunctionalDescription>
   <properties>
     <property>
        <name>Bitrate</name>
        <value>1.5</value>
        <relation>eq</relation>
        <unit>Mb/s</unit>
        <weight>1</weight>
     </property>
     <property>
        <name>Cost</name>
        <value>300</value>
        <relation>lt</relation>
```

```
            <unit>PLN</unit>
            <weight>1</weight>
        </property>
    </properties>
</nonFunctionalDescription>
```

This allows the introduction of domain-specific parameters that could be defined according to the current needs with only one constraint—they should be also represented in the domain ontology, which assures the interoperability and semantic consistency of our framework.

3.2 Service Composition Methods

This section presents a general composition scenario and introduces the Service Composer as a specialized service that performs service composition based on the SLA agreement and contains both functional and non-functional requirements. The service composer is a configurable composite service itself, and therefore offers a flexible approach in which each of the stages of the composition process may be performed using different methods of service selection and optimization provided as services. This approach can be described as being utilized to ensure efficient service composition and provisioning while addressing the problem of QoS-aware communication between the components of composite services.

3.2.1 Overview

The composition of Web services permits the building of complex workflows and applications on the top of the SOA model [7]. There are two standard approaches to service composition, namely, workflow composition and Artificial Intelligence planning (AI planning). Besides the obvious software and message compatibility issues, good service composition should be carried out with respect to the Quality of Service (QoS) requirements; these, in turn, can be split into functional and non-functional [26] components. The first are considered to be of key importance; however, preserving the non-functional requirements (availability, performance, and security—to name only the most important) is a key factor as well as its importance rapidly grows in distributed environments where complex services are composed of atomic components [25, 33, 40].

3.2.2 SSDL-Based Service Composition

There are two standard approaches to service composition, namely, workflow composition and AI planning. In both approaches, the emphasis is placed on identifying control and data flow. Generally speaking, functionalities of a composite service are represented as a Directed Acyclic Graph (DAG), also referred to as a functionalities graph, in which arcs denote interactions among functionalities and nodes determine functionalities (note that a service may be executed periodically or in a feedback loop—this mode is provided by the execution engine described in Sect. 4). Thus, the service composition task can be seen as a determination of a DAG structure, e.g., using AI methods and then a selection of atomic or composite services that provide required functionalities. The service selection is formulated as an optimization task in which an objective function reflects QoS attributes aggregation, e.g. response time or price, and constraints concerning structural dependencies in DAG. There are several known approaches to QoS optimization such as graph-based methods and mathematical programming (integer programming, stochastic programming). Recent trends also include semantic modeling and optimization via meta-heuristics, e.g. genetic algorithms.

The result of service composition is a composite service execution graph, which is, similarly to functionalities graph, a DAG, but its nodes are constituted not by functionalities but by services stored in the repositories. In fact, the composite service execution graph should have the same structure as the functionalities graph, and its nodes should contain services with the functionality described in the corresponding nodes of the functionality graph.

Composite Services Structure Composition

In general, service composition process consists of two steps. First, the required functionalities and their interactions, i.e., control and data flow, are identified. Second, for the functionalities graph set, the appropriate candidate services are selected, and as a result a composite service execution plan is established. Candidate services are services that share a specific functionality defined in the functionality graph; in other words, they fulfill the functional requirement but are different with respect to the non-functional properties. Only one of the candidate services for each functionality within the functionality graph will be selected to replace this functionality in the composite service execution plan. The service selection is accomplished based on non-functional properties which, generally speaking, can be referred to as Quality of Service (QoS) attributes: service execution time, service execution cost, service availability, service execution success rate, service reputation, or service execution frequency.

Depending on the structure of the functionality graph, the non-functional parameters of the composite service will be calculated differently with respect to the functional parameter type: i.e., the cost of a composite service is a simple sum of all services that build it, whereas service execution time is a sum of the execution times whenever services are executed one after another, and a maximum of those composite service parts where services were executed in parallel. For each of the non-functional service parameters, the user places a constraint, determining the sufficient conditions for the service selection. This defines the goal to select such services (from sets of candidate services providing all of the requested functionalities) that their aggregated non-functional parameters—constituting the non-functional composite service parameters—meet the user-defined constraints.

Service selection consists of several stages:

- Inverted service indexing—inverted indexing of services, using a domain ontology describing the input and output service parameters annotated with service metadata, allowed for faster selection of valid services, independent of the semantic filter being used (although semantic filters had to be redesigned to fit the approach).
- Semantic filtering—a method for service and requirement comparison that goes through a service requirement consisting of a set of required input and output parameters. Depending on the type of semantic filter and distance measure used, the method finds concepts that are identical to the requirements—or semantically close. Note that the definition of semantic closeness depends on the measure being used.
- Preliminary candidate service sorting—at this stage, a set of candidate services that fulfill the functional requirements (selected by semantic filters) can be sorted using various semantic filters. More strict or computation-heavy semantic filters could be applied to the candidate services sets to establish how well the functionalities were fulfilled. Some non-functional requirements could be checked locally at this stage as well.
- Composite service execution plan generation—this stage is defined as an optimization task from which only one service for each functionality is selected from sets of candidate services corresponding to nodes in the functionality graph. This occurs together with other selected services making up the composite service execution plan. Additionally, the inputs and outputs of the services included in the plan must be pair-wise compatible. The algorithm for this stage is the key component to service selection and is described in more detail below.

The service execution plan generation algorithm (Fig. 5) consists of several steps. First, based on a greedy approach, an initial plan is generated. If the initial plan does not meet the sufficient conditions then its neighbor plan is generated, then the neighbor plan is verified against sufficient conditions. If the neighbor plan is also not a valid solution, then it is put onto a taboo list and the next neighbor plan is generated. This is repeated until a valid solution is generated or the algorithm reaches the limit of iterations, which is a standard approach applied for service composition [61]. The taboo list ensures that the same composite service execution plan will not be generated repeatedly. Also, at certain times a simulated annealing mechanism accepts a neighbor plan that does not meet the requirements so that new solution spaces could be explored. This is more likely to happen due to the more iterations performed by the algorithm. The algorithm presented above is a general heuristic hybrid procedure based on simulated annealing and tabu search. It triggers more specialized methods such as the generation of an initial or neighbor service execution plan. Those methods are described below in more detail.

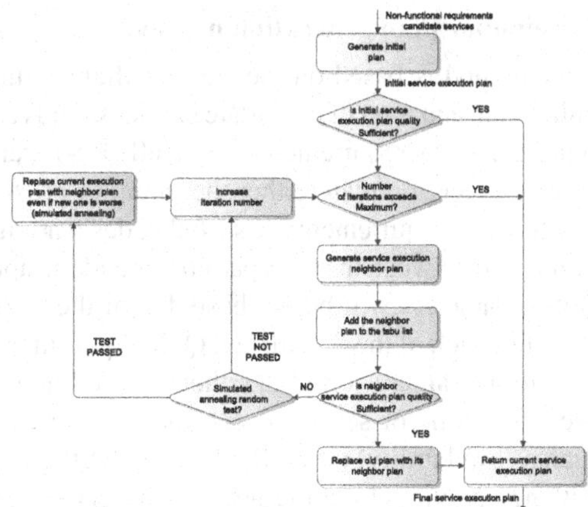

Fig. 5 Algorithm for service execution plan generation

1. Method for generating composite service initial execution plan

The method is a greedy approach in that each functionality requested by the user selects one service from a set of candidates with the same functionality. Within each set of candidate services, a service with the highest fitness score is selected. This method does not take into account the structure of the functionalities graph but locally selects the best candidate service with a requested functionality.

2. Method for generating local fitness function

The method establishes a local fitness function score according to the weighted QoS parameters of each service. The fitness function gives an aggregated, weighted value to all QoS parameters of a service, which is normalized relative to other QoS values of all candidate services with the functionality.

3. Method for calculating the composite service execution plan quality

The method calculates the quality of the composite service based on its aggregated QoS parameters compared to the user's non-functional requirements. The QoS parameters are calculated based on the composite service execution plan in the aggregation method below.

4. QoS parameters aggregation method for composite service quality estimation

The method establishes a vector of QoS parameters for any composite service provided with a composite service execution plan and with QoS parameters for each of atomic services in that plan. The method analyzes the functionalities graph structure: the sequence and parallel structures within the service execution plan and generate a vector of formulas for calculating the QoS parameters for a given functionalities graph. For any composite service execution plan with the structure of the functionalities graph, the generated formulas can be used to quickly estimate its QoS parameters. This is true because the generated formula is based on the functionalities graph and not any particular composite service execution plan.

5. Neighbor plan generation method

This method is based on the simulated annealing algorithm and, provided with an initial plan and a set of candidate services, finds a neighbor plan that is closer to satisfying the user's requirements. First, all QoS parameters from the previous composite service execution plan (either the initial or previous neighbor plan) that did not satisfy the user requirement are sorted in descending order according to the normalized difference between the QoS parameter and the appropriate requirement. The selected QoS parameters will be the basis for further modification of the service execution plan. For each of the dissatisfied QoS requirements, the atomic services in the service execution plan are sorted in either descending (if those QoS parameters are service execution time or service execution cost) or ascending (for other QoS parameters such as availability) order. Replace the first two atomic services from that sorted list with new services from the appropriate service candidate sets (services with the same functionality) in such a manner that a newly-selected service is not on the taboo list for that functionality; it is a service with the highest fitness function score among the other service candidates for that functionality. Then, add the selected services to the taboo lists for the appropriate functionalities. The composite service execution plan resulting in those service replacements is a neighbor plan to the previous one with respect to the minor changes made towards replacing the services with the most dissatisfying QoS parameters.

As the domain knowledge is represented in the form of ontology and the services are semantically described (in terms of associating their inputs, outputs, and functionalities with the concepts—or sets of concepts from the ontology), the composition is a process of transforming user requirements defined in the SLA into a fully-defined composite service that fulfills them. In the first stage of the composition process, user requirements are analyzed and assembled into a single structure that represents a final composite service structure by a graph where nodes contain user requirements and edges connect those requirements, determining the order in which a final composite service will be executed. Then, with the use of knowledge engineering, this structure is enhanced so it defines an execution scenario for particular atomic services in a composite service. In this stage, no requirement can be left disconnected from the other requirements. However, the services may differ in non-functional properties (as in execution time or cost), so in the last stage for each functional requirement, a single atomic service is selected so that all services that build a composite service jointly fulfill the non-functional requirements.

The key feature of the PlaTel's service composer is that it is a composite service itself; each of the components of the composition scheme (SSDL GUI, service matching, and service selection) are performed by the services. PlaTel's composition service is described in SSDL and may be freely reconfigured if there are more or better-suited services that could be used for a specific problem. This approach opens vast possibilities—each of the stages could be performed using different strategies. We may use different semantic discovery methods when searching for services that fulfill each of the requirements (or propose different distance measures for concepts in the ontology), or various optimization techniques could be used to produce the composite service, fulfilling the non-functional requirements, not to mention that a variety of non-functional parameters could be requested and optimized by this.

Equipped with the above mechanisms, our framework allows composition service designers to incorporate various approaches, test them, and deploy—in a form of service—enabled composition tools well-suited to different domains and problems. The tool consists of two main parts: one is the front end of the application which allows a business client to define his/her domain by connecting to external service and knowledge repositories (here: ontologies) that will be used to construct composite services; the second part is the layer of specialized services which is extensible and can provide a variety of services but mainly composition services.

The PlaTel service composer execution plan may be passed directly to the Execution Engine, but in this work we want to refine our approach even further and introduce the communication (ICT) services, incorporating them to our framework. This is a unique feature of the PlaTel which will be described in detail in the following section.

Composite Services Provisioning

Recently, enterprises tend to collaborate and integrate their business cores in Web markets in order to maximize both clients' satisfaction and their own profits [45]. Therefore, there is an increasing need to develop methods for combining existing services together to enrich functionalities and decrease execution costs. Hence, the old-fashioned way of developing composite services, i.e., human-based service designing, becomes very insufficient due to the enormous number of available services and a lack of computational resources. That is why the automatic or semi-automatic service composition method is a crucial element of any system that implements the service-oriented architecture (SOA) [45]. In general, service composition process consists of two steps [57]. First, the required functionalities and their interactions, i.e., control and data flow, are identified. Second, the execution plan is established, i.e., for sets of functionalities, an appropriate service version is selected. The service selection is accomplished basing on non-functional properties which, in general, can be referred to as Quality-of-Service (QoS) attributes.

In most of the proposed approaches for service composition, the domain services are considered apart from information and communication services (ICT services), which are understood as physical means of supporting the execution of domain services [9, 60, 63]. However, in order to allow Service-Oriented Architecture [45] to be applied properly, most of requirements need to be mapped to ICT services. Otherwise, it is ambiguous how the domain services should be executed physically. For example, a requirement for a building monitoring service consists of the following operations: signal acquisition, coding, sending, and decoding. All of these services are, in fact, the ICT services. Thus, the ICT service mapping is a process of mapping requirements to a combination of atomic ICT services which facilitates the delivery of requested domain-specific functionalities. In both approaches, the emphasis is put on identifying control and data flow. Generally speaking, functionalities of a composite service are represented as a directed acyclic graph (DAG) [19, 65, 67] in which arcs denote interactions among functionalities and nodes determine functionalities. As a result, the service composition task can be seen as a determination of a DAG structure, e.g., using AI methods [48] and then as a selection of atomic or composite services that provide the required functionalities. The service selection is formulated as an optimization task in which an objective function reflects QoS attributes aggregation, e.g., response time, price, and constraints concerning structural dependencies in DAG [57].

There are several known approaches to QoS optimization such as graph-based methods [18, 19, 65, 58] and mathematical programming (integer programming [67], stochastic programming. Recent trends also include semantic modeling and optimization via meta-heuristics, e.g., genetic algorithms. However, the problem of ICT service mapping is usually omitted, and lately only a few papers have pointed out the necessity of considering ICT services in the process of service composition [17, 64]. Hitherto, due to the authors' knowledge, the problem of ICT service mapping has not been fully stated, and hence, no solution has been proposed.

The service composition in the telecommunications domain requires one to aggregate, update, maintain, and process knowledge about telecommunication services in order to analyze, plan, and manage—in an optimal way—telecommunication resources. Moreover, the frequent behavioral patterns in business processes must be discovered for the purpose of optimization of telecommunication service delivery scenarios. The workflow of the service composition and mapping process is presented in Fig. 6.

To compose the telecommunication services, we use three main modules in our framework:

- *Domain service composition*—based on the request a domain service is composed;
- *Service mapping*—the domain scenario is mapped with ICT services
- *Service composition*—in a telecommunication scenario, ICT services are composed and a final execution plan is returned.

and repositories:

- domain service ontology;

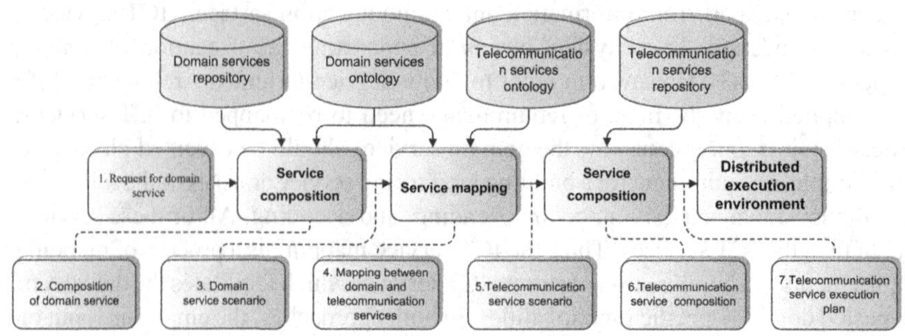

Fig. 6 Workflow of the services composition and mapping process

- ICT service ontology;
- domain service repository;
- ICT service repository.

Ontologies are representations for a seamless cooperation and knowledge flow between business processes and formal service descriptions in service-oriented systems. They play a crucial role in service mapping in translating domain service scenarios to ICT service scenarios.

Furthermore, two types of services could be distinguished, namely, computational and communication services. Ontologies, which are implemented in Protégé, are based on SIMS Ontology developed within a project of the Sixth Framework Programme [64]. SIMS has been extended for wireline telecommunication systems, QoS concepts, and parameters.

Generally speaking, computational services are used to process data streams, while communication services facilitate data transmission and end-to-end QoS guarantees. In the proposed system, there are multiple atomic or composite computational services, e.g.:

- signal merging and splitting;
- data compression;
- signal encoding/decoding;
- video/audio stream encoding;
- change detection in data stream.

The task is to compose composite ICT services which can deliver functionalities given in a business process describing all the required use-case scenarios for a particular business domain. The resulting composite ICT services are composed based on the available atomic ICT services (computational and communication) provided by ICT service providers and/or operators. Additionally, the above composite services must conform to the users' non-functional requirements concerning—among others—quality, security, and availability of the assumed business process.

The process of the composite ICT service composition consists of four stages: transformation of the business process into a set of composite services, composite domain service composition, domain and ICT service mapping, and composite ICT service optimization (see Fig. 6).

In the first stage, the business process provided by the user is transformed into a set of composite services which fully cover all use-case scenarios defined by the business process [59]. Each composite service is defined by the functionalities (atomic domain services) that need to be performed in order to fulfill the corresponding use-case scenario. Moreover, a composite service description includes non-functional requirements (e.g.: QoS, security, etc.) concerning the entire composite service and/or some of its functionalities. Each composite service is defined by a service-level agreement (SLA) including—among others—functionalities and values of non-functional parameters required to be fulfilled in the corresponding use-case scenario.

Next, each composite service defined by functional and non-functional requirements is composed of the available atomic domain services. The result of this stage is a set of atomic domain services and the order in which they must be executed in order to provide the user with the requested functionality and conforming to the non functional requirements. Note that in order to deliver the requested composite functionality, some of the atomic functionalities may have to be delivered in parallel and/or in sequence. Therefore, the order in which the atomic domain services are generally performed is defined as a directed graph, the nodes of which represent atomic domain services, and the edges represent the precedence relationship between them. Such a graph, called a composite service scenario, is passed to the next stage of the service composition.

Each node of the composite service scenario graph represents a single atomic domain service that provides a well-defined functionality. In order for this functionality to be delivered, certain ICT (i.e. communication and/or computational) services must be executed. The task of the third (services mapping) stage of service composition is to substitute each of the nodes (atomic domain services) of the composite service scenario graph with a sub-graph representing the execution order of the atomic ICT services necessary to deliver the requested atomic domain functionality.

As an example, let us consider the atomic domain functionality consisting of sending a billing information record for a certain time period being the response to a query to a CRM application. This atomic domain functionality is delivered by execution of three atomic ICT services (see Fig. 7a), i.e.: transmitting a query to the database (communication service), execution of the query in a database (computational service), and returning the results to the user (communication service). Moreover, if a user requires a secure data transfer, two additional computational services (cipher and decipher) for each communication service should be executed (see Fig. 7b).

Note, that in general, two communication services, taking part in the delivery of the above exemplary atomic domain service, are functionally and non-functionally different. The first one consists in sending small message (e.g. SQL query) while the other may require the transmission of a large volume of data.

The result of the service mapping stage is a composite ICT service scenario graph where the nodes represent atomic ICT services and the edges define the precedence relationship between the services.

There are multiple approaches to accomplishing the mapping stage. First of all, the mapping for each atomic business service may be predefined, which is a simple

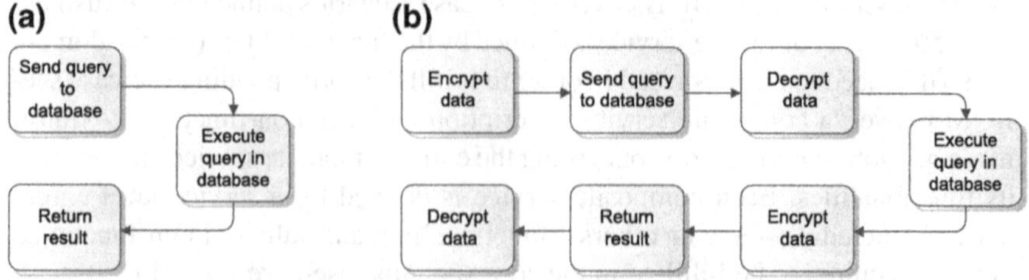

Fig. 7 Exemplary mapping of a billing information acquisition atomic business service: **a** unsecure and **b** secure data transmission

and efficient—but not flexible—solution. On the other hand, the mapping task can be treated as a special case of a composition task. In this case, some known methods (e.g. semantic composition or GraphFold algorithm [50]) can be used. Another approach is to use pre-defined mapping rules and a reasoning engine to find the best possible mapping conforming to the non-functional requirements. The last composition stage is an optimization stage. The task performed at this point consists in finding such a version of the atomic ICT service available within the system: that non-functional requirements for the entire composite service are met. Additionally, certain composite service optimization tasks may be performed at this stage. These tasks may include, among others, finding the least expensive composite ICT service that satisfies the non-functional requirements or finding such a composition for which the usage of the ICT resources is balanced. The first optimization task may be stated by the user in the composite service request, while the latter is rather the concern of the ICT resource operator or service provider. The result of this stage, which is passed to the execution environment, is a composite service execution plan defining exactly which versions of ICT services will be used to fulfill each particular composite service request.

The first stage of service composition has to be accomplished during the ICT infrastructure design phase for each business process. The following stages of the services composition and delivery may be performed once along with the first stage during the ICT system development (off-line composition), or they can be performed on-demand each time when a certain use-case scenario is launched (on-line composition).

In the first approach (i.e. off-line composition), all composite services are predefined in the design phase and only these services can be executed during the lifetime of the system. This approach is very conservative; it does not make it possible to change or add new composite services and is very inefficient in terms of resource consumption, service reusability, and composition flexibility. It allows, however, for a very quick response to new incoming composite services requests, since the time-consuming process of composite service composition is performed only once.

In the second approach (i.e. online composition), on the other hand, each new incoming service request is handled separately, namely the best composite service conforming with the non-functional requirements is composed, taking into account the current state of the system (e.g. usage of communication and computational resources). This approach fully draws from the service-oriented architecture paradigm and allows for efficient resource usage, service reusability, and on demand composition of new services, scalability and flexibility. Additionally, it is possible to store pre-defined composite service execution plans or execution plans resulting from the compositions performed during the system's lifetime and apply them to the incoming composite service request when the response time is critical.

The composition process of an exemplary composite ICT service is presented in Fig. 8. The considered example concerns handling a physical alert in a business process of laboratory monitoring.

The business process of monitoring a laboratory consists of four use cases (see Fig. 8a): two concerning the monitoring of environmental security and two concerning the monitoring of physical security. Each use case is handled by the execution of

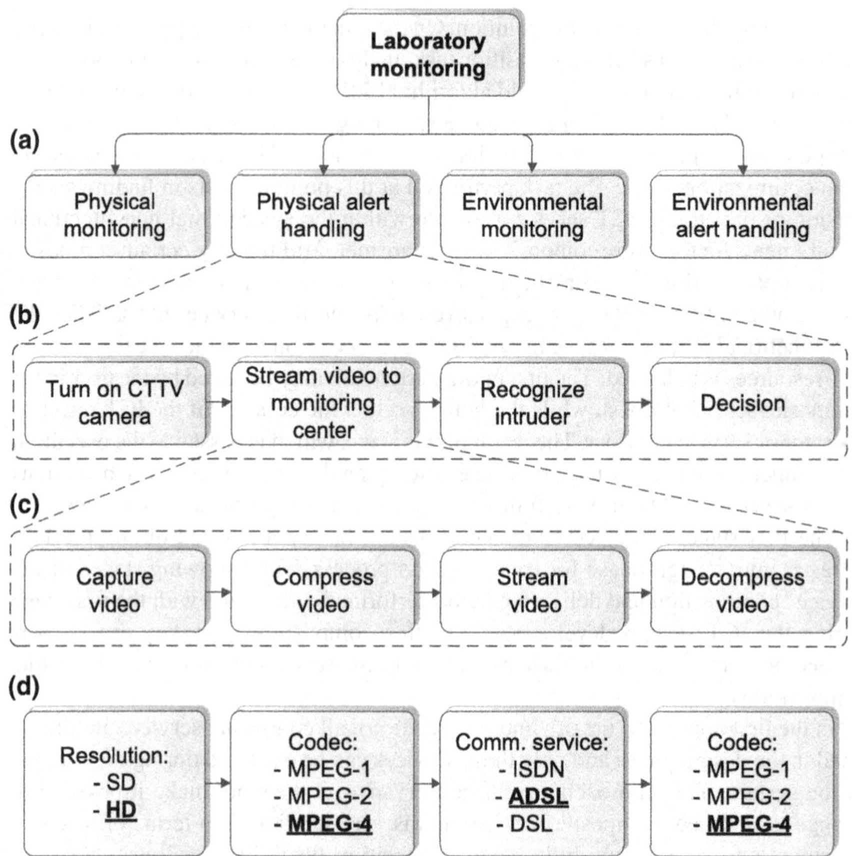

Fig. 8 Exemplary composition of telecommunication service

a separate composite domain service that has been predefined in a business process. In Fig. 8b, an exemplary composite domain service for a use-case scenario concerning the handling of a physical alert is presented. The definition of such a service is a result of the first stage of composite service composition, consisting of translating the functional and non-functional requirements of a business sub-process for physical alert handling into a sequence of atomic domain tasks which have to be performed to meet these requirements.

In the next stage (see Fig. 8c), each atomic domain service is mapped to a composite ICT service which guarantees the fulfillment of all the required functionalities. In this case, the atomic domain service stream video to monitoring center consists of a sequence of four atomic ICT services: video capturing, video compression, transmission of the compressed video stream, and video decompression.

In the last (optimization) stage (see Fig. 8d), particular versions of atomic ICT services are chosen. The choice is made based on given non-functional requirements and service availability. In this example, the streamed video is passed to the face recognition service. Therefore, a high definition (HD) video quality is chosen. Moreover, the MPEG-4 codec with the highest compression rate was chosen for video encoding. Since HD video encoded with MPEG-4 results in 16 Mbps one-way data stream, Asymmetric DSL (ADSL) with 16 Mbps upstream guarantees was chosen for a communication service. Finally, for the video decompression, the same codec as for compression must be used.

Composition of ICT Services

In this section, we indicate the problem of ICT service mapping in the process of service composition and propose the solution. The presented solution applies the idea of decision tables [31] as an ICT service mapping tool. We utilize the decision table to associate ICT services with the corresponding requirements. The decision table for ICT service mapping consists of two columns. In the first column, a requirement is given and in the second one, a DAG of ICT services. In the service mapping step, the service selection through QoS optimization is performed. Our approach allows one to perform the execution plans on physical machines seamlessly because ICT services describe the computational and communication aspects of SOC. Furthermore, it is worth mentioning that the presented approach is domain independent and can be applied in any business area that requires utilization of ICT resources.

In general, the task of service mapping can be stated as follows. Find functionalities that are matched with ICT services and select such ICT services that perform matched functionalities in the best way, according to QoS attributes.

Thus, the mapping problem can be divided into three sub-issues. First, the mapping functionalities must be found. Usually, not all functionalities need to be mapped to ICT services. Second, the found functionalities must be matched with available ICT services. Third, one complex ICT service is selected.

Hence, the ICT service mapping can be seen as a service composition but performed only for the functionalities that are identified to be mapped. However, the crucial step is how to determine which functionalities are supposed to be mapped.

In order to solve this issue, the following two assumptions are made. First, functionalities and services are described in the same language in order to be identified by the same keys. For example, functionalities, as well as ICT services, are described by XML-based language and by the same tags that are used for their content comparison. Second, the mapping between functionality and a complex ICT service is given by an expert. For such assumptions we propose to apply decision tables for ICT service mapping.

It is worth noting that the problem of ICT service mapping is strongly connected with the service composition process. Namely, functionalities for matching are checked whether to be mapped to ICT services, or to be matched directly with services in a service repository. Therefore, the augmented service composition process can be represented by a sequence diagram presented in Fig. 9. The ICT service mapping consists of two steps. In the first step, all functionalities which are supposed

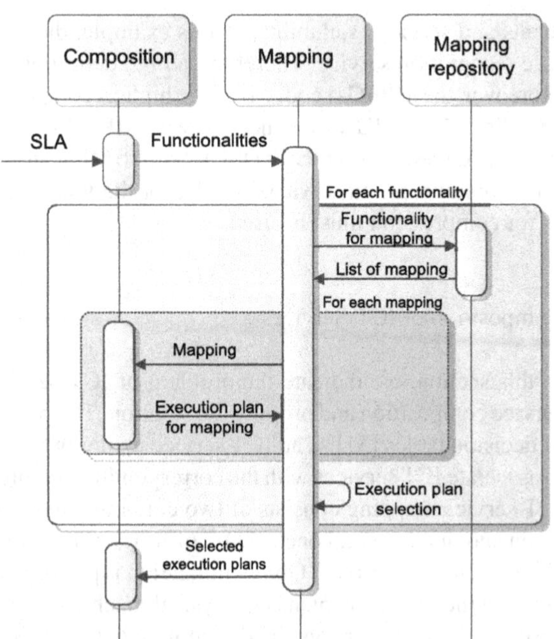

Fig. 9 Sequence diagram of interactions between ICT service mapping and service composition. First, for all functionalities, only those are found which are supposed to be mapped. Next, for each mapping an execution plan is prepared. At the end, for each of the functionalities, an execution plan is chosen

to be mapped are found (the light grey rectangle in Fig. 9). Next, for each of the functionalities identified to be mapped, a list of mappings is retrieved from a repository of mappings. For each mapping in the list, an execution plan is proposed by a service composition method according to given quality criterion (the dark rectangle in Fig. 9). At the end of the entire mapping process, one execution plan is selected according to QoS attributes.

There are two critical steps in the above process for ICT services mapping. The first step is how the repository is constructed; secondly, how the final execution plan is selected. Here, we focus only on the first problem as the second one can be solved using one of the methods known in the literature [48].

We take advantage of the earlier-mentioned assumptions, i.e., the mappings are given a priori and functionalities and mappings can be matched using universal keys. That is why, in this work, we decided to apply decision tables to represent the repository of mappings.

The decision table consists of two columns, i.e., the first column defines the condition and the second one—the decision. In the considered application, the condition is the functionality description, e.g., XML-based description, and the decision is the mapping. It is worth stressing that usually there are several possible mappings for one function. An exemplary decision table is presented in Fig. 10.

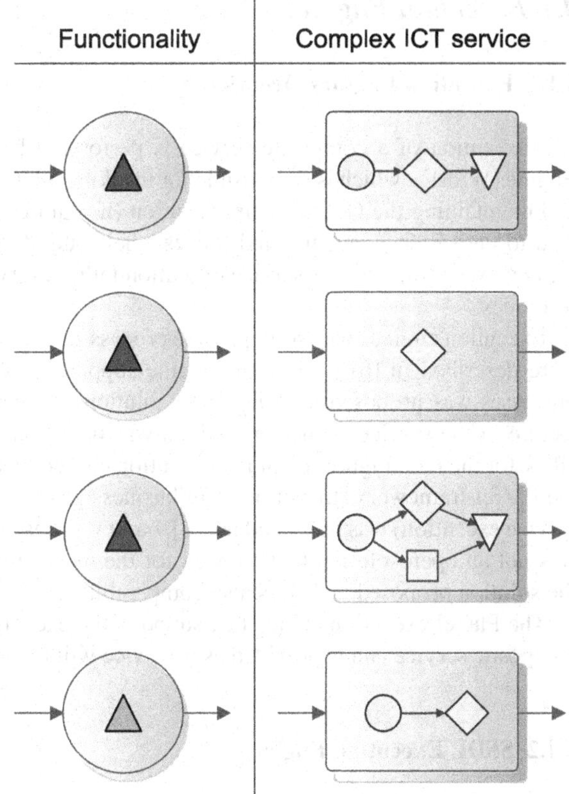

Fig. 10 An exemplary repository as a decision table. *Dark* and *light gray triangles* in circles represent functionalities. *Circles*, *diamonds*, and *triangles* denote atomic ICT services and rectangles—complex ICT services. Notice that one function (*dark gray triangle*) corresponds to several complex ICT services

4 Composite Services Execution

Distribution of the request inside the Service Oriented Architecture yields a number of issues which involve virtualization management related to the management of virtual machines that represent the services offered. Automation of such a process requires well-defined handling requirements, policies, and a description of service features to be properly matched with the request and to be integrated with the service execution engine functionalities [PSS25]. As it was pointed out in this work, PlaTel is making such a capability real. Increasing the level of service awareness already at the lower (hardware and virtual resources) level can make the system more flexible and more adaptable to the changes in customers' behavior.

4.1 Execution Engines

4.1.1 Execution Engines Overview

The execution of a composite service is performed by a Web Service Execution Engine (WSEE), which is responsible for invoking all its component atomic services and maintaining the QoS of their execution (if required). It needs to pass the input data to the services, acquire and process their output, and provide the procedures for any exceptions such as service execution failure, service blocking, or the lack of resources.

Execution engines which support the process-driven composite service execution were described in [61], while the engine support for execution of BPEL-defined processes was proposed in [20]. These solutions, however, assumed a fixed architecture of the service framework, which was based on certain devices and did not allow for the reconfiguration of the execution engine. An approach similar to that of the PlaTel framework (transition from business process definition to the composite service execution) was presented in [23], but it was based on the REST approach and was not an open solution that allowed for the use of multiple service repositories, the solution proposed in [23] is also comparable.

The PlaTel execution engine that supports the execution of any SSDL-described composite service and is provided as-a-service will be presented in the next section.

4.1.2 SSDL Execution Engine

After the composition and mapping of the composite service are completed, the service is executed by the SSDL execution engine. It assumes an 'engine-as-a-service approach' and offers an execution engine as a configurable composite service, which acts as a SSDL language interpreter and supports dynamic interpretation of service description files, service configuration, and execution control. The SSDL execution engine is implemented as a lightweight virtual machine that may be duplicated and migrated upon the decision taken at the SOA infrastructure level. The core feature distinguishing the SSDL execution engine from other execution engines is its focus on composition mechanism and, together with the expressive nature of SSDL language it interprets, its ability to configure its own behavior.

The main role of the execution engine is to govern the process of services execution (and dynamic interpretation of requirements leading to service execution); however the configuration of the engine is in fact also something to distinguish it among other engines. We designed its core to be composed of several elements, quite like a composite service consists of atomic services. One of the elements is service interpretation and execution but we can add other elements or, as we call them, phases that introduce instructions like pre-processing, validation, full composition, and other actions such as a classification-based transformation of requirements to the composite service based on previous compositions. Actions focused on the

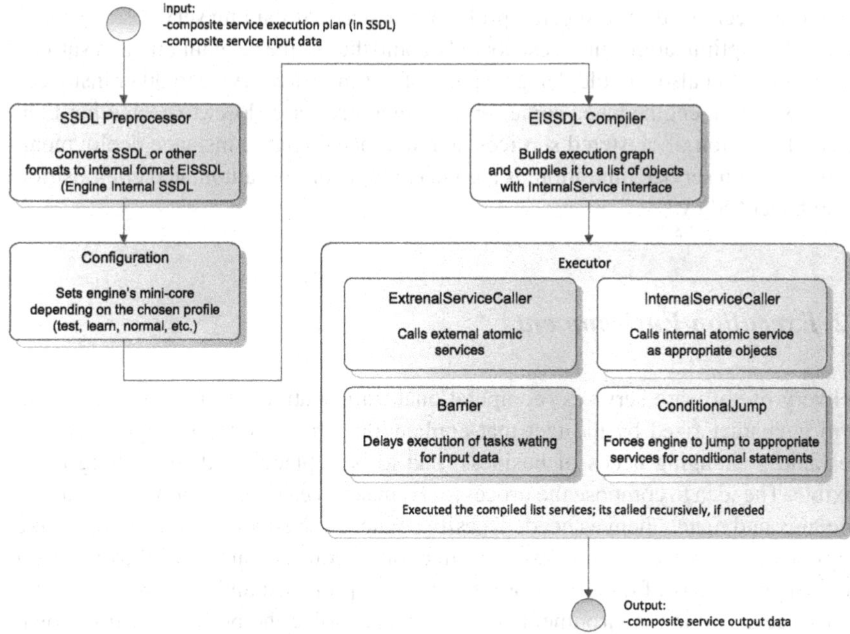

Fig. 11 The Execution engine scheme and basic functionalities

SSDL performed in the preprocessing phase could be internal methods or, in the configuration-driven approach, composite web services. Such a situation is depicted in Fig. 11, where we distinguish a two-phase engine with its configuration managed in an external web application. Based on this fundamental application model, other phases could be added to further personalize the behavior and expand the engine's functionalities. Such phases could be final reporting (to specific applications via web services or a general report hub) or, as shown in Fig. 11, an even more complex approach where a pre-processor to the internal format is added and various work profiles are supported. This could ultimately lead to further flexibility of the engine, supporting various composite service definitions as modules and various behaviors, depending on identified situations (like debug, testing, return from halt, learning, normal, speed-low, etc.).

To better exploit the benefits of cloud computing, a presented execution engine was designed to work as a service. It was implemented in Java and supports multi-threading to process multiple service execution plans at the same time. It is capable of executing composite services defined in SSDL, but it can also automatically generate web interfaces to composite services stored inside the engine, broadcasting as those services. In this mode, the Execution Engine can emulate any composite service defined in SSDL. As a result of multi-threading, a single execution engine could act as multiple services or, in extreme cases, multiple engines could act as atomic services—hiding composite services behind a layer of abstraction. As a service, the proposed execution engine could be replaced by a different execution engine, e.g. more specialized for a specific problem, such as a streaming services execution engine. For optimization purposes, not only could the engine be maintained in various localizations but also it could delegate parts of a composite service to other instances of the execution engine when some services in a series are closer to each other than others. Detection of clustered services and automated engine instance deployment, combined with service migration, is a good example of how automated composition could benefit SOA.

4.2 Execution Environment

Delivery of software services (computational, information, etc.) in the traditional form is characterized by the fact that applications are not open enough to follow the rapidly changing needs of business, had to be replaced with something more flexible. The idea to compose the processes from services publicly or privately available, mix and match them as needed, easily connect to business partners, seems like the best way to solve it. This involves functional requirements as well as assuring the changing needs of quality of processing. The proposed and discussed composite service execution environment is an attempt to solve the problem of traditional software's lack of flexibility. The environment takes into account the entire range of SOA advantages.

4.2.1 Infrastructure as a Service

Flexible and efficient service delivery is a very important and difficult challenge in SOA-based systems. The quality of service delivery is also a very complex problem in such systems.

The new concept of effective and quality-aware infrastructures built in accordance with the SOA paradigm relies on the idea of a virtual service delivery system. The service delivery system consists of two layers (Fig. 12). The client of System calls a service visible to him/her at the Virtual Service Layer (VSL). The VSL virtualizes real services available at given service execution systems that are hidden from the client's point of view. Moreover, service execution systems can also be virtualized by service providers in a real execution system. As a result, the client deals with a virtual service that is mapped to a virtual server. Additionally, a virtual service can be instantiated in multiple virtual servers.

The composite service is put together with basic atomic services, i.e. the ones that cannot be partitioned afterward. The atomic services can be localized in different execution systems and executed by service instances running in these systems. The set of execution systems, atomic services, composite services, and service instances are constant for some time, but generally over longer periods, can change. The implementation of the Virtual Service Layer and Virtual Resource Layer must support the dynamic updates of these entities.

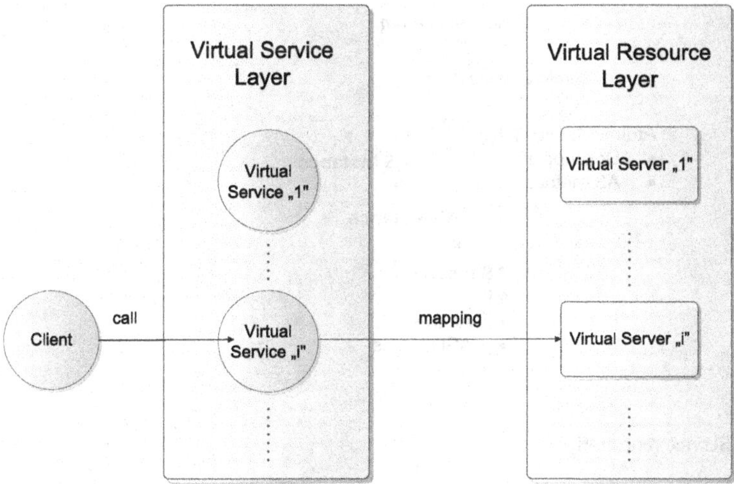

Fig. 12 Two-layer service delivery system

Both layers form the Service Delivery and Request Distribution System (SDRDS). The Virtual Service Layer is implemented by a network service broker (hereinafter—'Broker') with an integrated Virtual Service and Resource Manager (VSRM) module that supports the virtualization of network services. Broker B maintains information about resources and handles atomic services client requests (Fig. 13). It hides request processing in the delivery system.

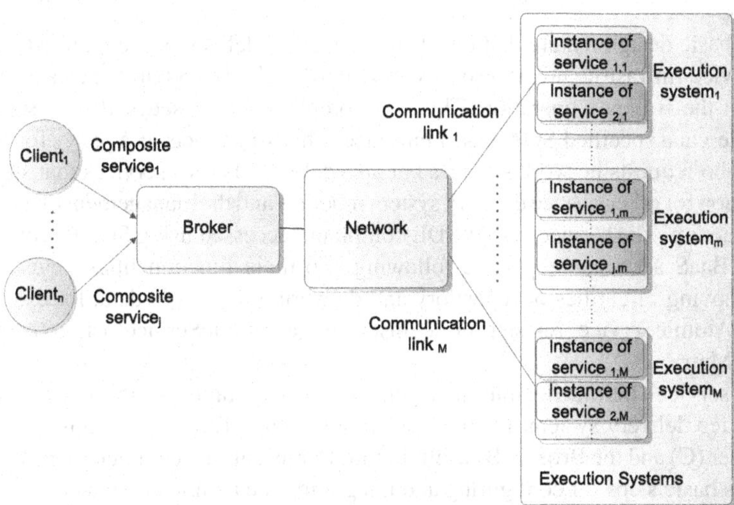

Fig. 13 General infrastructure of Service Delivery and Request Distribution System

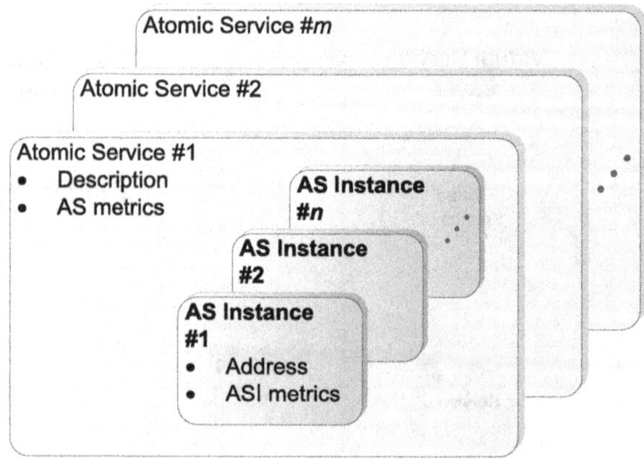

Fig. 14 Service repositories

From the Broker's point of view, every component requesting services is a client—both the ordinary client application and any system component (e.g. service composer). The Broker acts as service delivery component, while in fact distributes requests for services to known service processing resources. Distribution is performed on the basis of formulated service quality criteria and values of service instance non-functional parameters (metrics), which are monitored.

The Broker maintains a repository of all known services and components that support them (Fig. 14), i.e.: the set of atomic services (AS) available (and seen) for clients, the set of service instances localized at given execution systems, and the set of execution systems. It also maintains information about the available composite services.

The basic design feature that results in the service delivery system infrastructure as a service (infrastructure as a service, for short) is the BaaS (Broker as a Service) mode of the Broker's operation. This is accomplished by a set of Broker services available via a specified SOA-based interface. The Broker acts as a proxy for every client who requests network services but also delivers its own services that support integration for other service delivery system modules and the management of a virtual infrastructure. The services are WSDL compliant, accessed using SOAP protocol.

The BaaS services supply the following two main functionalities: registration and removing of entities of repository and delivering registered data, for example: registerAtomicService, registerServiceInstance, getAtomicServiceList, getAtomicServiceMetrics.

The services permit flexible management and integration of the main units of the service delivery system, i.e. the execution systems (ES), the composite service composer (C) and the Broker (B) itself. Figure 15 presents the complete scenario that includes basic steps for configuring and integrating the mentioned units.

After service installation in the execution system, the basic procedure that makes the network atomic service available includes three steps: registration of the execution

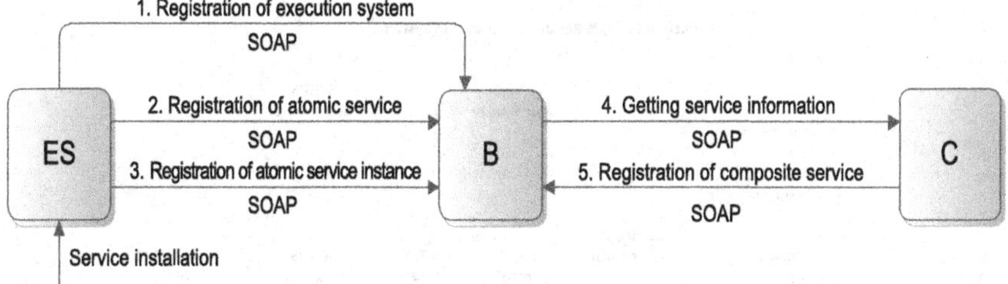

Fig. 15 Basic steps of virtual service layer instantiation

system by the Broker (if it is not registered yet), registration of atomic service (if it is already installed, it will be signaled to register), and registration of service instances. The order must follow the numeration of the presented steps.

If the client (e.g. service composer who wants to compose a composite service) does not know the available services, this information can be obtained with the use of a proper Broker service (Step 4). After that (or without the previous step, if the client is sure it knows the available atomic services), it can register a composite service for monitoring purposes (Step 5). The client can also get detailed information (e.g. monitored metrics) about each service, instance, or composite service.

Service Delivery and Request Distribution System Architecture

The architecture of the Service Delivery & Request Distribution System (SDRDS) is presented in Fig. 16. It is composed of a number of independent modules that provide separate functionalities and interact with each other using specified interfaces. The main components are:

- SDRDS-Broker—handles user requests and distributes them to proper instances of virtualized execution resources. Decision making is based on specified criteria of the request distribution: in turn, based on non-functional requirements. It also performs internal requests to coordinate the operation of the system components as well as obtain some necessary information.
- SDRDS-Facade—separates the rest of components, supports communication with them, and collects necessary information for the Broker. It also provides special services for the Broker to test the current state of a processing environment (e.g. characteristics of communication links).
- SDRDS-Controller—manages all components behind the Facade. It is responsible for control processing according to the capabilities of the environment and its current state. It can also route the requests to the services (capsules) independently, taking into account computational resource utilization and performing decisions to start/stop another instance of service.
- SDRDS-Virtualizer—offers access to hypervisor commands. Uses libvirt to execute commands that give the project independence from a particular hypervisor.

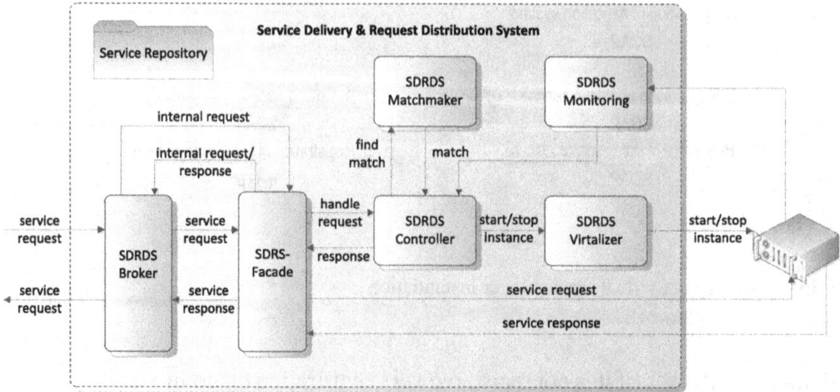

Fig. 16 Service delivery and request distribution system architecture

- SDRDS-Monitoring—collects information about particular physical servers as well as running virtual instances.
- SDRDS-Matchmaker—module responsible for properly matching the requirements of the request with capabilities of the environment and current state of it.

The instances of a given atomic service are functionally the same and differ only in the values of non-functional parameters such as completion time of execution or availability. SDRDS modules interact using two interfaces. Internal communication is XML-RPC based for components behind the Facade and uses SOAP messages for communication between the SDRDS-Broker and SDRDS-Facade. This allows for the flexibly to manage the distributed computational resources as well. Interaction with external components is based on the SDRDS-Broker services with SOAP messages. SDRDS delivers for clients' access to network services hidden behind it. Clients see network services at the Broker localization and do not know that services may be multiplied. From the client's point of view, services are described at the Service Repository using the WSDL standard and are accessible using standard SOAP calls. However, the Broker distinguishes individual execution resources, simultaneously coordinates activities with the SDRDS-Controller, and passes requests on to others. Each and every request is redirected to a proper instance based on the values of non-functional parameters of the requested service. Proper instance of service is either found from the ones that are working and available or the new one is started to serve the request. Such an approach gives the possibility to serve clients requests and manage resource virtualization and utilization automatically with minimal manual interaction.

The resource management is considered in the context of effective resource utilization and the delivery of network services with QoS. The effective resource utilization concerns performance issues as well as cost-related issues that embrace the financial cost of resource employment. The performance issues can concern such problems as load control (particularly load balancing) and the selection of suitable resources to support given tasks. The cost of use of resources can be supervised by the control of their consumption with respect to time and capacity. Resource management is related to the quality of services. In fact, most of decisions related to resource utilization concern the quality of services. Both issues demand online monitoring as well as resource usage and resource conditions as values of the parameters of delivered services. Finally, we propose to distinguish two main policies for managing resources in the Service Delivery & Request Distribution System:

- control of resource allocation, taking into account supporting the quality of delivered services with the use of service virtualization, requesting distribution algorithms, and monitoring service execution as well as the execution environment;
- management of resource instances, capacity, and behavior with the use of resource virtualization.

Resource allocation can be performed in two primary areas: service process execution resources and communication resources. Communication resources management can be achieved on different levels with the use of different methods. With full control of communication links on a data link and network layer level within the entire network being used, the control can include establishing the route, new route creation, control of communication link saturation, link throughput, etc. The last two examples also refer to full control of dedicated links. However, on the Internet this is usually not the case. After all, the resource allocation at the application layer level, performed as resource selection (allocation of given set of resource), can be applied. The last case is considered in this work. The resource allocation that takes into account the quality of services is performed with a request distribution policy by the Broker.

Communication Resources Management

Communication resources management is performed as a resource selection for every service request, i.e. allocation of a given communication link for the given request. At the same time, it works in tandem with the selection (optional instantiation) of computational resources. All is accomplished on the basis of the non-functional parameters of network service processing.

The request for atomic services can be labeled with non-functional attributes corresponding to the quality requirements of the request, such as response time, reliability, price, etc. In [52], a service selection was proposed in a way that maximizes user satisfaction expressed as utility functions over QoS attributes. However, this approach assumes that all service parameters are constant. In practice, it may be distinguished by two kinds of parameters: static—constant over a long period of time (e.g. price), and dynamic—variable in a short period of time (e.g. response time of service or throughput of service data transfer). For this case, we propose request control with a dynamic estimation of values of parameters characterizing network service instances. We distinguish two separate cases for considered approaches to request control:

- request distribution at the atomic service level only;
- request distribution at the composite service level.

The instance of a given atomic service is characterized by the values of non-functional parameters. The quality-aware service delivery can be performed using different approaches according to the constitution of requirements and the infrastructure of the delivery system. Taking into account communication infrastructure, we distinguish two general approaches:

- satisfying quality requirements for the request by direct control of consumption of resource (communication link) capacity for dedicated links from the Broker to execution systems;
- examining quality requirements for the request versus the values of non-functional parameters of service instances and the selection of proper service instances for service execution.

These two approaches essentially do the same thing: assign (or reject) the request to the communication link; however, they differ in the point of interest. The first one focuses on communication link parameters. The second focuses on network service parameters.

The Broker performs resource allocation decisions based on the actual values of the service instance parameters and chosen distribution strategy. Making control at the atomic service level, the problem of service request distribution can be expressed using criterion function on service instance non-functional parameters.

The criterion function can be expressed as different combination of more than one parameter and a formulated condition to fulfill the function requirements can be different as well, depending on the system's design assumptions. It may be formulated as guaranteeing the exact values of service parameters or, in this particular case, finding the extreme of the function and use the best effort, strategy, etc. An example of such distribution criterion can be minimizing the sum of data transfer time for given instance of atomic service and the completion time of its execution. It is a single request response time minimization. Assuming one communication link to one execution system, the selection of service instances is also the selection of communication link and selection of a virtual server.

To support distribution using dynamic parameters, its current values must be used. For the best effort, a distribution strategy following estimations of values of current parameters can be used:

- best last—the best (minimum) last monitored value of estimated parameter for all instances of service;
- best mean—the best (minimum) average value in k-window for all instances of service;
- best max—the best (minimum) maximum value of in estimated parameter for all instances of service;
- best prediction—the forecasted value derived using any forecasting method, e.g. artificial intelligence approach.

Considering something other than a best effort strategy, the same estimations can be used.

For most important dynamic parameter, request response time, the Broker monitors and registers transfer time and execution time of each request and response. These times are used to estimate response time.

For forecasting transfer times and execution times of the request for atomic service served by the given instance, the fuzzy-neural controller [15] is used. Each service instance and each communication link for each service instance are modeled as a two stage fuzzy-neural network with two inputs characterizing the current conditions of modeled entity and the estimated time as an output.

For execution time, the model refers to the execution server. The two inputs are processor loads and the number of requests actually served. For communication links, the two parameters can be chosen from the following:

- the estimated (using time series analysis) latency (the exact TCP Connect time is proposed);
- the estimated link load derived from a regular test of the link—Şdownload of test portion of data;
- the aggregated intense of transfer of all requests directed to a given link.

Making control on the composite service level is more complex. The composite service scenario (composite service plan) can be described as a directed acyclic graph (see example in Fig. 17 where nodes represent atomic services to be executed and the edges define dependencies between atomic services—Şprecedence relationships).

For the composite service scenario graph of the service requested at the given moment, there is an equivalent graph in which nodes are sets of instances of atomic

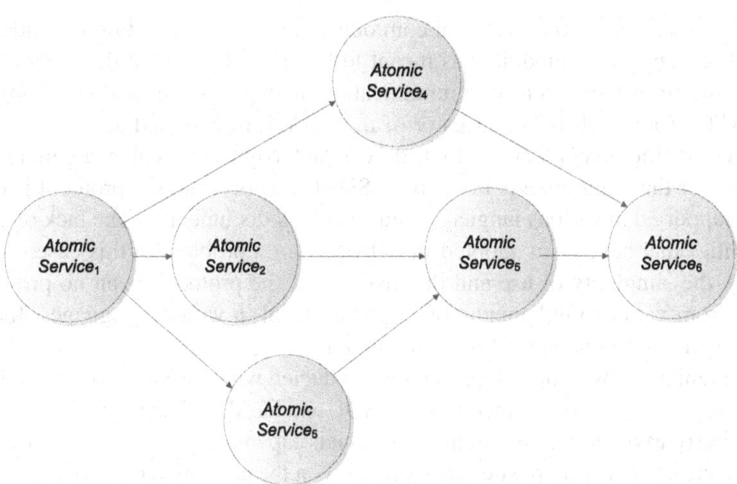

Fig. 17 Composite service scenario graph—an example

services that constitute composite services. This equivalent graph defines different execution graphs (i.e. graphs that refer to concrete instances of atomic services).

The request for composite service at the given moment can be characterized with requirements on values of the non-functional service parameters. Every non-functional parameter of composite service is a function on the non-functional parameters of atomic service instances composing this service. The form of the function depends on the kind of parameter as well as the structure of the composite service implementation graph.

Knowing all non-functional parameters of instances, it is possible to perform different ways to fulfill the composite service non-functional requirements. If we consider assuring requirements for each request for composite service separately, the goal of service request distribution can be expressed with criterion functioning similarly; as for atomic services, however, it is formulated with use of a set of parameter vectors instead of a set of parameters and is more complex.

The task to select such instances so that the criterion function is satisfied (minimized in this particular case) is equivalent to the determination of the composite service proper execution graph. The determination of proper instances is performed in two ways. The first approach is to specify all instances the moment the request for composite services is received. The dynamic approach assumes that every execution of a subsequent atomic service that makes-up the composite one runs the procedure for the determination of the service instance that fulfills the specified criterion. The SDRDS broker supports queries for the composite services execution plan registered with the Broker.

Computational Resources Management

The main driver of SDRDS' internal communication was the condition of independence. Each and every module was meant to be operable without the others. This leads to the implementation of communication among internal modules using the XML-RPC protocol. It is the ancestor of the SOAP protocol and as such has limited functionalities in comparison to its descendant. Nonetheless, they are more than sufficient for the requirements implied by SDRDS. Moreover, the protocol itself is widely supported in various languages plus it is well documented. The lack of some possibilities that have been adopted in SOAP is not a problem in this case; on the contrary, the simplicity of use and the maturity of the protocol mean no problems coming from various implementation of the protocol in various languages, further promoting the independency of internal modules.

Communication with hosted services is conducted with the SOAP protocol. Each of these services is also described using WSDL and has the advantage for the execution engine (part of the service composer) that it is capable of dynamically generating requests and creating composite services based on the existing atomic ones.

The WSDL document describing the service is linked with a URL address that allows one to call among a number of other resources available on the Web. Yet, due to the dynamic nature of the entire process—in this case, during the registration phase—it is modified in such a way that the URL is no longer connected with the original URL (e.g. a local address if a document was generated in this manner), but rather with the Broker that serves as a proxy to access the service. This hides the true service situation from its customer who does not need (or even should not) know how many instances there are of a given service and to which one his/her request would be directed.

The independence of all of the modules actually led to some overhead in the registration process which is now conducted in three phases. Those were realized automatically and come from the logical consequence of the actions. First, a new service is being registered in the SDRDS. It is responsible for managing virtual machines; thus, nothing can carry on without informing this system that there is a new service that it will incorporate into this process. During that step, some information concerning the execution details (e.g. required RAM, number of processors) is given. The next step consists of the registration of the service in the SDRDS-Broker and the beginning of the first instance of the service. During this phase, the provided WSDL is modified by the SDRDS-Broker and made available publicly as if it was his/her own. The final phase is connected with the registration of the service in the external service repository that belongs to the service composer. The last of the steps requires some additional parameters that cannot be found in the WSDL to be provided. Those are non-functional parameters to be used later by the service composer in the composition process.

Hiding the real address of the service has deep meaning in a system like this. It allows SDRDS to manage the services and their instances in an independent manner for the execution engine in which only the address is important and that it remains constant. On the other hand, information about services might be used by the SDRDS-Broker. It is a semi-internal module of SDRDS that presents the access point to the services. The SDRDS-Broker possesses information about the instances and the realization times regarding them, and can therefore suggest where the request should be directed. Its role in this process is basically limited to forwarding the request and appending the suggestion of where should it go. There is also an alternative path for processing the request that takes into consideration more conditions, including all that might have been already enclosed in the request. In that case, the SDRDS-Matchmaker module is used to find the best match.

The SDRDS-Broker module is an element that remains outside the others—in the way that is publicly available for service clients to access services. It provides an API called Broker as a Service (BaaS) that, in basic terms, is used to register and execute the services. The Broker is not, however, only a mere proxy between the service client and the provider but also influences the internal processing of the message by giving extra information regarding the proposed instance in order to process a particular request with the goal of reducing time spent. The most important functionalities of the Broker are: Execution system registration, Service registration in broker, Service instance registration with Broker.

During the composite service registration, SDRDS predicts the service execution time and the necessary resources needed for service execution. The time estimation made is the SDRDS based on earlier realizations of the requests. This allows for the prediction of the realization time either simply by averaging the times or by an approximation based on just the last requests. In both instances, we first need to register the composite service with the Broker to be able to trace not only a single request, but also a composite one as a whole. The time of the composite service realization is computed based on historical data from previous runs of the services. Those times are simply totaled to give the overall time of the composite service realization. In the case of time, it is important to note that the algorithm performing the computation assumes a full sequential of the composite service. It does not take into account the fact that some of the computation could be done in parallel, for example, and thus reducing the total time of the composite service.

The estimation of the resources required by the composite service execution is not a trivial task. Currently, it is done with few simplifications. It is assumed that the request is realized with the worst case scenario, in which all of the resources assigned to the service will be used while performing the request. Used resources are estimated assuming this negative scenario. For each atomic service involved in the composite service, it is assumed that all of the resources available for the instance will be utilized.

4.3 Execution Engine-Execution Environment Interaction

Communication between the Execution Engine and Execution Environment was done with independence in mind. In any event, neither of those two will depend on the other. Furthermore, the Execution Environment will preserve transparency for other modules, including the Service Composer and Execution Engine. On the other hand, composition and execution tasks performed by the Execution Engine will not depend on the location of a particular service or how it is hosted but solely on the provided functional and nonfunctional description. This leads to the usage of generic registration mechanisms offered by the Service Composer in order to communicate the presence of new services to this module.

The only dependency here comes from the requirement to obtain the address of the repository and to adjust communication to the interface offered accordingly. Registration is offered again using the XML-RPC protocol. Registration is simply a call to the AddService method that takes a list of information: name, class, address, inputs, outputs, location, cost, response time, and availability. After registration, the service can be immediately used by the Service Composer in its processes.

Realization of atomic service means usage of the SOAP protocol just like it is used when there is no interaction between the provider and client. Although there are no extensions to the regular message, they might be introduced to fine tune the realization of a particular request.

Extra parameters, which are then processed by the Execution Environment, are described with more detail further. Usage of those parameters is one of available paths of processing a request. The latter is connected with the Broker who can extend the

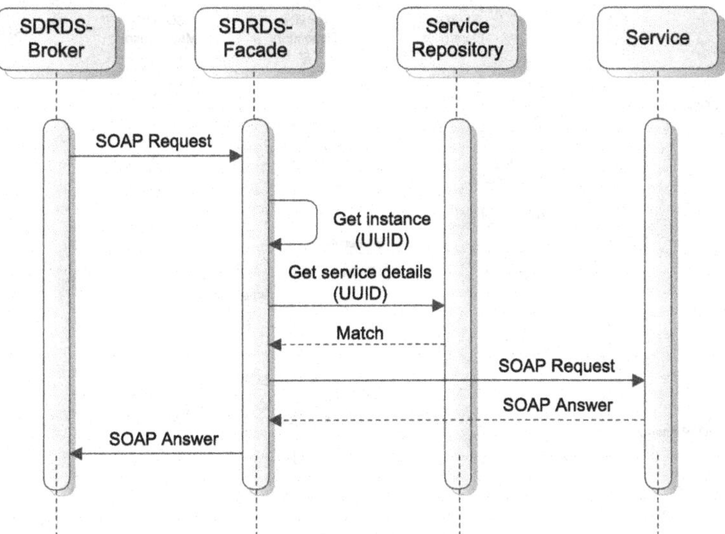

Fig. 18 Sequence diagram—request processing with broker suggestion

message with explicit information about the instance to the one who will process the incoming request.

The Broker, as it was already mentioned before, gathers some data on request performance in particular instances. Having those data, the Broker is able to suggest which of the available instances shall be used to perform a certain request. Such information is simply placed in a SOAP message header, and then the modified message goes to the SDRDS-Facade module. It is then the responsibility of the SDRDS-Facade form to get the details about the instance (IP address) and forward the message there. The answer is then sent back to the Broker and back to the client (in this case to the Service Composer). Figure 18 presents the simplified scenario for service request processing.

In cases where the Broker does not send the information about the instance upon which the request shall be directed, a longer path would be used to find the destination address. The SDRDS-Matchmaker module would then be fed the data describing the request and system status. First of all, the request is described by the name of the service: secondly, by all of the constraints passed in the header of the SOAP message. It is specified in XML and then translated into the form understood by the SDRDS-Matchmaker. Next, the matching process has two steps. In the first step, matching is done only among the already running instances, and only in the case when no matching instance is found the second step happens. The second step means checking the capabilities of the servers in order to begin a new instance that would handle the request according to the request submitted. The entire process is described in Figs. 19 and 20. Figure 19 shows a request processing when there is an active service instance

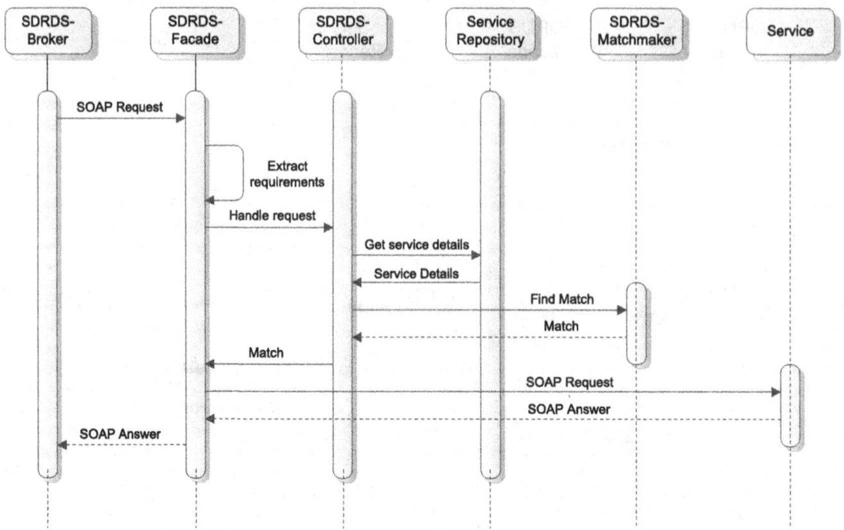

Fig. 19 Sequence diagram—request processing with successful matchmaking

that satisfied the request requirements and Fig. 20 when there is a need to activate the new service instance to fulfill the request requirements.

In some cases, access to internal data is required to work with an execution environment. Those rare situations are exhibited to the outside world using SOAP services—just like all of the services made publicly available. Each of the exposed actions extends the address of the system with certain keywords. The list of available actions is as follows:

- system_load—access to information about the number of requests finished and being processed;
- link_ load—allows measurement of the communication with SDRDS and as a response returns the requested amount of data;
- echo—simply sends back received message;
- monitoring—access to system logs; access can be filtered out using following parameters:
  - system.requests_ age—age of requests to be filtered;
  - system.requests_ since—exact point in time when the requests will be given;
  - system.requests_ until—complementary to the parameters above and making it easy to mark the range in time from which requests will be given;
  - system.requests_ to—provides name of the service to get the logs from.

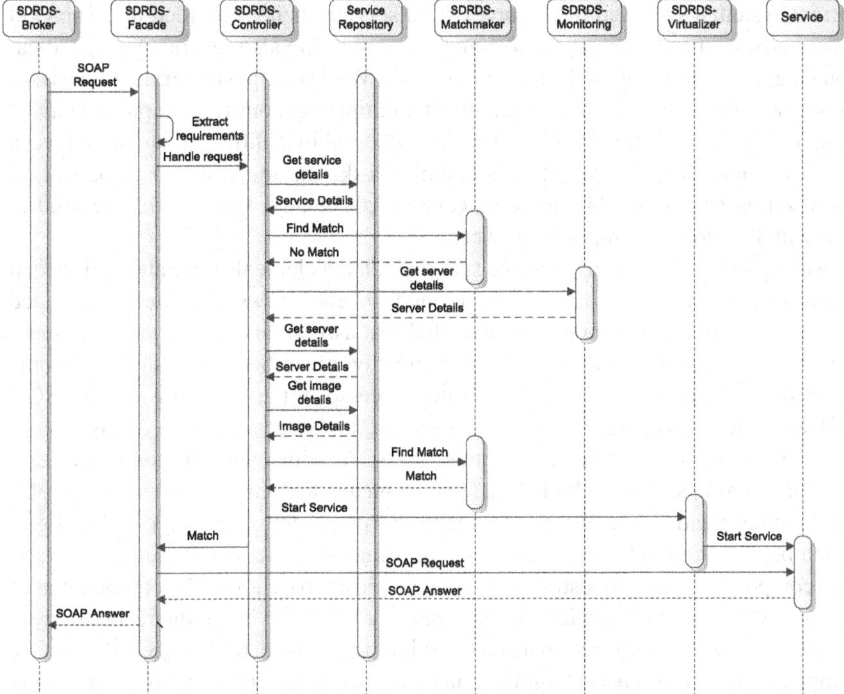

Fig. 20 Sequence diagram—request processing with unsuccessful matchmaking

5 Composite Services Validation

This section focuses on the method of composite service validation in SOA-based systems. The proposed method to measure the security level of services uses a layered approach derived from the SOA governance model. Based on the security level estimates of the atomic services, an information fusion scheme—utilizing Subjective Logic formalism—is proposed to evaluate the security level of a composite service.

The composition of Web services, which has been presented in detail in earlier sections of this chapter, allows the building of composite workflows and applications on top of the SOA model [62]. The main aim of service composition is providing a new functionality to the end-users but apart from this, composition should be also done with respect to the Quality of Service (QoS) requirements. Preserving the non-functional requirements defined by QoS parameters is a key factor in which its importance rapidly grows in distributed environments where composite services are generated from atomic components [49]

In order to fulfill the non-functional requirements, the Service Level Agreements (SLAs) are defined or negotiated between service providers and their clients. The SLAs must be precise and unambiguous; they should also be obeyed regardless of the current system's state and the complexity of services being provided [4]. The QoS characteristics may be expressed in many ways; there are also several QoS specification languages including QML and WSLA [28]. While composite services consist of atomic services which are often available in multiple versions (e.g. implementations using different programming languages or supported by a different hardware layer), the assessment of QoS properties is complex task, and there appear to be several optimization problems [24]. In some cases, sophisticated frameworks are used to deal with the QoS of composite services.

Among other QoS proprieties, the security-related elements have attained special importance as the number of service-oriented systems grows. The literature related to SOA security focuses on problems with threat assessment, techniques, and functions for authentication, encryption, or verification of services [13], and [42]. Some other works focus on a high level modeling processes for engineering a secure SOA [54] with trust modeling, identity management, and access control. Many studies focus on secure software design practices for SOA, with a special interest in architectural or engineering methodologies as the means to create secure services [30]. On the other hand, one could notice a rapid development of semantic technologies which benefit from ontologies and semantic similarity assessment methods for service composition and orchestration. The example approaches to the composition of semantically-described services are presented in [30]. The key issue resulting from the use of semantics is the automation of service management tasks, especially service composition. In most cases, building composite services converts into a satisfaction problem regarding constraint—there can be many candidates (atomic services) for building blocks of a composite service (process), and it is necessary to select the optimal execution plan. So, the constraints defined in the optimization process are selected to satisfy the chosen QoS parameters. There are many approaches to the task of QoS parameter assessment for composite services, such as service-oriented systems. However, all they miss the general approach to the composite service security level estimation problem. The proposed solution in this section fills this gap by defining a formal model of composite security level service evaluation.

5.1 SOA Security Governance

The general flexibility of SOA based systems results in problems related to information assurance. These types of problems, which are quite natural in most distributed environments that have been designed to provide inter-operability and availability, become more challenging in a context of a service-oriented paradigm. SOA-based systems assume a scenario with multiple service consumers interacting with multiple providers, all potentially without prior human confirmation as to the validity of either the consumer request or the provider response. According to application of knowledge-based systems supporting the process of a security level evaluation, it is possible to reduce direct human intervention or interpretation to seek or acquire "valid" information about the QoS of composite and atomic services. This idea is based on the automatic processing of well-defined resource description standards and the high availability of such resources.

The specific problems of service-oriented systems validation related to the system environment are as follows:

- **Identity management**—the identity needs to be decoupled from the services. All entities like users, services, etc. may have their own identities, and these identities need to be properly identified so that appropriate security controls can be applied.
- **Seamless connection to other organizations on a real-time basis**—organizations are providing services which should be integrated during the composition process. To enable execution of composite services where atomic services come from different providers, the connection between organizations must be seamless.
- **Proper security controls management**—there is a need to ensure that, for composite services, proper security controls are enacted for each atomic service and when services are used in combination.
- **Security management sovereignty**—SOA needs to manage security across a range of systems and services that are implemented in a diverse mix of new and old technologies.
- **Protection of data in transit and at rest**—execution of composite services is associated with complex data flow. Because of the weakest link security paradigm, this data flow must be secured at each intermediate stage.
- **Compliance with a growing set of corporate, industry, and regulatory standards**

These security issues arise in the context of the following SOA characteristic features:

- high heterogeneity of resources (both software and hardware),
- distribution of resources,
- the intensity and variety of communication,
- the different level of resources and tasks granularity,
- resource sharing,
- competition for resources,
- the dynamics of resources,
- the business level requirements related to security.

There are several standards and mechanisms that have been elaborated to provide and to maintain the high quality level of SOA-based systems. The basic solutions address the problems of confidentiality and integrity of data processed by an SOA-based system. Because of the network context of SOA and the multi-level security risk related to the layers of ISO/OSI network model, there are several solutions that offer data protection mechanisms at the corresponding level to each network layer.

In this context, the most commonly used and described are standards and protocols coming from the application layer that are maintained by the OASIS consortium. These solutions have been worked out to support the development of Web services and so also SOA-based systems. The other type of protection methods, mechanisms,

Table 3 The validation requirements for SOA functional layers (selection)

SOA layer	Evaluate/verify/test
Business process	Policy consistency, policy completeness, trust management, identity management
Composite service	Identification of the composite services, authentication of the composite services, management of security of the composite services
Service description	Description completeness, availability, identification, authentication
Service communication	Confidentiality, integrity non-repudiation
Service execution	Availability, protection from attacks

and protocols are common for all network applications and can be used also in SOA-based systems as well in any other type of software.

As the SOA-based system can be defined by its five functional layers, the corresponding definition of SOA security requirements for validation of composite services should address the specific problems within each of the defined layers.

A selection of elements defining security requirements for the five layers of SOA systems has been presented in Table 3. The complete list can be found in [30].

The security level estimated with respect to the requirements, which have been enumerated in the Table 3, are typically constant. For example, the implementation of an atomic service does not change during operation, so its security level of authentication, trust management methods, etc. remains constant. However, a practical impact of a particular service on a security level of a composite service may vary in time. The fluctuations of the security level are caused by how the services are executed in a system. The security level should especially reflect the states of identified attacks against the services. Only then can the validation procedure bring up-to-date information about the security level of services that can be used during the composition phase.

Both elements required by composite services validation: the evaluation of service security properties for SOA layers and the service security level evaluation accordingly to an observed service execution history have been presented below.

5.2 Service Security Properties for SOA Layers

The proposed framework for composite service validation benefits from the model of trust and opinions and is called Subjective Logic [27]. According to this model, the security level of the particular layer of the SOA system is expressed using Subjective Logic opinions and Subjective Logic operators. The aim is to assign quantitative values to the measured security of the SOA architectural levels. Opinions which have been used are defined in Subjective Logic as tuples composed of three values

<belief, disbelief, uncertainty>, for which the following interpretation has been defined:

- The ***Belief*** component of the Subjective Logic's opinion which reflects trust in the security level of the service under consideration. Its value close to one literally means that one perceives the service as secure.
- The ***Disbelief*** component reflects the opinion that a service is not secure.
- The ***Uncertainty*** component is to express that the knowledge is partial or incomplete and the security assessment does not give a definite result.

5.2.1 Business Processes Layer

The Business Processes Layer defines the requirements for SOA-based systems that come directly from the business process definition. The most important elements pertaining to this layer and their influence for the process of services composition and execution have been discussed in detail in Sect. 2. One of the most important results obtained was a conclusion that the standard approaches to business modeling, such as ARIS or BPMN, have received too little expressiveness for services composition (e.g., they lack the ability to define the functional and non-functional requirements of composite service). As the security-related requirements are typically non-functional, this also limits the application of these approaches to the task of security level evaluation or generally to composite services validation. Introduced in Sect. 2, PEPC notation overcomes this problem and offers the well-defined framework that can be successfully used to express security-related requirements.

For example, PEPC can be used to define the service requirements for security policy, trust and identity management, etc. After that, during the service validation process, subjective opinions describing the level of how user requirements have been satisfied by the particular service implementation are calculated.

The Subjective Logic opinion about the Business Processes Layer is calculated as the combination of opinions regarding policy consistency, policy completeness, trust management, and identity management.

5.2.2 Composite Service Layer

For proper validation of composite services, two elements are required. The first one is a result of the corresponding atomic service validation. The second is the execution plan defining the composite service. The first element is performed using information provided by the Service Description Layer (described below). The results of the service validation are represented as Subjective Logic opinions calculated in analogous way as it has been described for the Business Process layer (a combination of opinions about the service authentication method, service authorization, identity management, etc.). The second element, which is required for the composite services validation, is provided in the form of an SSDL file. The SSDL file contains a description of a composite service which has been defined by a set of nodes connected together in a graph-like structure. The details of the proposed method for enabling composite services validation has been presented in the next sections.

5.2.3 Service Description Layer

The Service Description Layer in the context of the services validation process focuses on analyzing the functionality offered by services (e.g., about authentication, input and output provided by services, etc.). Corresponding information used during the Subjective Logic opinion calculation can be obtained from appropriate WSDL files, for example [30]. However, Sect. 3 of this chapter presents the fact that the WSDL file can describe only a single Web Service and its functionality. Additionally, this description offers a limited level of precision. The approach to service composition presented in earlier sections benefits also from the knowledge representation in a form of ontology. Ontology contains many of the domain-specific parameters which may be defined according to the current needs; therefore, during the service composition process, several specific domain ontologies (e.g. business service ontology, telecommunication service ontology, etc.) have been used. In this context, domain ontology has become an important source of information that is used while calculating Subjective Logic opinions about services. For example, at this layer, the required information about what is implemented is provided by service authentication or authorization methods.

5.2.4 Service Communication Layer

This layer addresses elements that are responsible for communication between services. This is a critical element when we consider the composite services (services must be able to send/receive data to/from each other), and this is also an important element to be taken into account during the composite service validation process. The calculation of an appropriate Subjective Logic opinion about security at the Service Communication Layer is a result of the analysis of several elements influencing communication security, e.g. communication confidentiality, integrity, or authentication.

5.2.5 Service Execution

The proposed method for a service's security level estimation using a service execution history actually benefits from an anomaly detection approach toward finding attacks and security breaches in ICT systems. The main assumption of this method is that attacks and security breaches are related to abnormal system behavior. While the typical behavior of most ICT systems shows some periodicity, e.g. number of processes executed during the daytime or data transferred, it can be assumed that also the data rate, volume, etc. characterizing the services executed will show this type of dependency. Therefore, the anomaly detection approach should be also valid to estimate not only the system but also the security level service. The purpose of anomaly detection is to identify so-called anomalous states of the systems. To achieve this, a tuple of characteristic values describing the system's state is observed. Next, using pre-defined methods and criteria, the current values of that tuple are classified into one of the two following classes: normal behavior and anomalous behavior. An observation might be an anomaly in a given context, but an identical data instance (in terms of behavioral attributes) could be considered normal in a different context. Contextual anomalies have been most commonly explored in time-series data. Other popular methods frequently implemented to detect anomalies in ICT systems are outlying detection methods which can be divided between univariate methods and multivariate methods or parametric (statistical) methods and nonparametric methods that are model-free.

The analysis performed to detect an anomalous state of services can be done in four steps. The first step is a feature selection. A selection algorithm chooses among a set of candidate feature functions. The second step is parameter estimation, and in a general approach, a new feature function is added to a model and the weights of all feature functions are updated. Collected historical data about service behavior (features values) are compared with the current feature value. After this step the model of service behavior is constructed. This model is built by iterating Steps 1 and 2 until a pre-defined stopping criterion is met. Finally, the last step is anomaly detection. A large difference between the distribution of the selected feature value and a baseline distribution derived from training data indicates an anomaly.

The anomaly detection methods usually classify the system's state into one of two well-known classes: anomalous and normal. However, there are also approaches to anomaly detection which uses fuzzy methods, allowing for the presentation of a wider scale of a system's states. These fuzzy methods make it possible to define a more fine-grained representation of the service's security level. In the proposed method of evaluation in this section, opinions about service security expressed in Subjective Logic correspond to the fuzzy approaches to anomaly detection.

Subjective Logic opinion about a service's security level is calculated using the service execution history. Data describing the history of the execution is provided by a monitoring subsystem, which is an integral part of the composite service execution environment described in detail in Sect. 4. This data allows for the calculation of the values of Subjective Logic opinions in a form of three previously-defined values < belief, disbelief, uncertainty>. The disbelieve value, in estimation of the security level for anomaly detection, is proportional to the value returned by a distance function, which calculates the distance between the current observation and the typical observation (e.g. in some implementations it may be a simple difference between the current value and the mean value of the observed parameter).

The disbelieve value varies from 0 to 1. When the detected anomaly is relatively small (near the average value), the value of the disbelief component of the Subjective Logic opinion will be near 0. While the high deviation from the earlier observed values has been observed, the disbelief value is equal to 1.

The uncertainty value in the Subjective Logic opinion about security level of the monitored service is related to the variance level of the observed parameter values. This means that when the observed parameter has a small variance, the uncertainty level is equal or near to 0. While the variance increases, the uncertainty level of the Subjective Logic opinion also increases until it reaches its maximum value.

5.3 Aggregation of Security Properties for Composite Services

A composite service execution plan is a graph that (apart from indicating which atomic services are used in a composite service) represents the structure of a composite service. Execution plans consist of serial, AND-parallel, or XOR-parallel execution substructures. After having assessed the security level of atomic services, the security of a composite service may be evaluated. In general, it is assumed that the security of a serial execution plan—where there is a chain of services executed one-by-one, is defined by the security of the "weakest point" (or "link") in the chain (indicating the possibility that any security breach may occur during the execution of composite service). The same thing concerns the AND-parallel execution plan. In the case of the XOR-parallel plan—where it is not determined which service will be executed, the security level of all parallel connections should be taken into account.

In the presented approach, an execution plan of composite service is defined by the SSDL file. After obtaining the graph structure from a corresponding SSDL file, a security-level estimate for each connection between services connected by one edge is defined in the form of a Subjective Logic opinion. According to the—œweakest point—paradigm, if for example, a non-secure service is followed by a set of services executed in parallel, then its insecurity will greatly influence each following security level estimate. To find a security level estimate of a composite service, all opinions about links between services are aggregated independently to find the— **IJ**weakest link-**İ** of the composite service—that is, a connection with the lowest security estimate. According to this, a composite service *general security level estimate* has been defined as a probability expectation that projects security level estimates onto a one-dimensional probability space.

As the information used for security level evaluation comes from the sources which are fundamentally different by origin and technical nature, there has been a second strategy also defined for producing opinions about composite services—the *layer-dependent security level estimate*. The *Layer-dependent security estimate* has been based on layer-specific opinions which have been described in Sect. 5.2 In this case, the rules for generating opinions about the security of the execution plan apply separately to opinions concerning specific SOA layers. Calculation of a composite service *layer-dependent security estimate* for Service Communication Layer is performed in the following steps:

- calculate the Subjective Logic opinion about the Service Communication Layer security level for each service constituting the composite service of a given execution plan;
- calculate the composite service security estimate for the Service Communication layer using the formula that defines a probability expectation that projects security level estimates onto a one-dimensional probability space (in the similar way that it is done while calculating *the general security level estimate*).

The third possible estimation of composite service security level also takes into account the results of that anomaly detection process as it has been described in Sect. 5.2.5. The estimation of this type of composite security level is performed in the following steps:

- calculate service security estimate for each service from composite service execution plan for each SOA layer, also including security estimates concerning anomalies detected during service execution (as it has been defined by the *layer dependent security estimate*),
- aggregate those estimates into one security level estimate of a particular service using the Subjective Logic consensus operator form ;
- calculate *the general security level estimate* as it has been defined earlier in this section.

This flexible strategy allows the generation of three types of opinions about a composite service security level with required granularity and the depth of security analysis of a composite service in service-oriented systems.

As it has been presented in the Table 3, a security layer interacts with all other SOA layers. The first type of this interaction is data acquisition for SOA system validation at the corresponding level of description (Business Policy, Composite Service, Service, etc.) The main idea of SOA-based system validation, with respect to elements defined by particular layers, has been presented in Sects. 5.2.1 through 5.2.5. While the atomic service security level has been evaluated, the composite service can be validated using Subjective Logic operators, the execution plan defined by the SSDL file, and Subjective Logic opinions about atomic services (Sect. 5.3) On the other hand, the obtained results of the services validation are available for corresponding elements from specific SOA system layers. This type of feedback can be used to reorganize or reconfigure SOA system to fit predefined security requirements. For example, the results of the Business Policy validation may be used to redefine some policy elements that are responsible for the final security level of an SOA-based system. Similarly, the information provided by the service validation module may be used by a module responsible for service composition to select services, providing the required security level of composited service.

6 User Interface for Services

The problem of design and implementation of the personalized and automatically generated user interface (UI) for services in SOA-oriented systems is quite new. We can try to find similar solutions in the area of user interface adaptation that have been addressed in many papers before, for example [53] and [46]. The user interface is usually adapted to the personal user's needs and/or environment settings.

The SOA paradigm was proposed because of the discontent of the software community over interoperability, reusability, and other issues of traditional software development best practices and standards [6]. The service-oriented best practices are shaped to deliver strategic solutions for enterprise or other types of organizations that are focused on overcoming different shortcomings appearing at the tactical level. The SOA framework delivers a general guide on how to conceptualize, analyze, design, and structure organizational service-oriented assets [6]. It is believed that one of the most important features of the SOA-based systems is the composition of services. The composite service is built from at least two different reusable services. The consequence of this feature is that the particular user interface differs depending on the composition (sequence of services) and the context of use of the particular service.

In today's implementations of the SOA-based system, the user interface for each composite service and for each user role is designed and then implemented manually or semi-automatically using CASE-tools, however then the precise information flow has to be specified. A consequence of this is considerable time and money consumption, which is rising with the number of different user roles and the system environments to be programmed. These problems may be overcome by the application of the automatic user interface authoring procedures, which may be additionally enhanced by the application of the ontologies.

The UI implementation for PlaTel required the implementation of a specialized graphic tool—a Web-based SSDL GUI interface that was developed for supporting the Service Composer configuration. As a composite service, the service composer has its own execution plan, which may be redefined according to the needs and available services. In this way, it is very simple to construct different composition schemes just by dragging services from the repository and setting the necessary parameters.

In the following sections, we present composite service UI formal definition, UI automatic generation, user interface adaptation, and finally UI quality assurance. As an example, we take the delivery of UI for the applications of the PlaTel platform.

6.1 User Interface Definition

Automatic user interface generation for services in the SOA-based systems is possible according to the formal service description and the user model. The formal description given in the SSDL defines functional and non-functional description,

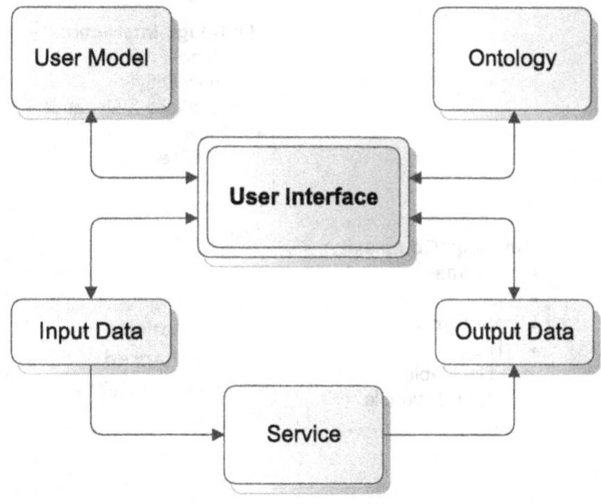

Fig. 21 General architecture for UI generation in SOA system

service input, and service output. The most important element that is necessary for UI construction is the service input attribute names and the type of values as well as output attribute names and values. However, to present the input and output properly, we should possess the proper user model that decides, for example, what language is appropriate for the given user and how the specified input values should be entered (i.e. directly or using a select box) and how the output values should be presented (i.e. using a table of values or charts). In some cases, we should also apply ontologies to resolve some complex concepts into simple ones (i.e. address could be divided into street, city, state and zip code—in the case of an address in the U.S.). The general framework for UI generation for the service is given in Fig. 21. The user interface first enables the user to enter the input data, the input data labels, as well as their form and is worked out according to the user model; the ontology as well as the output data is presented according to the same elements.

Ontologies are applied in many areas of information systems design and development, such as database integration, business logic or Graphical User Interfaces (GUI). An ontology-enhanced user interface is defined as a user interface, which visualization capabilities, interaction possibilities, or development processes are enabled or (at least) improved by the employment of one or more ontologies [46]. In the user interface enhancement, we may use different types of ontologies. We can distinguish the following ontologies that concern: the real world, the IT system, users, and roles [46]. The real world ontology characterizes its part, especially the area of the system application (i.e., e-commerce, travel, etc.), in order to identify the central concepts and relations among them. The IT system itself delivers formalized ontology by containing categories such as Software Module, Web Service, etc. We may further divide the IT ontology into hardware and software ontologies. Finally, users and roles ontology characterizes the users, their preferences, and their roles, which have an influence on the rights and possibilities they have in using a system. In the work [46] and also other categories of the ontology-enhanced user interfaces have

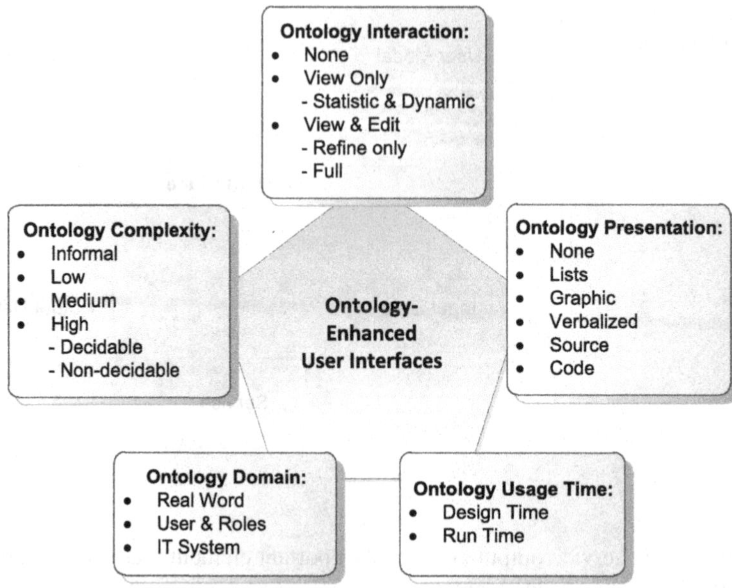

Fig. 22 Ontology-enhanced user interfaces characterization schema [46]

been distinguished, these are: ontology complexity, ontology interaction, ontology presentation, and ontology usage time (see Fig. 22).

Usually in order to formally define the UI, specialized language is used. Today in this area the most popular are languages based on XML; some of the most popular are XUL and XIML, but other UI description languages are also used such as: UsiXML, UIML, Maria, LZX, WAI ARIA, and XForms. It has been discussed, however, that the application of ontology will not replace the application of user interface description languages, but rather deliver its valuable enhancement.

In the implementation of the PlaTel user interface, we decided to choose the quite popular JavaScript library jQuery, which simplifies client-side scripting in HTML. We also applied different types of ontologies in the UI design and implementation. First, we adopted domain ontology, which is also used in the description of functional and non-functional parameter description of services.

In the work [68], we presented a method for Web-based user interface construction for SOA-oriented systems enhanced by ontology, where we postulate to extend the WSDL service file in order to define a user interface in SOA architecture for atomic as well as composite services. The proposed extension is based on populating WSDL nodes, which describe Web Service method parameters and child nodes describing the metadata.

6.2 User Interface Automatic Generation

User interface automatic generation is a process where the input is a formal definition of the UI, i.e. in the form of a file written in the specified user interface language (for example, XUL or XIML), and the output is usually in the form of a working GUI. PlaTel is a Web-based application that it is run in the Web browser, so in fact the final GUI rendering and control is made by the browser itself. The general idea of generating user interfaces for Web Services is based on the assumption that we can serialize user interaction in HTML form. Therefore, to generate a user interface for Web Services, we should be able to produce a Web form that would deliver all the necessary inputs from the user and make the interaction with a Web service possible [68].

The input of data in Web form should be validated by means of validation metadata. Some basic validation may be done using HTML5 Forms, which enable one to force enter a valid e-mail address or other pattern matching string. However, more complex validation, i.e. verification if given a credit card number, necessitates the application of more sophisticated methods needed.

There could be several approaches to the problem of input validation: one of which is the restriction of the user input to some pre-defined set of acceptable values, which are specified for each non-free text input, and practically limiting input into a select list. However, depending on the number of values and their types, we can distinguish the following restrictions [68]:

1. Size of the domain values set: if the select list contains more than 20 elements, applications of scroll list are very inconvenient.
2. Conditional nature of some inputs: if selecting one input value in one field determines set off acceptable values in another field, i.e. when in one field we select a certain country, then in another field the set of acceptable cities depends on the former choice.
3. Multilingual and internationalization support: the domain definitions should support multilingual and convention values, i.e. distances measured in kilometers or miles.
4. Multiple selection: Multiple selection: there are inputs that require a single value, some others might need entering a specified number (greater than one) of values entered from a defined domain, i.e. a single choice we have in the case of the student's faculty, and the multiple choice in the case of the list of enrolled courses.

The first problem could be solved using specialized input widgets that allow filtering domain values and shrinking the select list while the user enters the following characters contained in the value [68]. This problem could be also solved using an autocompleting mechanism.

Usually, the user interaction with a system is some kind of user-system conversation, and the accessible functionalities and possible user inputs depend on the current state of the system. As a result, some business logic has to be embedded in the user interface regarding user inputs in order to avoid invalid user input. For each set of domain values, we should be able to define the context in which these values are valid.

The following problem is natural language that is used during the interaction. It is obvious that most of the text presented to the user is language-dependent and also the input value domains are specific for each language. In the case of a simple text field type input, a label should be language sensitive, but in a more complex situation it should support localization and internationalization.

Finally, in the case of multiple selection support it is becoming more difficult the longer the selection list is. We can consider several different cases: when the values to enter are known to the user, then a simple text field with auto-complete may be sufficient; however, in the opposite case, a filtering application or a shortening of the list is needed.

In order to verify our method that is based on the WSDL extension, a UI generator has been built. It generates a jQuery widget Web-based user interface according to the description given in WSDL that is enriched with necessary metadata. The generated Web-based user interface may be localized for the specified language or date format, as well as pre-selected values in the select list, ratio, or checkboxes. So far, we have implemented several types of jQuery widgets, and the generator could be further extended by the application of graphical visualization, i.e. geo-data presentation using Google Maps API [68].

In the present stage of PlaTel user interface development, the user interface definition is specified manually. However, it is ready for the user interface adaptation that is described in the following section and also for dynamic generation in the event of compound services. In must be noted that HTML 5 gives its own solutions for generating user interfaces for Web-services; however, there is a problem in work flow of the user interface and service calls. This problem is being solved by HTML5 Web Workers, and thus in our future works we will concentrate on its application in PlaTel.

6.3 User Interface Adaptation

User interface adaptation is a process that results in delivering the personalized UI for the given user. The key element of UI adaptation is the proper user model [56]. The user model usually contains the user data, usage data, and environment data. The user data contains: demographic data and users' knowledge, their skills, interests and preferences, and also their plans and goals. The usage concern selective operations that express users' interests, unfamiliarity or preferences, temporal viewing behavior, ratings concerning the relevance of these elements, as well as different interactive events, such as opening a page, purchasing a product, sending feedback information to the system are stored. The environment data contains information about software and hardware platform data and data about usage conditions. The user model may be initialized and maintained in many different ways [56]. Additionally, in the case of PlaTel, we record many different usage data that can be used in UI adaptation.

Having the proper user model, we may utilize it by applying, for example, the methods developed within the recommender systems or more specific user interface adaptation [56]. They are based on the following filtering methods: demographic (DF), collaborative (CF), content-based (CB), and hybrid approach (HA).

To date, we have implemented a simple user interface adaptation based on demographic filtering; this implements stereotypical reasoning based mainly on the user's demographic data and is very easy. The adaptation concerns personalization of the user interface language, date format, some other default data, etc. More sophisticated adaptation methods may be applied when we gather more information about the user interface usage by a considerably large number of users.

6.4 User Interface Quality Assurance

The key notion of quality of user interface is usability. Usability, and also other connected concepts, are defined within many different international standards and guidelines such as: ISO 9126, ISO 9241 and ISO 13407 standards, W3C WAI (Web Accessibility Initiative), or design guidelines such as the Sun Java Look & Feel, Gnome 2.0, Apple, or MS Windows Guidelines. Considerable comprehensive usability definition is given in ISO 9241-11 standard: *Usability is an extent to which a product can be used by specified users to achieve specified goals with effectiveness, efficiency, and satisfaction in a specified context of use*. Where *effectiveness* is defined as: *accuracy and completeness with which users achieve specified goals*, *efficiency* is defined as: *resources expended in relation to the accuracy, and completeness with which users achieve goals*, *satisfaction* is defined as: *freedom from discomfort, and positive attitudes towards the use of the product* and *context of use* is defined as: *users, tasks, equipment (hardware, software and materials), and the physical and social environments in which a product is used.*

Many different methods of usability testing have been developed. They may be applied at a specified phase of the interactive system development: from the planning and feasibility studies to post release. In practical cases, we should apply at least several methods to achieve acceptable results. However, usability is not the only quality parameter we should verify in SOA systems quality assurance. We should also generate the following types of tests: functional, automatic, and sensitivity. Below, we present the quality assurance tests we have made for the PlaTel prototype verification. The functional tests verify if the application is able to return the results that have been assumed by the designers. We verify this by checking each single function, possibly with several sets of representative input data. When we verify the user interface we should test it manually, using the specified input devices such as: keyboard, mouse, touch-screen, etc. PlaTel is implemented using an agile programming methodology—SCRUM, which is iterative and incremental, and the following steps are made within so-called sprints. Each sprint is restricted to the particular length of time (i.e. one week) and brings specified requirements which should be implemented. These requirements must be and are recorded in the sprint backlogs, so then it is quite easy to verify if the specified functionalities are completed.

The second type of the verification tests that we have made is called automatic testing. PlaTel is a Web-based interactive application that is run in an Internet browser. Today, this type of application is tested automatically by a software tool that enables repeatable tests to be conducted. Test automation has many advantages over manual testing; they are mainly related to a test's repeatability and speed. There are many such tools available commercially and free of charge. In the automatic testing of PlaTel, we have used a free tool called Selenium IDE (Integrated Development Environment), which is a prototyping tool for building test scripts. This is a Firefox plugin, which provides an easy-to-use interface for developing automated tests with a recording feature of user interactive actions. These actions are exported to a reusable script that can be later executed.

The last type of test we have conducted is a sensitivity test, which verifies PlaTel behavior by using different software and hardware platforms. PlaTel is a Web-based application, so the most obvious verification is in using different Web browsers: Firefox, Opera, IE and Chrome while using standard a PC with Windows 7. As mobile apps are becoming more and more popular, we also verified PlaTel running on iOS using both the iPad and iPhone and also Android (versions 2.3 and 4.0) running on the smartphone and an emulator. The tests have shown that PlaTel is working pretty well on most PC browsers, iOS, and Android 4.0-based systems.

We have also conducted two types of quality tests: the first usability tests using heuristic evaluation and the second application code validation using different automatic validators. Heuristic evaluation is a usability testing method where usability experts judge whether the verified application follows a list of established usability heuristics (i.e. Nielsen's Ten Usability Heuristics). In this test, usually 3–5 experts are making independent judgments.

Automatic code validators are becoming more and more popular; they are easy to use and give quick results by locating specific problems in HTML or JavaScript codes. We have used the following validators: syntax validator W3C Validate by URI, CSS W3C validator, W3C link checker, JuicyStudio Readability test, and color visibility Vischeck.

However, making all the tests after completing each sprint would be very time consuming. We must admit that automatic code validators and automatic testing may be applied pretty often, even several times within the single sprint. Functional tests should be performed after each sprint. However, sensitivity and usability tests are time consuming because we should engage at least three experts for two working days; so, these tests should be performed only when we have available resources, but at least twice during the entire project.

7 Summary

The presented platform offers a set of unique applications, tools, and techniques which provide an integrated approach to composite service composition, service execution in a distributed environment, and quality of service monitoring. Originally developed service description language allows the inclusion of QoS requirements and composite service execution-related parameters in service description, which is directly used by the Workflow Engine. All the parameters are measured during service execution and may be used when needed by the Service Composer, which feature forms feedback between composition and service execution.

The invented framework is an extensible solution in terms of adding new methods, algorithms, and techniques for business processes analysis, composite services composition, as well as resources allocation, provisioning, and utilization monitoring purposes. The results of first experiments of the developed framework's components functionality, scalability, openness, and reusability with are very promising.

The presented platform is designed to gain support in decision-making tasks, which may be distinguished in the process of the matching required (business process) and available (computer-based information system) services. The business process and ICT systems' services integration scenario, as well as the required and available services matching process, is divided into several, well-defined steps which cover all activities that should be undertaken between requesting an arriving time until service delivery time (end-to-end working prototype). The scope of functionalities obtainable at distinguished steps may be easily extended by adding new procedures, methods, and algorithms.

The general concepts, implemented in the platform as a set of cooperating applications and tools, is based on the component-oriented software development idea; required business process information services may be delivered performing predefined (an available form services repository) or composed on-demand from service components (service on demand) distributed IT systems services. The presented platform is an attempt to offer various functionalities delivered as services: data processing as a service, composition as a service, security as a service, composition as a service, infrastructure as a service, monitoring as a service, etc. Such an assumption means also that the functionalities available in the presented platform above are reusable at different steps within the services matching process.

Functionalities of the presented modules are based on extensive data, information and knowledge gathering activities, and processing. It is evident, especially in the case of service composition, where ontology matching and prediction algorithms are used to select proper services or to obtain the optimal services composition of components available in various instances within a space-distributed environment.

The presented platform is still under development. The general framework idea assures that all of the presented functionalities may be easily extended by adding general-purpose and specific-oriented data in addition to information and knowledge gathering activities, and processing units. Moreover, it is assumed that both the available and planned processing capabilities are reusable in many parallel service delivery processes as well as in various steps of the same distinguished service delivery process.

The functionality of the described tool also presents a multi-step methodology for semantically annotating services regarding matching, selection, composition, and execution. There are several innovations in the presented attempt. First of all, the services matching (management) is available in three versions: service selection, service level agreement negotiation, and service composition. The second is that the services composition (service on demand) process is decoupled into stages that allow for the utilization of various kinds of knowledge to obtain optimal composite service and its execution.

Acknowledgments The research presented in this paper has been partially supported by the European Union within the European Regional Development Fund program no. POIG.01.03.01-00-008/08.

References

1. A guide to the business analysis body of knowledge, International Institute of Business Analysis. www.teiiba.com (2009)
2. Agarwal, V., Chafle, G., Dasgupta, K., Karnik, N., Kumar, A., Mittal, S., Srivastava, B.: Synthy: A system for end to end composition of web services. World Wide Web Conf. **3**(4), 311–339 (2005)
3. Aguilar-Saven, R.S.: Business process modeling: reviewand framework. Int. J. Prod. Econ. **90**, 129–149 (2004)
4. Anderson, S., Grau, A., Hughes, C.: Specification and satisfaction of SLAs in service oriented architectu res. In: 5th Annual DIRC Research Conference, pp. 141–150 (2005)
5. Badr, Y., Abraham, A., Biennier, F., Grosan, C.: Enhancing web service selection by user preferences of non-functional features. In: 4th International Conference on Next Generation Web Services Practices, IEEE Computer Society Washington (2008)
6. Bell, M.: Introduction to Service-Oriented Modeling. Service-Oriented Modeling: Service Analysis, Design, and Architecture. Wiley, Hoboken (2008)
7. Blanco, E.: Techniques to produce optimal web service compositions. In: IEEE Congress on Services, pp. 553–558 (2008)
8. Borzemski, L., Zatwarnicka, A., Zatwarnicki, K.: Global distribution of HTTP requests using the fuzzy-neural decision-making mechanism. In: Proceeding of 1st International Conference on Computational Collective Intelligence. Lecture Notes in AI, Springer (2009)
9. Brzostowski, K., Drapa ła, J., Świątek, J.: system analysis techniques in eHealth systems: a case study. Lecture Notes in Computer Science. Lect. Notes Artif. Intel. **7196**, 74–85 (2012)
10. Brzostowski, K., Tomczak, J.M., Rekuć, W., Sobecki, J.: Service discovery approach based on rough sets for SOA systems. In: Nguyen, N.T., Zgrzywa, A., Czy£ewski, A. (eds.) Advances in Multimedia and Network Information System Technologies, pp. 131–141. Springer, Heidelberg (2010)
11. Cardoso, J., van der Aalst, W.: Handbook of Research on Business Process Modeling, Information Science Reference, ISBN: 978-1-60566-288-6 (2009)
12. Chynał, P., Szymański, J.M., Sobecki, J.: Using eyetracking in a mobile applications usability testing. LNCS/LNAI **7198**, 178–186 (2012)
13. Department of Homeland Security: National Vulnerability Database of the National Cybersecurity Division. http://nvd.nist.gov (2009). Accessed 20 March 2009

14. Fraś, M.: The architecture of complex service requests broker. Grzech, A. (eds.) Information Systems Architecture and Technology: Networks and Networks' Services, pp. 369–379. Wroclaw University of Technology Publishing House, Wrocław (2010)
15. Fraś M., Zatwarnicka A, Zatwarnicki K.: Fuzzy-neural controller in service request distribution broker for SOA-based systems. In: Kwiecień, A., Gaj, P., Stera, P. (eds.) Proceeding of International Conference Computer Networks 2010, pp. 121–130. Springer, Berlin (2010)
16. Fraś, M., Grzech, A., Juszczyszyn, K., Kołaczek, G., Kwiatkowski, J., Prusiewicz, A., Sobecki, J., Świątek, P., Wasilewski, A.: Smart Work Workbench : integrated tool for IT services planning, management, execution and evaluation. LNCS/LNAI **6922**, 557–571 (2011)
17. Grzech, A., Świątek, P., Rygielski, P.: Dynamic Resources Allocation for Delivery of Personalized Services. In: I3E 2010, IFIP AICT 341, pp. 1728 (2010)
18. Grzech, A., Świątek, P.: Modeling and optimization of complex services in service-based systems. Cyb. Syst. Int. J. **40**, 706–723 (2009)
19. Grzech, A., Rygielski, P., Świątek, P.: Translations of service level agreement in systems based on service-oriented architectures. Cyb. Syst. Int. J. **41**, 610–627 (2010)
20. Hackmann, G., Haitjema, M., Gill, C., Roman G.: Sliver: A BPEL workflow process execution engine for mobile devices, LNCS 4294 pp. 503-508 (2006)
21. Havey, M.: Essential Business Process Modeling, O'Reilly, ISBN: 0-596-00843-0 (2005)
22. Hoyer, V., Bucherer, E., Schnabel, F.: Collaborative e-Business Process Modeling: Transforming Private EPC to Public BPMN Business Process Models, Business Process Management Workshops, pp. 185–196 (2008)
23. Brzeziński, J., Danilecki, A., Flotyński, J., Kobusińska, A., Stroiński, A.: ROsWell Workflow Language: A Declarative, Resource-oriented Approach. New Gener. Comput. **30**(2 & 3) (2012)
24. Jaeger, M.C., Rojec-Goldmann, G., Muhl. G.: QoS aggregation in web service compositions. In: IEEE International Conference on e-Technology, e-Commerce and e-Service, pp. 181Ű185 (2005)
25. Jinghai, R., Xiaomeng, S.: A Survey of Automated Web Service Composition Methods, Semantic Web Services and Web Process Composition. In: First International Workshop, SWSWPC, San Diego, CA, USA, pp. 43–54 (2004)
26. Jong Myoung, K., Chang Ouk, K., Ick-Hyun, K.: Quality-of-service oriented web service composition algorithm and planning architecture. J. Syst. Softw. **81**, 2079–2090 (2008)
27. Josang, A.: A Logic for uncertain probabilities. Int. J. Uncertainty Fuzziness Knowl Based Syst. **9**(3), 279-311 (2001)
28. Keller, A., Ludwig, H.: The WSLA framework: Specifying and monitoring service level agreements for web services. IBM Research Report, May 2002
29. Klush, M., Fries, B., Sycara, K.: OWLS-MX: A hybrid Semantic Web service matchmaker for OWL-S services. Web Seman. Sci. Serv. Agents World Wide Web **7**, 121–133 (2009)
30. Kołaczek, G. Opracowanie koncepcji specyfikacji metod i modeli szacowania poziomu bezpieczeństwa systemów SOA i SOKU, WUT, (in polish) (2009)
31. Kohavi, R.: The power of decision tables. Proc. ECML LNCS **912**, 174–189 (1995)
32. Korherr, B., List, B.: Extending the EPC and the BPMN with Business Process Goals and Performance Measures, 9th International Conference on Enterprise Information Systems, pp. 287–294 (2007)
33. Kozik, A., Rudek, R., Świątek, P., Grzech, A.: Resource allocation problems in network processors for the Future Internet. In: Grana, Manuel, et al. (eds.) Advances in knowledge-based and intelligent information and engineering systems /, pp. 1509–1520. IOS Press, Amsterdam (2012)
34. Kruczyński, K.: Business process modeling in the context of SOA Ú an empirical study of the acceptance between EPC and BPMN. World Rev. Sci. Technol. Sustain. Dev. **7**(1/2), 161–168 (2010)
35. Kwiatkowski, J., Fraś, M., Pawlik, M., Konieczny, D.: Request distribution in hybrid processing environments, Lecture Notes in Computer Science, vol. 6067. Springer, Berlin, pp. 246–255 (2010)

36. Kwiatkowski, J., Papkala, G.: Service aware virtualization management system. In: Grzech, A. (eds.)Information Systems Architecture and Technology: Service Oriented Networked Systems, pp. 317–326. Wroclaw University of Technology Publishing House, Wroc^{3}aw (2011).
37. Kwiatkowski, J., Pawlik, M., Konieczny, D.: Efficient Computational Resources Allocation for Service Request, Application of Systems Science. Academic Pub-lishing House EXIT, Warszawa (2010)
38. Kwiatkowski, J., Pawlik, M., Fraś, M., Konieczny, D., Wasilewski, A.: Design of SOA-Based Distribution System, SOA Infrastructure Tools. Concepts and Methods, pp. 263–288. Poznan University of Economics Publishing House, Poznan (2010)
39. Lodhi, A.: An extension of BPMN meta-model for evaluation of business processes. Sci. J. RTU 5. series *46*, 27–34 (2011)
40. Milanovic, N., Malek, M.: Current solutions for web service composition. IEEE Internet Comput. **8**(6), 51–59 (2004)
41. Minoli, D.: Enterprise Architecture A to Z, Frameworks, Business Process Modeling, SOA, and Infrastructure Technology. CRC Press, Boca Raton. ISBN: 978-0-8493-8517-9 (2008)
42. Nakamura, Y., Tatsubori, M., Imamura, T., Ono, K.: Model-driven security based on web services security architecture. IEEE Int. Conf. Serv. Comput. **1**, 7–15 (2005)
43. Nguyen, N.T., Sobecki, J.: Determination of user interfaces in adaptive systems using a rough classification-based method. New Gener. Comput. ltextbf24(4), 377–402 (2006)
44. Ovrien, L., Merson, P., Bass, L.: Quality Attributes for Service-Oriented Architec-tures. In: Proceeding of the International Workshop on Systems Development in SOA Environment. IEEE Computer Society, Washington, DC (2007)
45. Papazoglou, M.P., Georgakopoulos, D.: Service-oriented Computing. Commun. ACM **46**(10), 25–28 (2003)
46. Paulheim, H., Probst, F.: Ontology-enhanced user interfaces: a survey. Int. J. Semant. Web Inf. Syst. (IJSWIS) **6**(2), 36–59 (2010)
47. Ponnekanti, S.R., Fox, A.: SWORD: a developer toolkit for web service composition. In: 11th World Wide Web Conference, pp. 97–103 (2002)
48. Prusiewicz, A., Zięba, M.: The Proposal of Service Oriented Data Mining system for solving real-life classification and regression problems. Technol. Innov. Sustain. IFIF series **349**, 83–90 (2011)
49. Rao, J., Su X., A Survey of Automated Web Service Composition Methods, Semantic Web Services and Web Process Composition, SWSWPC, San Diego, CA, USA, pp. 43–54 (2004)
50. Rygielski, P., Świątek, P.: Graph-fold: an efficient method for complex service execution plan optimization. Syst. Sci. **36**(3), 25–32 (2010)
51. Rygielski, P., Tomczak, J.: Context change detection for resource allocation in service-oriented systems. Lecture Notes in Computer Science. Lect. Notes Artif. Intell. **6882**, 591–600 (2011)
52. Schmietendorf A., Dumke, R., Reitz, D.: SLA management—challenges in the context of web-service-based infrastructures. In: Proceeding of the IEEE International Conference on Web Services, San Diego, California (2004)
53. Shahzad, S.K.: Ontology-based user interface development: user experience elements patterns. J. Univers. Comput. Sci. **17**(7), 1078–1088 (2011)
54. SOA Reference Model Technical Committee. A Reference Model for Service Oriented Architecture, OASIS (2006)
55. Sobecki, J., Żatuchin, D.: Knowledge and data processing in a process of website quality evaluation. In: Nguyen, N.T., Katarzyniak, R.P., Janiak, A. (eds.) New challenges in computational collective intelligence, pp. 51–61. Springer, Heidelberg (2009)
56. Sobecki, J.: Ant colony metaphor applied in user interface recommendation. New Gener. Comput. **26**(3), 277–293 (2008)
57. Strunk, A.: QoS-Aware Service Composition: A Survey, In: Eighth IEEE European Conference on Web Services, pp. 67–74 (2010)
58. Szpala, A., Rutkowska-Kucharska, A., Drapała, J., Brzostowski, K., Zawadzki, J.: Asymmetry of electromechanical delay (EMD) and torque in the muscles stabilizing spinal column. Acta Bioeng. Biomech. **12**(4), 11–18 (2010)

59. Świątek, P., Juszczyszyn, K., Brzostowski, K., Drapała, J., Grzech, A.: Supporting content, context and user awareness in Future Internet applications. Lect. Notes Comput. Sci. **7281**, 154–165 (2012)
60. Świątek, P., Rygielski, P., Juszczyszyn, K., Grzech, A.: User assignment and movement prediction in wireless networks. Cybern. Syst. **43**(4), 340–353 (2012)
61. Świątek, P., Stelmach, P., Prusiewicz, A., Juszczyszyn, K.: Service composition in knowledge-based SOA systems. New Gener. Comput. **30**(2&3), 165–188 (2012)
62. Tari, Z., Bertok, P., Simic, D.: A dynamic label checking approach for information flow control in web services. Int. J. Web Serv. Res. **3**(1), 1–28 (2006)
63. Tomczak, J., Cieślińska, K., Pleszkun, M.: Development of service composition by applying ICT service mapping.In: Kwiecień, A., Gaj, P., Stera, P. (eds.) Computer Networks, pp. 45–54. Springer, Berlin (2012)
64. Xu, D., Wang, Y., Li, X., Qiu, X.S.: ICT Service Composition Method Based on Service Catalogue Model, In: Proceeing of AIAI'2010, pp. 324–328 (2010)
65. Yu, T., Lin, K.-J.: Service selection algorithms for Web services with end-to-end QoS constraints. Inf. Syst. E-Bus. Manage. **3**, 103–126 (2005)
66. Zeng, L., Benatallah, B., Ngu, A.H.H., Dumas, M., Kalagnanam, J., Chang, H.: QoS-aware middleware for Web services composition. IEEE Trans. Softw. Eng. **30**(5), 311–327 (2004)
67. Zeng, L., Benatallah, B., Ngu, A.H.H., Dumas, M., Kalagnanam, J., Chang. H.: QoS-aware middleware for web services composition. IEEE Trans. Soft. Eng. **30**(5), 311–327 (2004)
68. Kopel, M., Sobecki, J.: Web-based user interface for SOA systems enhanced by ontology. In: Aleksander, Z., Kazimierz, C., Andrzej, S. (eds.) Multimedia and Internet Systems: theory and practice, pp. 239–247. Springer, Heidelberg (2013)

Chapter 3
A Platform for Development of Electronic Markets of Sophisticated Business Services

Stanisław Ambroszkiewicz, Anna Ambroszkiewicz, Waldemar Bartyna, Mirosław Barański, Marek Faderewski, Piotr Kulma, Dariusz Mikułowski, Marek Pilski, Andrzej Ryżko, Marcin Stępniak, Grzegorz Terlikowski and Iosif Vojteshenko

Abstract The key idea of our approach is that the arrangement phase (request-quote) of business service can be realized in a generic way independent of application domain. On the basis of this idea a new information technology is proposed (in the form of specifications and protocols) and implemented as a prototype system for business process automation (including planning, composing of services, and execution). Augmented with social media it is the basis for creating a communication platform for developing electronic markets.

S. Ambroszkiewicz (✉) · A. Ambroszkiewicz · W. Bartyna · M. Faderewski · P. Kulma · M. Stępniak
Institute of Computer Science, Polish Academy of Sciences,
al. Jana Kazimierza 5, 01-248 Warsaw, Poland
e-mail: sambrosz@ipipan.waw.pl

W. Bartyna
e-mail: wbartyna@ipipan.waw.pl

M. Faderewski
e-mail: marekf@ipipan.waw.pl

P. Kulma
e-mail: pkulma@ipipan.waw.pl

S. Ambroszkiewicz · W. Bartyna · M. Barański · D. Mikułowski · M. Pilski · A. Ryżko · G. Terlikowski
Institute of Computer Science, Siedlce University of Natural Sciences and Humanities,
Al. Sienkiewicza 51, 08-110 Siedlce, Poland

I. Voiteshenko
Applied Mathematics and Computer Science, Belarusian State University,
4, Nezavisimosti Avenue, 220030 Minsk, Republic of Belarus
e-mail: voit@bsu.by

S. Ambroszkiewicz et al. (eds.), *Advanced SOA Tools and Applications*,
Studies in Computational Intelligence 499, DOI: 10.1007/978-3-642-38957-3_3,

1 Introduction

Economy consists of activities related to production, exchange, distribution of consumer goods and provision of services according to the needs of final consumers. Markets are meeting places for customers, producers, brokers, and service providers. New information technologies support the development of the economy and markets, create new opportunities and challenges at the same time. Automation of business processes (considered as composition and integration of loosely coupled services according to the SOA paradigm) is one of these challenges. The automation should involve not only service publication and discovery but also automatic planning, arrangement and execution of complex and often long-term business processes. Since modern markets are electronic, the business process automation is essential for their development.

Electronic markets for products are relatively well developed; they exist in the form of online stores, auction sites (e.g. Allegro, eBay), or portals for comparing product prices and specifications (e.g. ceneo.pl, pricerunner.co.uk).

Services (in comparison with products) are more difficult to categorize and standardize in terms of their functionality and quality. Processes of service publishing by service provider, discovering and invoking by client should be automated as much as possible. For this very purpose advanced information technologies are required, much more sophisticated than simple classifications of goods and services.

Existing electronic markets for services are usually restricted and dedicated to services provided on a mass scale, e.g. text translation, that are simple to use. Electronic markets should also support provision of more complex and diverse business services for business partners (mainly small and medium-sized enterprises) to be able to reflect closely their actual capabilities, and respond to the customer needs and expectations.

There are a lot of portals that support electronic markets for simple services (e.g. oferia.pl). The main problem of such sites is that offers and commissions are to be specified mainly in plain language, that is, the descriptions of the services related to these offers are not exact and formal enough to be processed automatically. As a result, these sites are usually used by users only to submit offers and commissions while transactions take place outside the site in the usual way.

Electronic markets of services (being a part of electronic economy) are also meeting venues for business partners. Hence, they should support cooperation between these partners, discussions, joint planning and realization of business ventures in the form of electronic business processes.

Generally, business processes comprise collaborative services that can provide much more complex functionalities, than the single services. For classic business process definitions see [1, 2], and perhaps more comprehensive one from IST CONTRACT Project [3]. There is large and extensive literature devoted to Web service composition and integration starting with the very beginning of XXI century, and inspired by the work done in Microsoft and IBM. Just to mention some of them (perhaps more important ones from our point of view) [4–8]. There are also important monographs and handbooks like [9–11]. Although a lot of work has been done in these research and technology subjects, still the problem of automatic service composition is a great challenge.

Related work to particular components of service composition (like ontology and planning) will be discussed in the subsequent sections describing these very components.

However amazing it may appear, it has usually been overlooked that the best (if not the only) way to proceed with automation of sophisticated electronic business processes is to begin with automating the use of a single business service.

Typical, not automatic way of using a business service by a customer consists of the following four phases:

1. The service requester sends a query to a service provider specifying roughly what she/he wants from the service, and then gets back a quotation (a proforma invoice) from the provider. The quotation specifies details of what (and how) the service can perform for the requester.
2. Having the quotation, the requester creates an order and sends it to the provider. Usually, the provider replies with an appropriate contract.
3. If the service is performed according to the contract, the provider sends expertise or carriage letter to the requester, whereas the requester sends back a delivery note (or acknowledgement of receipt) or a complaint.
4. Finally, the service provider sends invoice, and the requester realizes the payment for the provided service.

Sometimes, after sending the order, the service provider sends back an invoice, so that the payment must be done before service delivery.

In order to automate the use of a whole service, each of these interrelated four phases must be automated. For any specific type of service this may be done by a dedicated software application implemented, for example, in BPEL [12]. Note that in each of the aforementioned phases, there is a document flow, that is, query and quotation in the first phase, order and contract in the second phase, carriage letter, delivery note, and complaint in the third phase, and finally invoice and payment. These documents can be created in XML. For each of the phases, the document flow can be described as a separate operation in WSDL [13]. Since WSDL does not provide means to define the relations between the four phases, the complete process of using a single service can be implemented (hardcoded in, e.g. BPEL) separately for any service type in a dedicated way. However, our goal is to do so generically, that is, to construct tools that allow automation of service use for any service type.

Relations between the phases (operations) are crucial to the automation of business processes. They must be formally specified, so that they can be automatically processed. For this purpose, the following concepts are proposed (see also [14] and [15]):

1. XSD schemata of the XML documents (as service environment representation) processed in the second, third and fourth phase.
2. Formal language describing the XSD schemata.

3. Arrangement of business service, i.e., request and quote (in the first phase), is expressed universally as formulas in the formal language, and are related to the documents from the other phases.

The proposed language does describe service (like WSDL) and describes also the environment in which the service is executed, that is, the XSD schemata and corresponding XML documents. This description consists of a precondition and an effect. The precondition specifies the local (for this service) situation in the environment (specifying a collection of input XML documents) that must occur before the service can be executed. The effect specifies the situation in the environment caused by the service execution, i.e., a collection of appropriate output documents. This corresponds to IOPE (Input-Output-Precondition-Effect), i.e. service description in OWL-S [16]. However, in the proposed approach the service implementation is considered as a *black box*, while in OWL-S it is specified in a form of *Service Model* that describes the internal process of service performance.

In this way, only external view of service is described, i.e. the environment of the service. In other words, only changes in the environment (i.e., initial situation and effect of the service execution) are described. The environment is represented by XSD schemas of XML documents created and processed by clients and services. This representation provides a grounding (a concrete semantics) for service description language used to automate the first phase, that is, the request-offer phase of the business process. The information about relationship between phases is specified in the first phase in the formal language and can be automatically processed in the next phases.

Since a user (on the client side) may not (and often does not) know the specific requirements of the service in relation to its intention, it is necessary to query the service by a client in order to obtain all the parameters needed to send a proper order and agree on a contract. Hence, during the process of service arrangement (the first phase) all the parameters and their values necessary for the order and contract (documents in the second phase) creation are determined. These parameters (e.g. date and time of service performance, options, related scope, and price, etc.) are specified in a quote as a formula in the universal description language.

So that, the request and quote (exchanged in the first phase and defined as formulas in the description language) include all essential information about documents to be created and processed in the next phases.

Hence, the crucial point for the automation of the first phase (and then the next phases) is the concept of the service environment, its formal representation in XSD, and the description language used in the communication between services and clients. OWL, Entish [17–20], or any other simple language of first order logic without quantifiers can be used as the description language. The only requirement is that the services and clients must be able to *understand* the language, i.e. create and interpret messages defined in that language.

Actually, we propose an extensions of the business service architecture related to Web service and Semantic Web. In the architecture there are two ways of communication. The first way (during the arrangement phase) consists of exchanging messages (with contents defined as formulas in a description language). The second way of communication is a document flow and processing using WSDL during the order-contract, execution, and payment phases.

The concept of business service and the automation of the arrangement phase (using formal description language) give rise to automaton of the whole complex business process consisting of many services. It constitutes the core of the proposed new technology specified in the form of protocols (see Sect. 2) The corresponding prototype system for automation of business process (including planning, composing of services, and execution) has been implemented as a communication platform, and is available online in the form of a set of tools [21]. The platform (called *SOA-enT*) is generic and may be applied to any application domain when adjusted. The adjustment encompasses a definition of the appropriate XSD schema. The technology, and the platform is presented in detail in the subsequent sections.

A considerable work on automatic service composition (also within the framework of the IT-SOA project) based on the approach described above (however with different concept of ontology) was done the group of W. Penczek, see [22–28].

The proposed system architecture, its components and communication protocols (defining the order, format and semantics of message exchange between these components) are described in the following Sect. 2. In Sect. 3 the system components are described more precisely. In Sect. 4 a use case of the systems is presented. Section 5 describes the importance of introducing social media to modern electronic markets. The final Sect. 6 summarizes the chapter.

2 Overview of SOA-enT

SOA-enT is based on: the SOA paradigm, the principle that its components are loosely coupled, and the introducing a universal service description language to the system. A service is described by specifying its interface as a type of operation performed by this service (i.e., types of input documents, types of output documents, and the operation name), the precondition that has to be satisfied before the service is performed, as well as the final result (effect) of the operation execution. The system is designed to be scalable, open, and heterogeneous, that is, services implemented in different programming languages, on different platforms, by different providers may be joined to the system in plug and play manner if the service interface is already published. The service is like a *black box* for the client that knows only the description of its interface (the way how to use it), however does not know how the service accomplishes its operations. The client is interested only in the result described in the service interface.

2.1 The General Architecture

Services are joined to the system via *Service Manager* (SM), independent component for managing services of the same type. For different service types there may be different Service Managers. SM is an interface for a service provider to introduce its service to the system by providing the information about the method of communication with the service (its network address), its type, the precondition and the effect of the service.

The client side is represented by *Task Manager* (TM). It is responsible for the interaction with the user, generating the plan to realize the task submitted by the user, its arrangement and execution (by calling services).

According to the SOA paradigm it is also possible to introduce a broker between client and service providers. This component operates also as a service registry, in which service providers publish descriptions of their services (i.e. service interfaces), and clients can search for information about services. In our system, this functionality is delegated to *Service Registry* (SR) that is a part of *Service Broker* (SB). It is responsible not only for service discovery, but also for the arrangement phase. General architecture of SOA-enT is presented in Fig. 1.

Dictionary is the system component responsible for defining representation of service environment (that may also be called *ontology*) for specific application domains. It is a web application used by domain experts for defining types of documents and types of services specific for these domains.

Planner is the component responsible for automatic (or semi automatic) abstract plan constructions on the basis of information stored in the Dictionary. A plan consists of a list of service types, links between the input and output documents of appropriate services, as well as partial order of execution of the services.

The SOA-enT architecture including these two additional components is presented in Fig. 2.

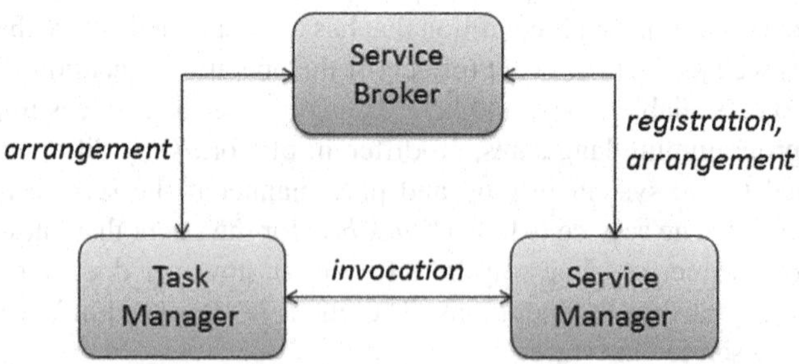

Fig. 1 The concept of the system—diagram

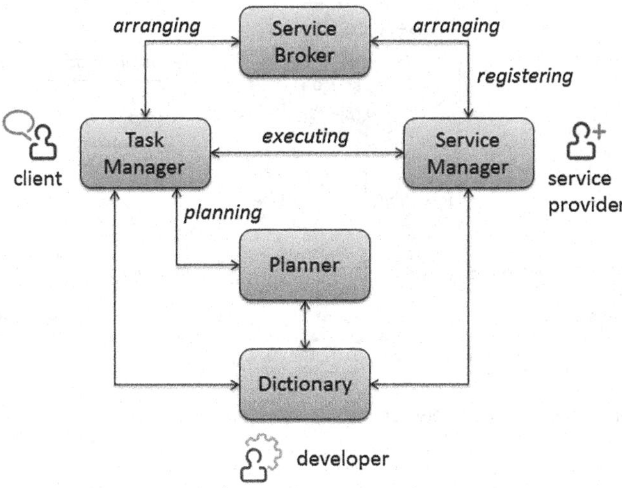

Fig. 2 System architecture

2.2 Service Arrangement

Note that an abstract plan indicates only what types of services should be used, and the partial order of execution of services corresponding to these types. However, these *corresponding* services should be discovered and then arranged (in the arrangement phase) according to the specific requirements, that is, preconditions and effects.

The arrangement phase consists in sending query (in the form of client intention) to a concrete service, and getting back a reply as a quote in the form of a service commitment to realize the intention. The intention partially specifies output documents (desired by the client), whereas the commitment specifies precisely the input documents needed by the service in order to produce the output. All these specifications are expressed in the common universal service description language. The arrangement phase is realized by *Service Agent* (SA) which is a part of Service Broker (see Fig. 3). In the figure also the basic communication protocols between the system components are shown.

2.3 Communication Protocols

The basic protocols are listed below.

1. Service registration protocol.
2. Intention submission protocol.
3. Service discovery protocol.
4. Commitment query protocol.

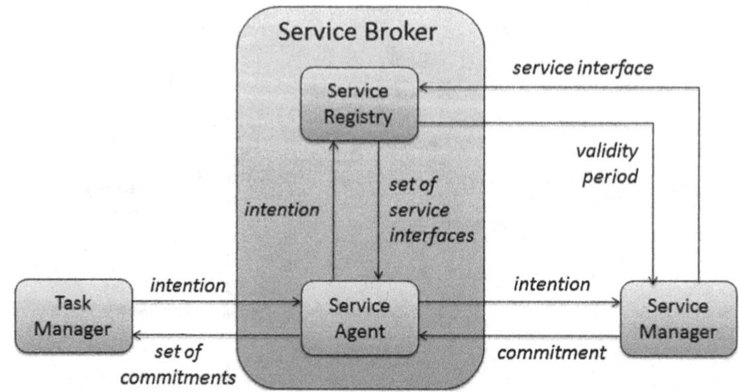

Fig. 3 Communication protocols between the system components

The first three protocols are simple, natural, and used (in many forms) in the systems based on SOA. The fourth, Commitment query protocol, is also simple and is the basis of our approach to business process automation.

Intuitive meaning of the protocols is as follows. Service providers publish (using appropriate Service Manager) their services to the Service Registry via the Service registration protocol. A user submits (via Task Manager) its task (in the form of a partial specification of the desired final document) to be accomplished by the system. The Task Manager gets abstract plans from Planner; this is not specified as a protocol. The Planner may be considered as internal part of TM. Having an abstract plan, for each service type in the plan, TM sends (starting with the last type in the execution order and then consecutively previous ones) a request to the Service Agent by using Intention submission protocol. The request is in the form of an intention corresponding to a subtask to be accomplished by a service of that type. SA discovers appropriate service interfaces in Service Registry using Service discovery protocols, and then tries to contact the services and arrange with them realization of the intention using Commitment query protocol. If the SA succeeds, the commitments (from the services) to realize the intentions are returned as a response to TM via Intention submission protocol. Intentions correspond to queries, whereas commitments to quotes in the first phase of using a business service.

Having arranged the whole plan, TM presents it to the user. If the user accepts the concrete plan, it is executed by calling subsequent services in the partial order defined in the plan. In the case of failure, the mechanism for handling transaction is started, see Sect. 3.5 for details.

The meaning of the message content is grounded in the representation of the service environment, that is, in XSD schemata of the XML documents processed by services. Input and output of each service are specified as appropriate types of documents (actually the XSD schemata).

A service commitment is a complete specification of the input documents, necessary to produce the output document partially specified in the client intention.

The specifications (intention and commitment) are not directly consistent with XSD schemata of corresponding types of documents defined in the Dictionary, so that they must be validated.

During an execution of a concrete plan, commitments and intentions are replaced by complete input and output XML documents. Input documents are created by the client (actually TM), and sent to initial services arranged in the concrete plan. The services produce output that is the required input for the subsequent services, and so on, step by step. Finally, the document desired by the user is delivered to the TM and presented to the user.

The protocols are specified in the usual way (as in the distributed systems), i.e. by defining message format in XML, types of exchanged messages, their contents, and actions taken by senders and recipients. The message header contains the following obligatory information (in the listed order):

1. sender's address (*sender*),
2. recipient's address (*recipient*),
3. protocol name (*protocolName*),
4. version of the protocol (*protocolVersion*),
5. message type (*messageType*),
6. session identifier (*sessionId*).

Apart from these elements, the header may contain a list of additional parameters (name/value), e.g. for handling transaction.

A type of message determines the way it is interpreted, and the way its content is processed.

Generally, the following 8 types of messages are used:

1. For the service registration protocol:
    - request to register (register-request),
    - registry response (register-response).
2. For the intention submission protocol:
    - request to collect offers for a step of the process (task-request),
    - Service Broker response (task-response).
3. For the service discovery protocol:
    - request to return information about matching services (info-request),
    - Service Broker response (info-response).
4. For the commitment query protocol (arrangement phase):
    - query (service-request),
    - service commitment (service-response).

A session identifier is defined by the component that initiates the conversation. The following messages have the same session identifier:

- messages sent during service registration(*register-request* and *register-response*),

- messages sent during discovery of services and collecting offers for the client's intention (*task-request*, *info-request*, *info-response*, *service-request*, *service-response* and *task-response*).

The name of the protocol (in the message header) indicates language used to express the messages content. For example, in the messages of type (service-request), and (service-response) it may be OWL, Entish or any other service description formal language. As a result, the arrangement phase may be realized in an arbitrary language provided that the sender as well as recipient do understand (can process) this language. To introduce a new service description language, it is necessary to implement translation from Dictionary into this language and vice versa.

Since arrangement phase is independent of the description language, further explanations will be also independent of language syntax, and based on the intended meaning of a message content.

The following information may be sent within the messages:

- a service interface,
- a validity period,
- an intention,
- a set of services interfaces,
- a service commitment,
- a set of service commitments.

Service Registry is an autonomous component of the system responsible for:

- gathering information about services available in the system,
- finding services of a particular type,
- preliminary (static) filtering,
- service registration protocol handling,
- information request protocol handling.

Exchanging messages according to these protocols requires, usually, a revision/ change of sender internal state and recipient internal state.

Depending on the type of received message, Service Registry calls appropriate function. Each of these functions is responsible for creating and sending a reply related to the request as well as updating knowledge stored in the SR database. In case of a request to register, a new entry is created or an existing entry is updated. In case of a request to unregister, appropriate entry is deleted. In case of a request to provide information, the SR state remains unchanged.

Service Agent receives an intention (specified in a *task-request* message) from the client (Task Manager), and obtains, form Service Registry, information about services currently available in the system that may realize the intention. Once SA gets a response (as a set of service interfaces), it sends the intention (in a *service-request* message) to each of the services, and waits for a reply (*service-response* message) for a specified period of time.

A state of Service Agent consists of:

- submitted intention,

- client's network address,
- session identifier,
- timeout (time period the SA waits for a service response),
- entries for each service consisting of:
  - service network address,
  - commitment (obtained from the service in response to the sent intention).

If Service Agent receives a response from a service, then it performs the following actions:

- a commitment (from the response) is saved in the entry related to the service,
- deletes entry related to the service if the response contains refusal,
- deletes entry related to the service if an error occurred during communication.

If a service does not send a response in a specified period of time, SA deletes entry corresponding to this service.

Service Agent sends a response to the client (Task Manager) if:

- Responses from all services has been received. If there are no entries in the Agent's state (i.e. all responses were refusals) it sends to the client a *task-response* message with the status set to *false*. Otherwise, it sends the collection of commitments as the response.
- The timeout is over. If there are no commitments, the SA sends back a response with the status set to *false*. Otherwise, the response contains all commitments collected so far.

2.4 Service Registration

In order to join a service to the system, it must be registered, and published to the Service Registry. Service provider registers its service using Service registration protocol. A message that initiates communication (sent by the service provider to Service Manager and then automatically by SM to the Service Registry) contains a description of the service interface. The type of this message is set to *register-request*, and the unique session identifier is determined by SM.

The register message contains the following information:

- Type of service (description of the type depends on the used language).
- Service network address used to communicate with the service in the arrangement and execution phase.
- Static restrictions for the service—conditions that must be satisfied by the output documents. Only intentions consistent with the conditions may be accepted in the arrangement phase. Static filtering is based on them.

It is reasonable to define (by a service provider) additional conditions during service registration if they are related to attributes (fields) of documents:

1. Conditions that do not depend on attributes of input documents, i.e. they are constant (at least for a certain period of time) for the service.
2. Conditions that may appear in client's intention. They may only be used to check if a service may realize an intention provided that the client enters the values for these attributes.

Information about registered service interfaces is stored by the Service Registry as a set of entries. Each entry consists of the following elements:

1. Information obtained directly from service interface:
   - service type,
   - service network address,
   - restrictions (limitations) on the output documents of the service.
2. Additional information:
   - date and time of service registration,
   - expiry date and time for the entry.

Upon registration of a service, the Service Registry specifies the expiry date and time for the new entry; it depends on the configuration parameters of Service Registry. The expiry date is sent to the address specified in the header of the registration message. The address may be different than the network address of the registered service. Usually it is the address of the SM that is the mediator in the registration process.

If the Service Registry cannot register a service, a *register-response* message with the status set to *false* is sent back.

A service may renew or update its entry in SR using the same registration protocol. If the entry has already existed, the new expiry date is set and returned.

Service restrictions may also be taken into consideration when checking if an entry already exists. If the values of attributes in the restriction have been changed in the service interface, or when the service provider decides to add new conditions, (actually it is a new interface of the service), Service Registry updates and changes both the expiry date of the entry and restrictions.

If there is a failure to notify the Service Registry by the service provider about changes of the service restrictions, then outdated and wrong information (about some attributes values) may be sent to Service Agent. However, during the arrangement phase, such service will not return commitments because of the wrong attributes values. It does not influence the final outcome of the arrangement phase, however it makes additional unnecessary workload for the system.

If a service provider decides to withdraw its service, which has already been added to the system by publishing its interface to the Service Registry, it may unregister the service. If the provider does not do it, this service is removed from the SR because of the timeout.

In order to unregister a service, it is necessary to send the Service Registry information that precisely identifies the service in an *unregister-request* message (request to unregister a service). This information consists of service interface along with its network address, and is sent to SR when registering the service; restrictions are not necessary. Then, Service Registry sends back the response as a *register-response* message having the same session identifier as the one in the request. In case of successful removal of the appropriate entry, the message contains the status set to *true*; otherwise, it is set to *false*.

2.5 Intention Submission

The basic function of the Service Broker is handling the arrangement phase for a single step of the business process accomplishing the user task, i.e. arranging a single service corresponding to a single service type in the abstract plan. Task Manager is responsible for the realization of the entire business process, in particular, in creating concrete plan, executing it and monitoring. SB starts its activity when it receives a *task-request* message from TM with an intention. Intention contains the following information:

- type of service that may realize a step of the process,
- partial specification of the output documents for this type of service,
- expiry date for the intention.

In the next two subsections the process of service discovery and querying for a commitment is described in detail.

2.6 Service Discovery

If the intention from the client is received by Service Agent, the agent passes it to the Service Registry sending an *info-request* message. The task of SR consists in finding out services (in its knowledge base), which possibly may realize the intention, that is:

1. the service type matches the type of the operation described in the intention,
2. the service restrictions do not contradict the intention.

The first step of selecting services for an intention is based on the service type. So that a set of interfaces of services that realize appropriate type of service is returned. Then, to limit the cardinality of the set, restrictions are taken into account; this is called *static filtering*.

The general idea of the algorithm for static filtering is based on the description of attribute values of the output document described in the intention, and the service interfaces stored in SR. If there are no restrictions for an attribute, then any attribute values are permitted. If there are restrictions for an attribute, then the sub-formula of the intention, describing this attribute, must follow from these restrictions.

Even if a service interface is selected by the filtering algorithm, it does not mean that the service (when requested in the arrangement phase), will return a commitment as the positive answer to the sent intention. Internal business logic of the service may not agree to realize the intention due to dynamic current restrictions on the service resources.

Once the services are selected, SR sends an *info-response* message (with the session identifier the same as in the request) with the interfaces and network addresses of the all selected services. If the set is empty, the message contains the status set to *false*.

2.7 Commitment Querying

If Service Agent receives a set of selected service interfaces and their network addresses as response from Service Registry, then the intention is sent (in the *service-request* message) to each of the services. A service responds (or does not) by sending *service-response* message with a commitment if a service is able to realize the intention, otherwise the reply message contains status set to *false*.

At this step, *dynamic filtering* of services is realized, i.e. a service that is not able to realize the client intention does not send a commitment because of its dynamic restrictions, which, as opposed to static restrictions, could not be included during the service registration.

A service commitment consists of:

- service type,
- complete specification needed by the client to produce the input documents for the service,
- validity period of the commitment.

For an intention, the commitments from all the requested services are sent by SA to Task Manager. In the execution phase, the arranged service must be supplied (by an appropriate client) with input documents containing all the parameter values specified in the commitment of that service. Only then the service is able to produce the required output document that was earlier partially described in the intention. Actually the commitment may be considered as an offer or quote expressed in the universal way in a chosen description language. The best offer is selected and the service corresponding to that offer is included in the concrete plan. Actually, this is done by Task Manager that acts as the client for all services arranged in a concrete plan.

The crucial point of the proposed technology consists in the way TM composes a concrete plan. The last (in the execution order) service type in an abstract plan is taken as the first one in the composition process. Then, TM constructs the intention consisting of this type and the original user task in the form of the partial specification of the output documents desired by the user (see Sect. 2.5 for intention definition). The intention is sent to SA that returns a set of commitments from services of this type.

Each commitment specifies input documents that must be sent to the corresponding service necessary for the intention realization. TM divides the specification into separate specifications concerning each single input document. Each of the separated specification becomes a subtask associated with one of the previous (in the execution order of the abstract plan) service types. Along with the type of this service, the subtask becomes a new intention to be sent to SA. Service Agent discovers appropriate concrete services, sends the new intention to them getting back (eventually) commitments. The commitments are sent back to TM by SA, and the next step of the composition is done iteratively, up to the final step where all the resulting (in this final step) input documents could be created and delivered by the user via TM, that is, no more external services are needed. More details how Task Manager works are given in Sect. 3.5.

The protocols presented in this section constitute the basis of the proposed technology. Three of them, that is, Service registration, Intention submission, and Service discovery are simple and obvious and are applied in the most of the existing systems for automation of electronic business processes. The fourth protocol, Commitment query, is also simple, especially in the form presented here, where it concerns only arrangement of a single service. More complex protocol was introduced in 2003 in [29] for composing services into concrete plan, where the composition was the integral part of the protocol.

In the version presented in this work, composition is undefined in the protocol, and is left to be implemented as a part of functionality of the Task Manager as a hardcoded application responsible for creating concrete plan and transaction on the basis of the simple Commitment query protocol. This solution seems to be better, and more scalable for open and heterogeneous distributed systems.

3 The Components of SOA-enT

The architecture of SOA-enT consists of the following components: Dictionary, Planner, Service Manager, Service Broker, Task Manager. Below these components are described more precisely except Service Broker that is strongly related to the protocols and was presented in the previous section.

3.1 Dictionary

Upper ontology (also called a top-level ontology or foundation ontology) describes very general concepts that are the same across all knowledge domains to support very broad semantic interoperability between domain specific ontologies. Upper ontologies are usually expressed in languages of the first order logic (predicate calculus), and can be exported to OWL that is de facto regarded as the standard now.

There is a considerable work done here. More important upper ontologies are listed below.

- Cyc [30] is a project that attempts to codify, in machine-usable form, millions of pieces of knowledge that compose human common sense. Knowledge Base of Cyc contains over one million human-defined assertions, rules or common sense ideas. These are formulated in the language CycL [31], which is based on the predicate calculus and has a syntax similar to that of the Lisp programming language.
- WonderWeb project and the WorderWeb Foundational Ontologies Library (WFOL) consisting of: DOLCE [32], DnS (Descriptions and Situations) [33], and OCHRE (Object-Centered High-level Reference Ontology) [34].
- WordNet [35] developed within the Framework of the project called Semantic Network. It is a lexical database for the English language for grouping words into sets of synonyms, providing short, general definitions, and define the various semantic relations between these synonym sets.
- Suggested Upper Merged Ontology (SUMO) is one of the most important ones. It is the candidate for the *standard upper ontology* considered by IEEE P1600.1 Standard Upper Ontology Working Group (SUO WG) (http://suo.ieee.org/). SUMO is intended to be a general foundation ontology for information processing systems, and is concerned with meta-level concepts not belonging only to a specific problem domain.
- General Formal Ontology (GFO) [36] is yet another upper ontology expressed in OWL.

It must be stressed that all foundation ontologies are only formal theories in formal languages. In mathematical logic, semantic for these theories are only model theoretic and, in fact, have no concrete groundings. As formal theories, their *semantics* consists in axioms and fixed derivation rules. Our approach is different. Generally, Dictionary is for introducing, editing, and storing upper ontology (fundamental notions) in SOA-enT, and then to develop representations of service environment for concrete application domains. The key feature of the Dictionary is its independence of specific service description languages, like WSDL, OWL, Entish. It means that it contains the concrete semantics (the representation of service environment) so that these languages can be grounded in this representation. Once a description language is joined to the Dictionary (by incorporating its syntax), formulas of that language describing the representation may be generated from the Dictionary.

In that sense Dictionary is more than yet another upper ontology; it is the grounding (the concrete semantics) for general concepts (belonging to upper ontology), and for specific concepts for particular application domains.

Technically, Dictionary is a web application implemented according to the three-layer architecture: the user interface layer, business layer, and data access layer. Business layer consists of tools for using the information stored in the database in order to produce XML schemata representing concepts defined by the user (see Fig. 4). Documents produced by these tools are the following:

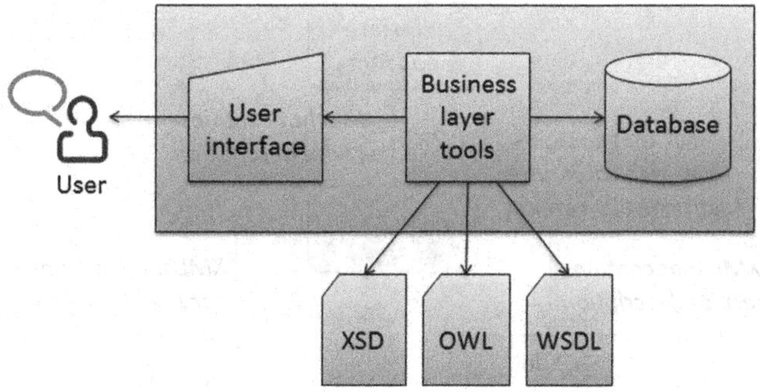

Fig. 4 Dictionary architecture

- XSD schemata to validate the XML documents produced by services,
- WSDL documents for defining service interfaces,
- OWL ontology documents generated on the basis of the concepts defined in Dictionary.

Concepts stored in the dictionary are expressed in the form of appropriate XML tags. Parameters of documents are translated into simple elements. In the following example, `ProfileID` is an element, `cbn` name space, whereas `urn:ipipan:waw:pl:docs:application` is a value of the element.

```
<cbc:ProfileID>
  urn:ipipan:waw:pl:docs:application
</cbc:ProfileID>
```

Types (grouping related parameters of documents) are described using complex XML elements, e.g. a type describing payment information contains payment description as a simple element, as well as payer and payee account details as a sub-type (nested complex element):

```
<iac:PaymentInfo>
  <cbc:PaymentDueDate>2009-07-21</cbc:PaymentDueDate>
  <cbc:InstructionNote>
    Fee for land development design
  </cbc:InstructionNote>
  <iac:PayerAccount>
    ...
  </iac:PayerAccount>
  <iac:PayeeAccount>
    ...
  <iac:PayeeAccount>
</iac:PaymentInfo>
```

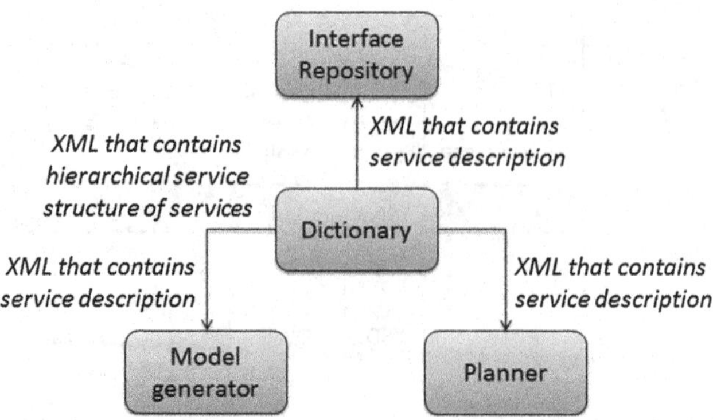

Fig. 5 Integration of the Dictionary with other tools

Universal Business Language (UBL) [37] standard is used to define the structure of documents. Since a lot of important elements are not present in the standard (especially for new application domains), and in order to adjust documents to Polish standards, it was necessary to define additional parameters and types in Dictionary for application domains.

Dictionary is the key component of SOA-enT, and used by the other components as the source of concepts, document types, and service types that together form representation of service environment, that is XSD schemata of XML documents processed by services. Relations of Dictionary to other system components are shown in Fig. 5. These components will be described in detail in the subsequent sections. Here only their relations to Dictionary are presented.

Interface Repository stores models, scripts, and configuration used while displaying and processing documents. It contains a creator, which generates models based on the concepts defined in the Dictionary, and automatically updates data in the registered Service Managers.

Planner is an autonomous component of the SOA-enT system, used to generate an abstract plan for user's task accomplishment. When the Planner receives a request from the Task Manager, it queries the Dictionary about available service types, and (on the basis of input document types and output document types of the service types) generates an abstract plan.

Model Generator produces the source code of models on the basis of concepts the Dictionary contains. A model is a class in Java for storing data contained in documents and document elements. Unlike the two tools described above, Model Generator is not a Web application. It works as a plug-in for the *Eclipse* programming environment.

Model Generator sends a request with selected service type names. A description of the types, and the entire structure of all documents related to that service type is generated by Dictionary and returned. An example of such XML document (a list of service interfaces) is presented below.

```
<?xml version="1.0" encoding="UTF-8"?><servicesTypes>
  <serviceType name="ArchitectCompany_Building
                              ProjectAdaptation">
    <inputDocumentsTypes>
      <documentType name="ProjectPurposeMap" />
      <documentType name="BoroughPlanStatement" />
      <documentType name="BuildingProject" />
      <documentType name="Order" />
    </inputDocumentsTypes>
    <outputDocumentsTypes>
      <documentType name="Invoice" />
      <documentType name="AdaptedBuildingProject" />
    </outputDocumentsTypes>
    <faultDocumentsTypes>
      <documentType name="Fault" />
    </faultDocumentsTypes>
  </serviceType>
    ...
  <serviceType name="WaterPidCompany">
    <inputDocumentsTypes>
      <documentType name="CertificateOf
                    Completion_Waterpid" />
      <documentType name="Order" />
    </inputDocumentsTypes>
    <outputDocumentsTypes>
      <documentType name="Invoice" />
      <documentType name="CertificateOf
                    Completion_Waterpid" />
    </outputDocumentsTypes>
    <faultDocumentsTypes>
      <documentType name="Fault" />
    </faultDocumentsTypes>
  </serviceType>
</servicesTypes>
```

The structure of *documentType* could be nested and quite complex, an example of a corresponding XML document is presented below:

```
<documentType name="xxx:HelloDictionaryDocument">
  <documentType min="1" max="1" inherited="0"
              name="cac:Country">
    <attributeDescription
              name="cbc:IdentificationCode"
              min="0" max="1" baseType="string">
    </attributeDescription>
  </documentType>
```

```
    <attributeDescription name="cbc:CountrySubentity"
                     min="1" max="1" baseType="string">
    </attributeDescription>
    <attributeDescription name="cbc:PostalZone"
                   min="1" max="1" baseType="string">
    </attributeDescription>
    <attributeDescription name="cbc:CityName"
                   min="1" max="1" baseType="string">
    </attributeDescription>
    <attributeDescription name="cbc:BuildingNumber"
                   min="1" max="1" baseType="string">
    </attributeDescription>
    <attributeDescription name="cbc:StreetName"
                   min="1" max="1" baseType="string">
    </attributeDescription>
  </documentType>
```

The XML code listing presented above gives only intuitive picture of the technical details of the Dictionary. As the conclusion it should be stressed again that the key idea of Dictionary is the generic representation of service environment (i.e. XSD schemata of documents processed by services) that can be introduced, edited and stored for specific application domains. This concept along with the arrangement protocol constitutes the foundations for our approach to an automation of business processes.

3.2 Planner

Planner is an autonomous component of SOA-enT used to automatically generate an abstract plan for user task accomplishment. After receiving a request from Task Manager, Planner queries the Dictionary in order to obtain descriptions of available types of services. Then, it creates an abstract plan (for the task) based on matching input and output types of document processed by the services. Abstract plan is a partial order of service types that, when arranged in a concrete plan (work flow or rather document flow) and executed (according to that order), will produce a document specified in the task. The plan is sent back to the Task Manager. Planner may also use process models (e.g. in BPMN notation) supplied by domain experts.

Plan Generation Algorithm

The plan generation is based on the algorithm presented below on a simple example. User task is represented by `DesiredDT` and `OwnedDTs` that denote respectively the formula specifying the desired output document, and the formula specifying the input documents that can be generated by the user via TM. Planner extracts the types of input documents from `OwnedDTs` (suppose that they are `DT5, DT1`, and extracts the types of output documents from `DesiredDT` (suppose that it is `DT7`).

In order to generate an abstract plan, Planner needs the simple descriptions of all the service types (in the form the names of input document types and output document types) currently available in Dictionary in the form of an XML document or an OWL ontology. The document returned by Dictionary is shown in Example 1.

Example 1 XML document with the description of service types provided by Dictionary to Planner

```
<servicesTypes>
  <serviceType name="ST1">
    <inputDocumentsTypes>
      <documentType name="DT1"/>
      <documentType name="DT2"/>
    </inputDocumentsTypes>
  <outputDocumentsTypes>
    <documentType name="DT3"/>
      <documentType name="DT4"/>
    </outputDocumentsTypes>
</serviceType>
  <serviceType name="ST2">
    <inputDocumentsTypes>
      <documentType name="DT5"/>
    </inputDocumentsTypes>
    <outputDocumentsTypes>
      <documentType name="DT4"/>
    </outputDocumentsTypes>
  </serviceType>
  <serviceType name="ST3">
    <inputDocumentsTypes>
      <documentType name="DT4"/>
      <documentType name="DT6"/>
    </inputDocumentsTypes>
    <outputDocumentsTypes>
      <documentType name="DT7"/>
    </outputDocumentsTypes>
  </serviceType>
  <serviceType name="ST4">
    <inputDocumentsTypes>
      <documentType name="DT3"/>
    </inputDocumentsTypes>
    <outputDocumentsTypes>
      <documentType name="DT3"/>
```

Table 1 Relationships between types of services and types of documents

	DT1	DT2	DT3	DT4	DT5	DT6	DT7	DT8
ST1	1	1	2	2	0	0	0	0
ST2	0	0	0	2	1	0	0	0
ST3	0	0	0	1	0	1	2	0
ST4	0	0	3	0	0	0	0	2

```
      <documentType name="DT8"/>
    </outputDocumentsTypes>
  </serviceType>
</servicesTypes>
```

Having this description, the first step of the algorithm consists in construction of the list of all service types (in this case: ST1, ST2, ST3, ST4, and the list of all document types, that is, (DT1, DT2, DT3, DT4, DT5, DT6, DT7, DT8).

The next step is to create a table of all possible connections between the service types and documents types based on the definitions of the service interfaces (see Table 1). Rows of the table correspond to the service types whereas columns to the document types. The number in the intersection of a row (service type) and a column (document type) has the following meaning: 0—there is no relation between given document type and the service type, 1—document type is an input of the service type, 2—document type is an output of the service type, 3—document type is an input and at the same time an output of the service type.

Based on the table, the directed graph (representing a partial order) is constructed. Nodes of the graph represent document types (denoted as rectangular notes), and service types denoted as ovals. The edges represent relationships between the nodes (*is input of*, *is output of*). The direction of the edges is determined by the flow of document to be processed by services of the corresponding types. For example, if document type DT1 is one of the input documents of service type ST1 the edge is directed from node DT1 to node ST1. The graph for the Table 1 is shown in Fig. 6. The algorithm for selecting sub-graphs (that will correspond to abstract plans) proceeds as follows. First, the root node is determined, it is the type of the final desired (by user) document in the task; it is DT7. Starting with this node and going up (in the opposite direction of the edge directions) the upper nodes of the graph are reached. The paths leading to that upper nodes determine the sub-graph shown in Fig. 7. Let the sub-graph be denoted by SG.

In the next step of the algorithm, Planner checks whether SG contains plan variants, i.e. whether a service type can receive one of its input documents from more than one source. In the example, service type ST3 requires an input document of type DT4 that can be provided by service type ST1 as well as by service type ST2. Variants in SG are detected by checking nodes representing document types. If more than one edge is pointed to a given document type node, the sub-graph has at least as many variants as the number of the pointing edges. A variant plan is a maximum

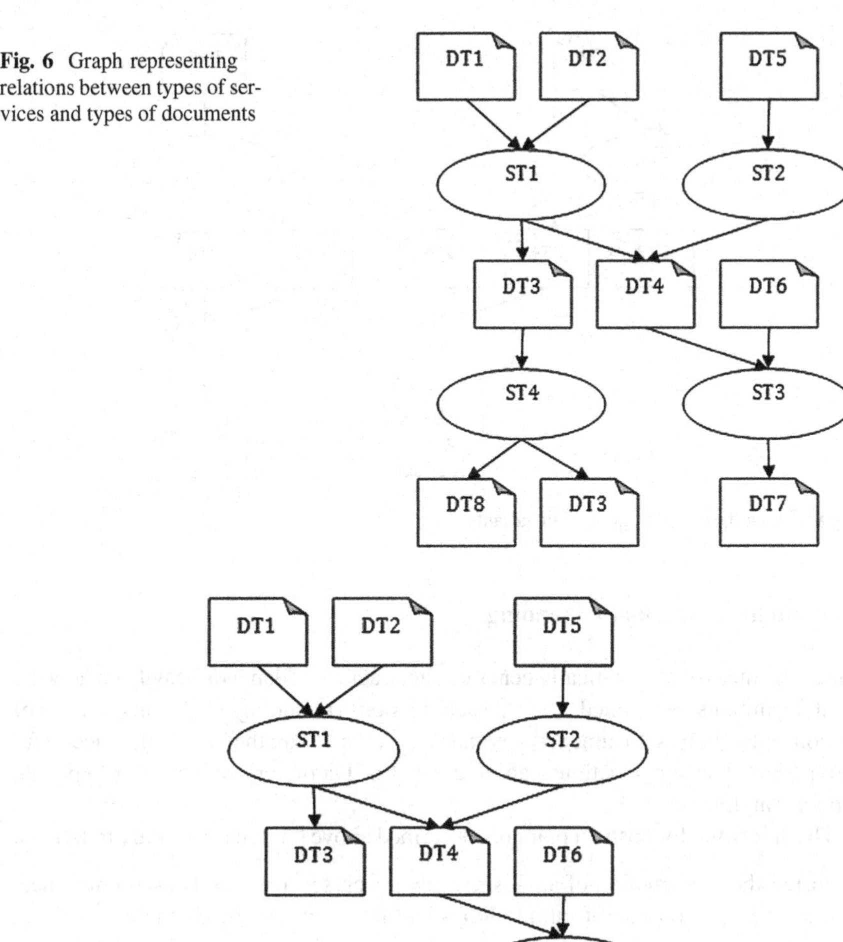

Fig. 6 Graph representing relations between types of services and types of documents

Fig. 7 Graph representing the plan with variants

sub-graph of SG that has no variants. The variant plans for our example are shown in Fig. 8.

In the last step of the algorithm, Planner transforms the variant plans into a list of XML documents and sends it back to Task Manager.

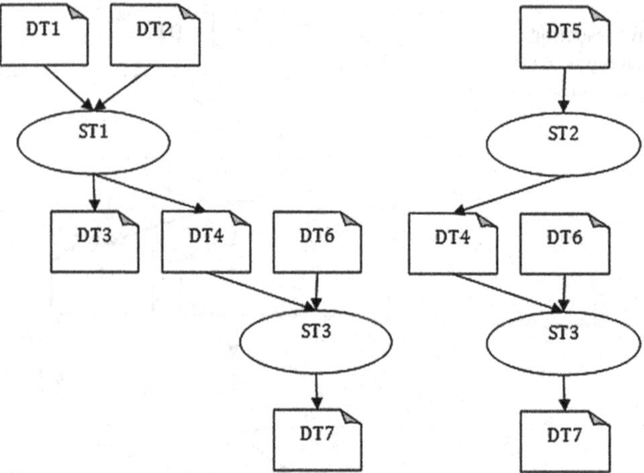

Fig. 8 Two graphs representing plan variants

Limitations of Automatic Planning

The correctness of automatically generated abstract plans depends heavily on how the input documents are defined. Formal specifications (including XML namespaces) of the concepts, such as document types and service types, are the key for the successful automation. The specifications can be created in Dictionary by experts of specific application domains.

The algorithm for abstract planning described above has the following limitations.

- During the construction of an abstract plan each service type is used only once. This prevents creation of infinite loops and large number of alternative plans. A complex process in real world scenarios may use the same type of service many times. Planner will not generate such a plan. Multiple uses of the same service type in all possible matching would lead to "effect of explosion" of number and length of the alternative plans.
- Planner does not consider service types for which the type of input and the type of output are the same. That document is passed to the next service type in the plan. This may also lead to infinite loops during the process of plan generation.
- Planner may not generate a plan if the basic graph SG has unconnected components. The continuous flow of documents between services is the basis for matching types of services and therefore for automatic abstract plan generation.

Manual Abstract Planning

In order to overcome the limitations (described above) manual planning must be used to create partial abstract plans that cannot be generated automatically. These partial

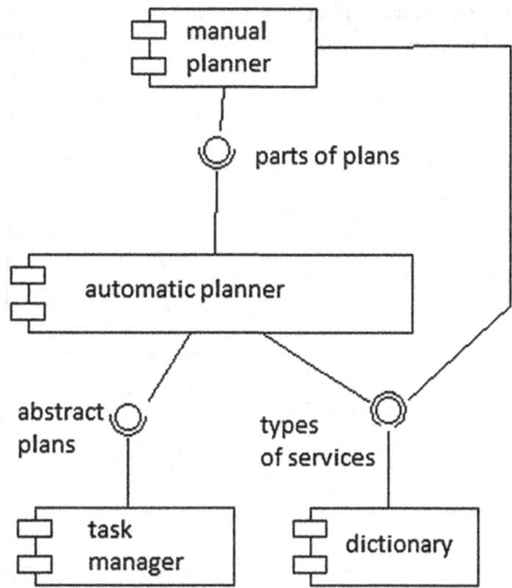

Fig. 9 The relations between elements of the abstract planning system

plans can be used as ordinary plans or as *complex service types* (explained below) in automatic plan generation.

In SOA-enT a dedicated module for storing manual plans is created. It can also provide tools supporting design and management of manual plans by domain experts. Manual plans may be also prepared by using existing business process modeling techniques (e.g. BPMN [38]) and then transformed into format used by the repository of manual plans. The relationships between elements of the abstract planning system are illustrated in Fig. 9.

Manual plans defined and stored in the manual planning module are used during automatic plan generation in the same way as service types received by Planner from Dictionary. They can be seen as *complex service types*. In the graph of manual plan, the uppermost nodes are the input document types of the complex service type, whereas the lowest nodes are output documents of this complex service type. An example of a complex service type is presented in Fig. 10.

During plan generation, automatic planner uses the complex service types in the same way as any other type of service, i.e. it checks if given type fits into plan by matching the names of types of input documents and the names of types of output documents. The difference is that while transforming a graph (representing a generated plan) into XML documents, the complex service types are unfolded, i.e. they are replaced by the corresponding manual plans, see Fig. 11.

The Concept of Process Shadow

To automate the process of adapting business process models from the existing standard tools (e.g. BPMN [38]), the concept of *process shadow* is introduced. The

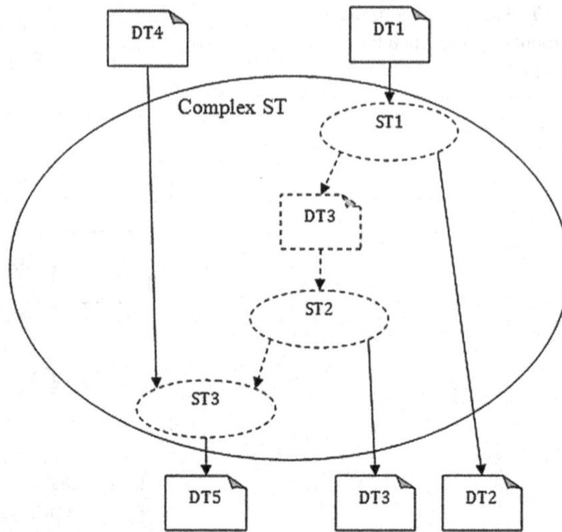

Fig. 10 Manual plan as a complex service type

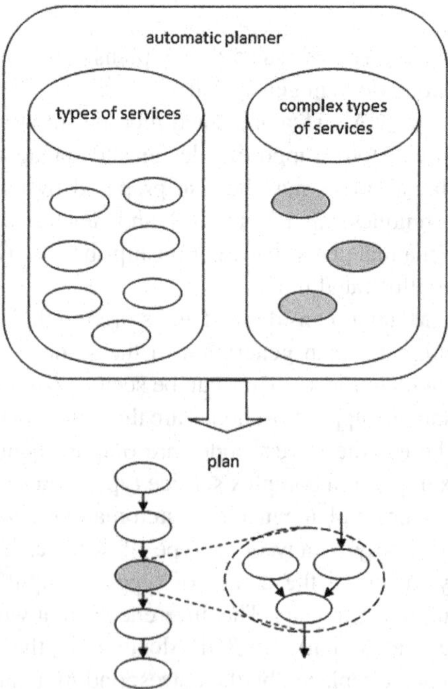

Fig. 11 Use of manual plans within the automatic planner

shadow is derived from a (complete or partial) process definition created by a domain expert using a visual modeling (e.g., TIBCO Business Studio Community Edition [39], and BizAgi Process Modeler [40]), and exported into a standard format (e.g.

BPMN notation). This process shadow can be converted to an abstract plan, and stored in the repository of manual plans, so that it may used by automatic planner.

The procedure of converting a model expressed in BPMN to an abstract plan is shown in Fig. 12.

A model of processes designed in BPMN standard is usually in XPDL format. It is loaded to Planner and converted into internal format. Then, the shadow of a process model is created by removing unnecessary elements such as partitions, paths, gates. Remaining elements (activities, artifacts and transitions) can be directly translated into service types, document types and relations.

A model of a business process defined in the BPMN notation is presented in Fig. 13. A corresponding version of the model with all the removed elements not needed for the process shadow is presented in Fig. 14. The BPMN model is *trimmed* during its conversion to data structures representing the process shadow. Any activities and artifacts that do not have definitions in Dictionary are also omitted.

The abstract plan generated from the process shadow (see Fig. 14) is shown in Fig. 15.

A process shadow is represented in XML format, where the main element is an *abstractPlan* that includes three main groups of elements:

1. *activities*—they correspond to activities in XPDL and are mapped to service types;
2. *artefacts*—they correspond to the *Data Object* artifacts in XPDL, and are mapped to document types;
3. *relations*—they correspond to the following elements:
    - *inputs*—input data corresponding to the associations in XPDL;
    - *outputs*—output data corresponding to the associations in XPDL;
    - *dataFlows*—a data flows that corresponds to a pair of association in XPDL: the first one in the pair connects a source activity with an artifact, whereas the second one connects the artifact with a target activity;
    - *sequences*—a sequence is a pair of activities that should be executed one after another but are not connected by a flow of documents; a sequence may also be a result of removing activities that do not have their service type equivalents in the process model.

Relationships between the elements of the shadow process language are illustrated in Fig. 16.

There is an extensive work done concerning abstract planning not only in the context of Web services, SOA and business processes. The following table (see Table 2) collects the most important approaches (i.e., Golog [6], PDDL [41], APPL [42], ISP [7], CAT [5], Synthy [4], SHOP2 [43], MBP [8]), and compares their main features. The columns denote particular approaches whereas the rows the most important feature in abstract planning. The symbols in the table have the following meaning:

- Y—the approach has this feature,
- N—the approach does not have this feature,

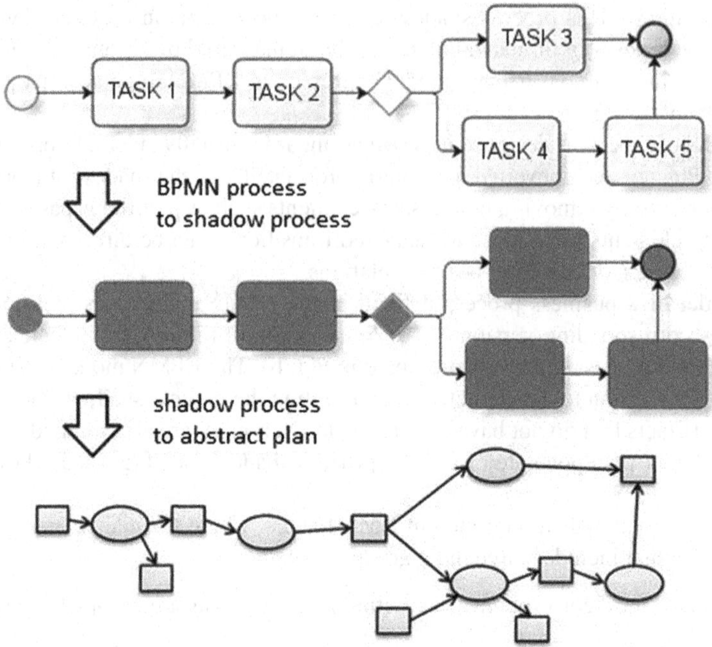

Fig. 12 Conversion of a BPMN model into an abstract plan

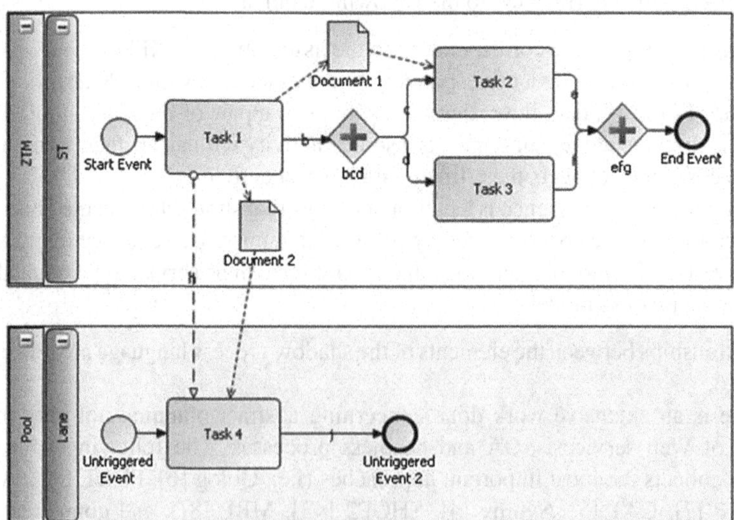

Fig. 13 A model of a business process defined in the BPMN notation

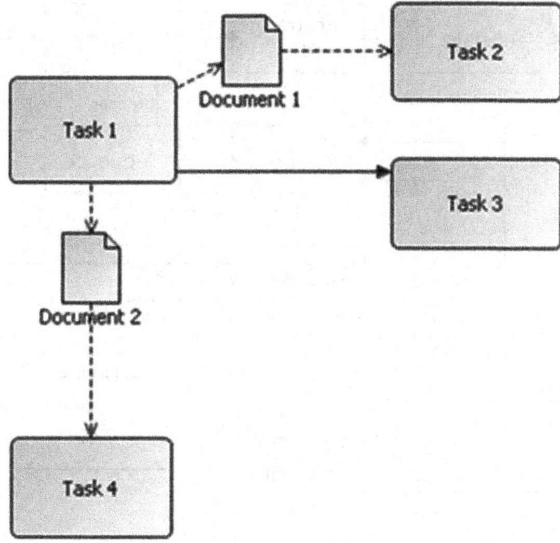

Fig. 14 Trimmed version of the process model

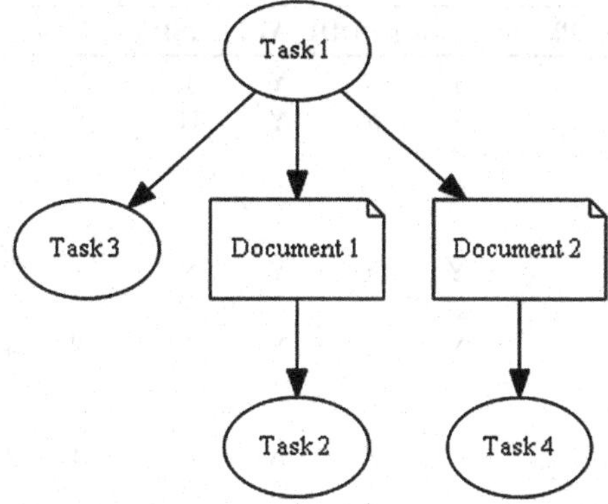

Fig. 15 Graph representing the trimmed version of the process model

- * (star)—partially supports this feature,
- ?—unknown,
- G—good scalability,
- P—partial automation,
- A—full automation.

All these planners (except our Planner) use advanced reasoning for matching (in a plan) input-output of the services, and backward or forward methods. The reasoning is necessary if the whole service description is taken into account, i.e. IOPE that includes the structures of the input and the output documents, as well as precondition and effect being formulas in a formal description language. A service description

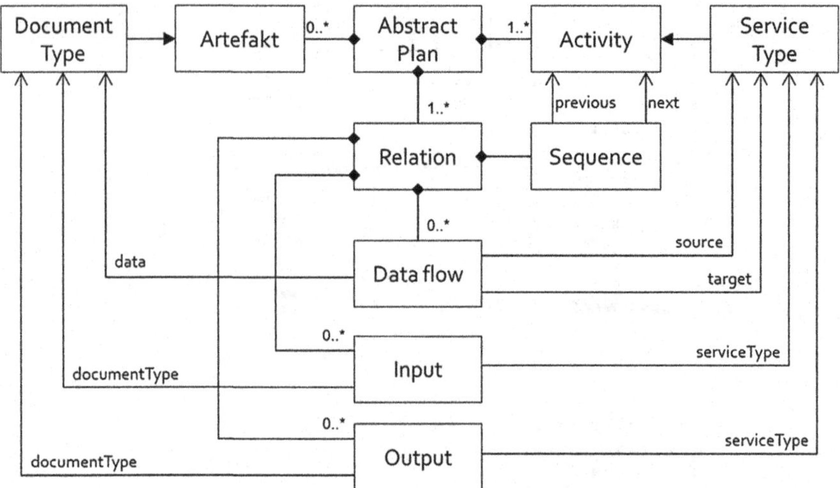

Fig. 16 Relationships between elements of the shadow process language

Table 2 Systems for abstract planning and their features—a comparison

	SOA-enT Planner	Golog	PDDL	APPL	ISP& E	CAT	Synthy	SHOP2	MBP
Use of standard	Y	Y	Y	Y	Y	Y	Y	Y	Y
Complex objects	N*	N	Y	Y	N	Y	Y	N	N
Abstraction, hierarchy	N	N	N	N	Y	Y	Y	Y	N
Non determin-ism	N*	Y	Y	Y	N	N	Y*	Y*	Y
Generation of non-linear plans	N	Y	Y*	Y	?	N	Y*	Y*	Y
Automation level	A,P	P	A	A	A	P	A,P	A	A
Plan selection	N	N	N	N	Y*	N	Y	N	N
Concurrency	Y	Y	N	Y	Y	?	?	N	Y
Scalability	G	?	?	?	?	?	G	G	G
Extended goals	N	Y	N	Y	?	?	?	Y	Y

(based on IOPE) is stored in a service registry and is static, i.e. does not takes into account the dynamically changing capabilities of the service.

Abstract plans generated by these planners are somewhere between our concrete plans and our abstract plans (that are nothing more than simple graphs of the names of service types and the names of corresponding output and input document types).

A concrete plan generated in IT-SOA is verified (using the simple *Commitment query protocol*) by services engaged in the plan via their commitments.

This verification means that the current service capabilities are taken into account in the plan. These very capabilities are dependent on the limited service resources, and cannot be included in the information contained in IOPE that is the basis for concrete plan construction in the most of the contemporary planners mentioned above.

In SOA-enT, Planner lacks of many of these advanced reasoning features, because they are not necessary. It is simple and efficient, and this should be viewed as its advantage. Actually, the reasoning in IT-SOA (necessary for matching services in a business process) is distributed between Service Registry and particular services. It consists of the *static filtering* done in Service Registry, and the *dynamic filtering* done by a concrete service in response to an intention sent by Service Agent. Some simple reasoning is also done by Task Manager to divide a service commitment into several intentions (corresponding to the input documents of the service).

The dynamic filtering (and the corresponding reasoning) is extremely distributed, and at the same time the most efficient because it is done locally in a service according to its business logic that *knows* best how to response to the client intention. In fact, it is *the business* of the service provider to do so in the best way to maximize his/her income. This seems to be a great advantage of our approach to the challenge of automation of business processes.

3.3 Interface Repository

The two main components of SOA-enT (i.e., Task Manager and Service Managers) use Interface repository very extensively in document exchange. Both components were implemented using Spring Framework MVC design pattern [44]. So that each document has a model (storing the document data), and a script presenting the data in the document. When several programmers work on implementing different services and simultaneously use the same document, it may happen that versions of the same document created by the programmers are inconsistent. For this very reason the Interface Repository was designed to store all data used by Service Managers and the Task Manager to process documents. The repository stores not only the model and the script but also additional data, e.g. internationalization files (used by scripts), relations between types, and default configuration. The Repository solves the problem of data inconsistence. Service Managers and Task Manager may update the data automatically, by requesting them from the Repository at any time. The data stored in the Repository have the following structure.

Document is the basic element, which is exchanged by Service Managers and the Task Manager. Each document consists of sections and (if necessary) single fields, which are grouped together as a separate section.

Section is a part of a document and may contain sub-sections.

Model is a class of the Java language. Its task is to store data contained in a document or a section. Each document (also section) has its own model. It is generated by Model Generator, provided that its description is in Dictionary.

Compiled model is in a “bytecode” form, which is sent to Service Managers and Task Manager.

Script is a jsp file, which is used to present data the model contains. Each document and each section has its own script. A validation mechanisms are coded in the scripts.

Internationalization files contain messages for a specific language. They are used by scripts to facilitate translation of messages into different language.

Type is a set of data for a specific document or document section, having the unique identifier. A type is associated obligatory with a model and a script. A type may have a default configuration (it may be automatically generated on the basis of the compiled model), which is then used while specifying the configuration for the service. A type may contain a set of internationalization files, and a set of its sub-types.

Default configuration is required to create the configuration for a specific service. It maps the model fields as well as the configuration parameters: “*ignored*”, “*read only*”, “*optional*”. By setting appropriate values of these parameters one can specify whether related field will be ignored, read only or optional in the default configuration. This is an initial setting and may be overwritten for a particular service.

Each type may contain sub-types as well as super-types, with the exception of types classified as documents, which do not have super-types. Defining a *type hierarchy* is important because it provides information which is needed during model compilation. If the hierarchy is defined incorrectly, the compilation fails.

Service configuration concerns the following four elements: intentions, commitments, input documents, output documents. These configurations are created on the basis of default configuration for corresponding documents, and the sections they consists of. Interface Repository is used by: Dictionary, Task Manager, Service Managers, Model Generator (internal module of SOA-enT), and Eclipse programming platform. Each of them uses, and updates (in different way) the data the Repository contains. Figure 17 shows the flow of data to and from the Repository.

A Service Manager receives packages with models, scripts, configurations, and translations only for document types processed or produced by service types the Service Manager handles. The Task Manager has access to all the data stored in Repository. These components may automatically or manually update (get new versions of) the Repository data (see Fig. 18).

To conclude, although the Repository is an internal (and technical) component of SOA-enT, it plays the crucial role to provide consistency of service interface and documents type definitions, and to provide the uniform document presentation in the entire system.

3.4 Service Managers

A Service Manager is the container for services of the same type; it is used by service providers to:

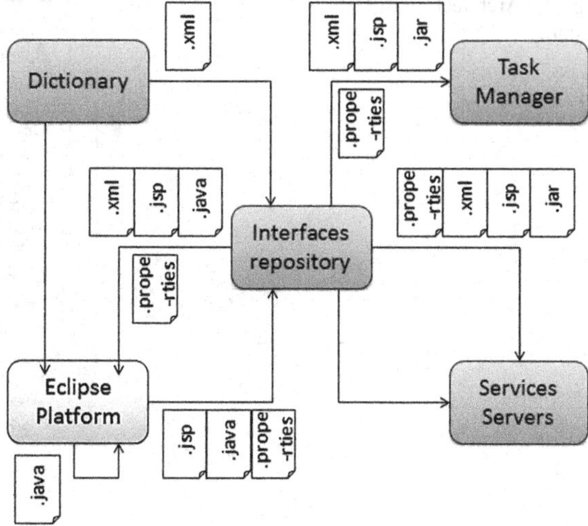

Fig. 17 Interface Repository integration with other components

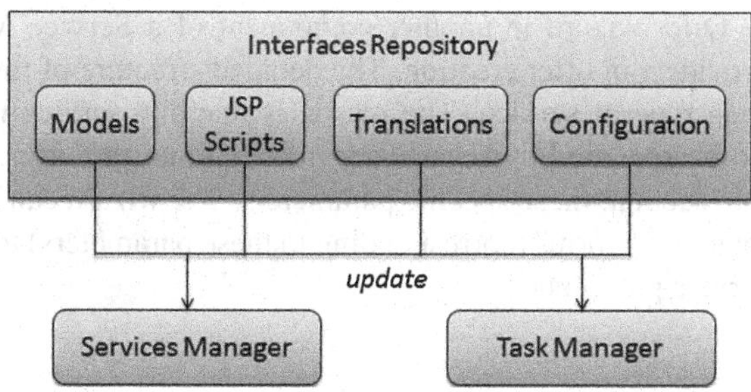

Fig. 18 Using the Interface Repository by Task Manager and Service Managers

1. create users accounts to identify the users in SOA-enT,
2. register services and submit offers related to these services,
3. manage documents during service execution phase.

To implement business logic of the application related to document processing, a programmer may use the Help system to provide the syntax of the documents, and to notify about errors. The syntax is tightly coupled with their definition stored in Dictionary, so that the help system is integrated with Dictionary, and its architecture (see Fig. 19) consists of the following components:

1. *Compatibility analyzer* is the gate for the programmer to use the help.
2. *Document change monitor* is for monitoring changes made by the programmer in editing the source code.
3. *Context menu assistant* prompts the structure of the document being edited currently.
4. *Dictionary data access object* is for getting from Dictionary the structure of documents currently used.

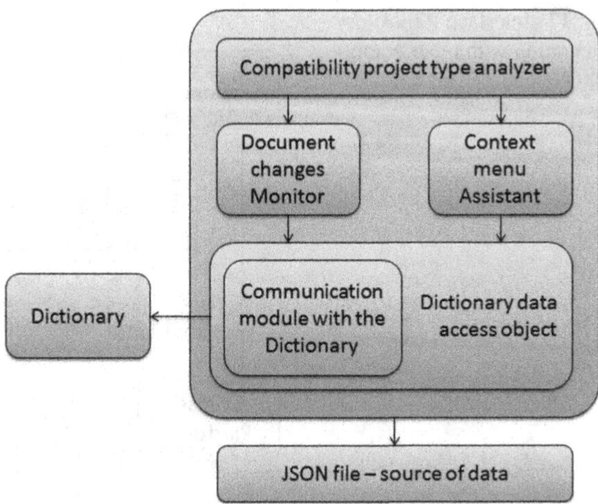

Fig. 19 Architecture of Help system

Offer wizard in another component of a Service Manager. It supports service providers in offer creation. The detailed structure of this component is different for each type of service. The reason is that any service type has its own parameters (to be specified by service provider) so that the offers for a given type must take into account these specific parameters. The wizard can provide sufficient number of interactive forms (corresponding to these parameters) to describe offers precisely by service providers.

3.5 *Task Manager*

Task Manager (TM) is a universal tool in SOA-enT for supporting task formulation by users. It is responsible to accomplish the task automatically by planning, arranging, and composing services into complex business processes. Finally TM executes the processes, and provides their monitoring and controlling.

TM consists of the following components that cooperate with each other using strictly defined interfaces (see Fig. 20).

- **Graphical user interface** is responsible for communication with the user.
- **Document manager** provides appropriate Spring MVC models for documents (Web user interface implementation is based on MVC paradigm), and allows users to view, create and edit documents.
- **Process access object** provides access to a persisted process state stored in a file (on a hard disk or in a database).
- **Process manager** reacts when certain events occur, and then changes the process state. The events may be triggered by the user (e.g. by Document manager) or by receiving messages from other components of the system:

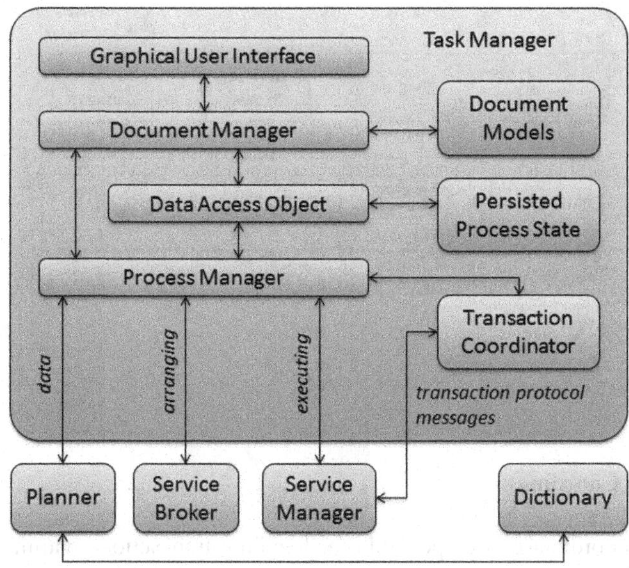

Fig. 20 Task Manager architecture

- Planner in the planning phase,
- Service Broker in the arrangement phase,
- Service Manager in the execution phase.

- **Transaction coordinator** handles transactions among services participating in the process.

The first TM functionality is to provide friendly communication with users. It involves support for the following activities:

- to register and login,
- to define new tasks,
- to monitor the dynamic state of a task accomplishing process,
- to visualize the process in the form of a diagram,
- to manage task accomplishment,
- to browse, selecting and specifying commitments,
- to notify about relevant events,
- to manage documents.

Each task is realized in four phases, during which interaction with the user may be needed in order to specify required data or to decide how to continue the process. Figure 21 presents task realization phases, as well as corresponding modules with which the Task Manager communicates in each of the phases. If there is a failure during the arrangement phase, the process returns to the planning phase. In case of failure in the execution phase, the process returns to the arrangement phase.

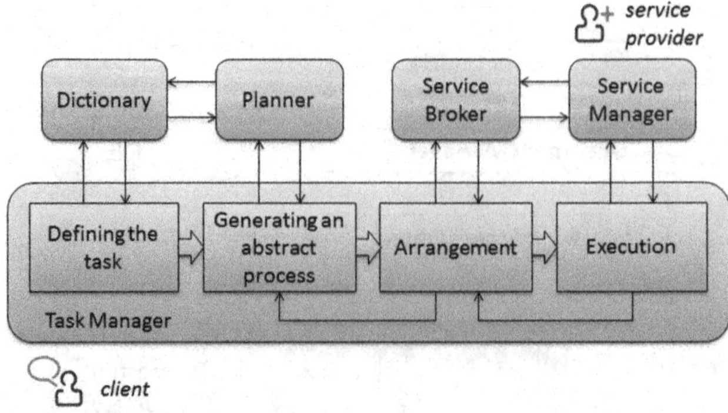

Fig. 21 Task realization phases

Transaction Coordinator

Transaction coordinator is responsible for handling transactions within a process. Transaction control in the execution phase is done according to the standard transaction protocol, i.e. *BusinessAgreementWithParticipantCompletion protocol handling* that conforms to the WS-BusinessActivity specification [45].

Although WS-Business Activity specification protocols were developed for implementing them in a coordinator that conforms to the WS Coordination specification, in SOA-enT our own coordinator was created consistent with the SOA-enT architecture. However, the general principles of the WS-Coordination were implemented also in our coordinator.

The creation of the transaction context is initiated inside the Task Manager by the special Process Manager Module. This context contains all elements that conform to the WS-Coordination. Registration module, according to the WS-Tx Specification, is responsible for informing the coordinator that a service will participate in the process, and providing the service with the address of the end point that handles the transaction protocol.

In the SOA-enT, services composed into a process are determined in the arrangement phase. For this reason, the coordinator does not have a registration module, and obtains necessary execution information from the Process Manager.

Initiation of transaction protocol for a given step starts if the input documents have been sent to a service. At this moment, a new participant (corresponding to the process step) is added to the context. Transaction coordinator supervises the service transaction according to the commands received from the Process Manager. When the coordinator receives a message from a service that its operation has failed, it relays the information to the Process Manager that replaces this service with another one, or generates a new process fragment that realizes the same sub-task as the service being replaced. If this replacement does succeed, the process coordinator, having arranged a new fragment of the process, continues to handle the transaction for the modified process. If it is not possible to replace a service that failed or to recompose

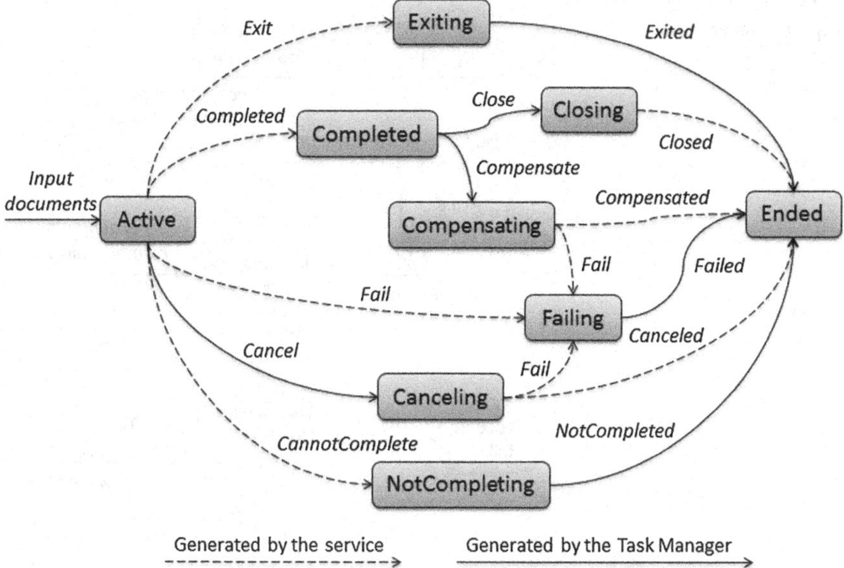

Fig. 22 State diagram of a single business service

a part of the process, the coordinator cancels the transaction and sends appropriate messages to all services taking part in the transaction (e.g. cancel, or compensate). Figure 22 shows all states transitions of a single service in the transaction protocol.

Unexpected Event Handling

Figure 23 presents how unexpected events are handled in the arrangement phase, and in the execution phase. Correct paths were omitted for the purpose of presentation. The algorithm starts the arrangement phase with the message "No commitments". It means that there is no service of the specified type in the Service Registry, and it must be replaced by another type. In the execution phase, when a service cannot perform the commissioned task, it sends an appropriate message (that conforms to the WS-BA specification protocol) to the Task Manager. The Manager makes an attempt to rearrange the plan, and tries to find other services of the same type. If this fails, the subsequent step of the algorithm is realized, which also starts handling failures in the arrangement phase. In this step, final state for the new fragment of the plan is also defined. It involves specifying appropriate types of documents that will be used in the subsequent services. If it is not possible to define a new final state (e.g. an attempt to generate a plan, that produces documents needed for the realization of the task, has been unsuccessful), the process ends in failure. Otherwise, an attempt to generate a new plan is made.

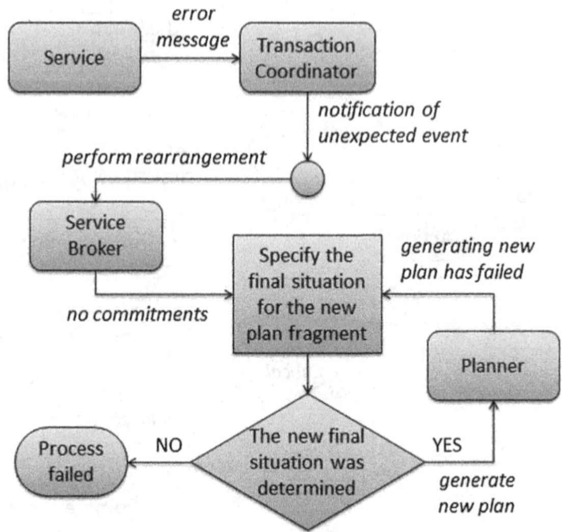

Fig. 23 Algorithm handling unexpected situations

The algorithm consists of the following steps (the next step is performed when the previous one fails):

1. *Replace a service with another one of the same type*—this is the simplest case and involves finding out (in the service registry) a service of the same type as the service that failed, and then replacing this service. This can be done if the commitment received from the new service agrees with the documents (already produced by the previous services in the concrete plan) that now are the input documents for the new service.
2. *Generate plan fragments, the execution of which would result in production of at least all the documents returned by the service being replaced*—if the service is not the first one in the branch (it needs documents generated by other services), the fragments, whose inputs are documents returned by services preceding the service being replaced, have to be taken into account first. If the inputs of the generated fragment of the plan match the inputs of the service that failed, then the process generated on the basis of this plan fragment, may be called a complex service. From the point of view of the task accomplishment, the modified process (where the failed service is replaces by this complex service) does not differ from the original process.
3. *Generate a new plan fragment, the execution of which would result in production of at least all the documents returned by the services that are next (in the execution order) to the failed service.* These services need (as their input) the documents that the flailed service was supposed to produce. These services have not been executed yet, so that the rollback mechanism can be applied to remove them from the process.
4. *Repeat step 3 until the correct plan is generated*—if a new fragment would need the same output as for the whole process, then the entire process must be

reconstructed what amounts to the total failure with no possibility of recovery. Then all services in the process are to be canceled by the rollback (if possible) or by a compensation.

When a new plan fragment is generated, a corresponding new fragment of the process is created by arranging the services in this plan, and the process fragment is linked to the main process so that the task accomplishment may be continued.

Building a new fragment of the process in steps 2 and 3, may result in a disconnection of some services (or even the whole "branches") whose outputs are not used in the modified process. Such services are removed from the process. Compensation mechanism is used for the already executed services whose output documents are needless for the modified process. Services in the execution phase are requested to terminate. For the services being in the arranging phase, their commitments are cancelled by the rollback.

If one or more services have been replaced (by a new fragment of a process), arranging is carried out for new elements of the process. The simplest case is when a new fragment of the process does not use the documents that have been produced so far in the process. In this situation, it is enough to perform arranging for this fragment and go to the execution phase. If the process uses documents produced in previous steps, then, after the arranging phase, appropriate output documents from previous services are compared with the commitments of services which will use them as the input documents. If such verification is successful, it means that the existing documents may be used. Otherwise, the service that produced incorrect (in the present situation) result must be rearranged (considering new requirements), and executed again. Such a situation is presented in Fig. 24. If service 2 fails, it is necessary to check if the document generated by service 1 is compatible with a commitment of service 10.

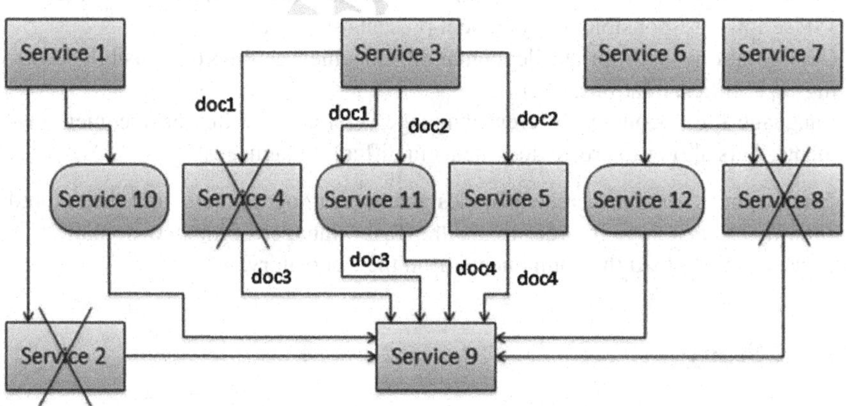

Fig. 24 An example of a business process after re-composition

Figure 24 presents a business process after re-composition. Services that failed are replaced with others. Services that failed and services being replaced are stroked out, and new services are in the oval form.

Service 2 was replaced with service 10. Both services are of the same type, so inputs and outputs are compatible. The services were replaced according to the first step of the algorithm presented above. Service 8 was replaced with service 12 according to the second step of the algorithm. Service 12 may replace service 8 because it has smaller requirements for the input than service 8 (it does not need the result generated by service 7). To deal with failed service 4, a new fragment of the process was generated, which, in this case, consists of service 11. In comparison with service 4, service 11 additionally generates document *doc4* and it needs document *doc2* as the input.

The presented diagram of the recomposed process also shows unnecessary services. Service 7 produces the result needed by service 8. Service 12, which replaced service 8, does not need this result. Therefore, service 7 may be removed from the process, and if it has already been executed, it is necessary to perform compensation for this service. Another service that may be removed is service 5. The result it produces is used by service 9. However, after re-composition of the process, service 11 has appeared, which produces the same document as service 5. In this situation, service 5 may be removed from the process, and compensation may be performed for this service if needed.

Applying TM to a New Application Domain

Task Manager is a universal element, and does not depend on a specific application domain. However, special components are needed to handle documents in a new domain. Such components include:

- Models of documents—Java class files representing document data structure.
- JSP scripts—responsible for document presentation layer.
- Configuration—appropriate files containing document settings (e.g. visibility, editing capabilities, relations, etc.).
- Language files—contain different language versions of different document elements. This allows to create documents in different languages.

These components are created by designers and programmers, and then stored in *Interface Repository*. In order to use the Task Manager in the new domain, it is necessary to download the components from the Repository.

4 Case Study

SOA-enT can be adjusted to a new application domain by:

1. introducing to the Dictionary new document types and new service types specific for that domain,

2. defining scripts and models of new documents to be available for the whole system through the Interface Repository,
3. implementing Service Managers for handling service of new types.

Types of services are abstract descriptions of operations performed by services. Each type of service is described by specifying the three elements listed below:

1. input document types that define what documents a given service requires to perform an operation,
2. output document types that specify what documents will be produced by the service as a result of the operation,
3. fault messages as output documents if:
   - a service cannot be correctly executed,
   - an unexpected event or failure has occurred.

A panel for viewing/editing service types is available in the *Types of services* tab in the Dictionary section, see Fig. 25. Below the tab there is the category element for grouping related services. By choosing a service type name in the panel, a structure

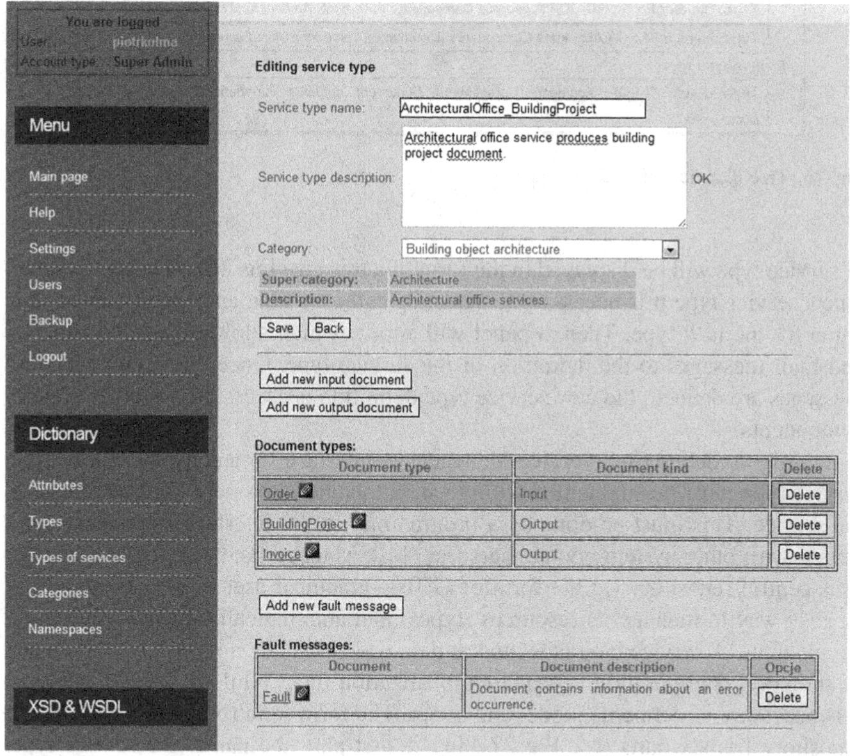

Fig. 25 Editing a service type definition in Dictionary

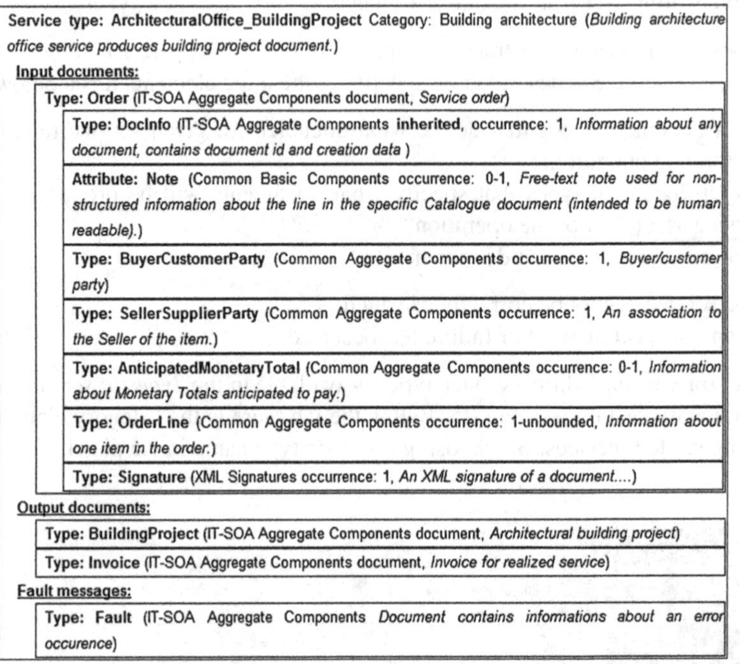

Fig. 26 Tree like view of a service type

of service type will be displayed in the form of a tree (see Fig. 26). In order to define a new service type it is necessary to select its category first, and specify an unique name for the new type. Then, a panel will appear which allows to add documents and fault messages to the definition of the service type. Once the documents and messages are defined, the new service type definition is accessible for other system components.

In order to define (in a Service Manager) an interface for a concrete service, the service type must be augmented with a validation and a business logic specific for the service. This must be done by a programmer in the Interface Repository, and shared with other system components (i.e. Task Manager and Service Managers). This central repository (in the form of GUIs—graphical user interfaces) provides an easy way to manage its resources (types) and automatically updates registered components. A service type is related to data which include a model, a script, a set of subtypes, configuration and internationalization files. While defining a subtype it is necessary to define its model and script. The form to introducing a type to the repository has six parts (see Fig. 27). In the first part, the name (the service type identifier) must be specified in the field. Each type may have an unlimited number of aliases, see the second part.

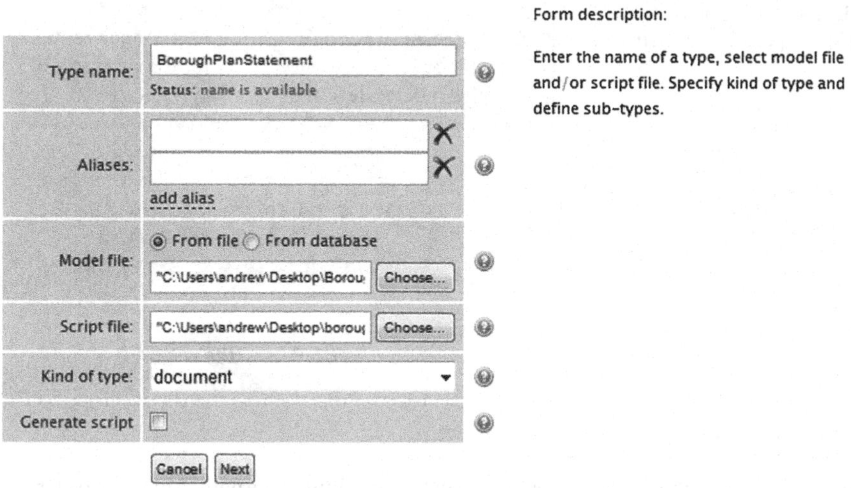

Form description:

Enter the name of a type, select model file and/or script file. Specify kind of type and define sub-types.

Type name: BoroughPlanStatement
Status: name is available

Aliases:
add alias

Model file: From file From database
"C:\Users\andrew\Desktop\Borou Choose...

Script file: "C:\Users\andrew\Desktop\boroug Choose...

Kind of type: document

Generate script

Cancel Next

Fig. 27 Adding a new type to the Interface Repository

In the third and fourth parts it is necessary to indicate the location for model and script that may be fetched either from files or from the database which stores the .java files produced by the model wizard. The fifth part contains the field to specify the type. Three values are available: *document, template, section*. The value *document* is selected for a type not having a super-type, *template* specifies that the type can be inherited by other types, whereas *section* specifies the it is a sub-type used in documents and templates. The last part of the form is for generating a script if the type being added is a document. The script is generated on the basis of templates defined while adding sections.

The most difficult step of adjusting the system to a new application domain is the process of implementing Service Managers that can handle a new service type. There are available tool packages that provide templates and code libraries that can greatly decrease the time and effort needed to implement a Service Manager. The packages handle communication with other system components according to the communication protocols.

Task Manager, Planner and Service Broker do not require any changes in new application domains. A user defines his/her task using the Task Manager on the basis of the names and format of document types defined in the Dictionary. The task is in the form of a formula that describes the document (desired by the user) by specifying values of some (not necessary all) attributes of this document. During the arrangement phase the rest of the attributes will be specified according to the commitments from the services composed into the concrete plan. In the beginning of the plan execution phase the user is asked to complete the necessary input documents that are sent to services arranged in the plan. Task Manager also handles exception and failures during plan arrangement and execution by rearrangement of a single

Fig. 28 An example of the business process in logistics

service or a number of services, or even by applying alternative plan in order to accomplish the client task successfully.

The business process presented below is only one of many examples realized in the SOA-enT testing environment. The main purpose of the tools created in the framework of SOA-enT was to apply them to construct communication platform for developing electronic markets called Neiberia presented in the subsequent section.

Ontology (as a collection of document types and service types) for a specific application domain is introduced to the system via Dictionary, for details see Sect. 3.1. In the quite similar way all the notions and corresponding ontology for realizing processes related to logistics were already introduced to the Dictionary. Based on the ontology, business processes can be planned, arranged, executed, monitored and controlled in SOA-enT.

The example of a business task related to logistics consists in purchasing by a company (called BestTea) of a large quantity of goods (say, tea from China and India) and transporting it to the company store in Bilgoraj city, see Fig. 28. Here, only a sketch of the example is presented; a detailed presentation is on www site http://www.itsoa.ipipan.eu/.

The business task is decomposed by Planner into two consecutive subtasks. The first one is the purchasing whereas the second one is the transportation. For the simplicity of the presentation, suppose that the first task was already done, however only to the arrangement stage, i.e. BestTea has decided (on the basis of the offers from several warehouses) that the goods will be purchased from the warehouse in Szczecin city.

An arrangement consists of the following consecutive steps:

1. Sending by the customer a query to vendors (service providers).
2. Receiving offers (pro forma invoices).

Then, the user (customer) selects the best offer and prepares an order.

In our use case, this is done by the company via Task Manager; the order for purchasing goods (along with their delivery to the nearby railway station) has been already prepared by the company. The transport to Bilgoraj must be arranged in the quite similar way by a carrier company, however, first a railway wagon must be rented. After successful delivery the wagon must be returned to the location specified by the rental company. All these subtasks (shown in the Fig. 28) are interrelated and their accomplishments must be synchronized and optimized. This is done by the company using Task Manager in the following way. The rental of a wagon (subtask 1.2) and its return (subtask 3.1) to Hrubiesz?w city was arranged with the company called PKP Wagon-Szczecin. The transport was arranged initially with carrier called PKP Cargo.

The execution of the arranged business process started with sending digitally signed order by BestTea to the warehouse and receiving digitally signed contract. The next order was sent by BestTea to PKP Wagon-Szczecin; the corresponding signed contract for wagon rental was sent back as the response. However, after sending the order to PKP Cargo it turns out that it cannot provide the arranged transport, so that the contract was not returned; the reason was that, say, due to the current license constrains, the PKP Cargo could transport the wagon only to Sławkowa city. This part of the plan must be rearranged taking into account the two contracts: one with the warehouse and the second one with PKP Wagon-Szczecin. To recover from this failure, first new partial plans must be generated by Planner that cover the transportation from Szczecin to Biłgoraj. In one of these abstract plans the transportation could be done in two steps: the first one is from Szczecin to Sławkowa by PKP Cargo, whereas the second step is from Sławkowa to Biłgoraj by the carrier called LHS. The plan is presented to the user (i.e., BestTea) that once chosen, the arrangements are performed with LHS and PKP Cargo that synchronize them with the purchasing and the wagon rental. If it is done successfully, the appropriate orders are sent first to PKP Cargo and then to LHS, getting back the signed contracts.

From the virtual point of view (on the basis of the signed orders and contracts) all is done except the payments. However, there is also the associated process to be realized in the real word, so that in this case, the execution phase consists in realizing all effects that are described in the contacts. During this execution phase possible failures are handled by clauses written in the contracts. However, if for example the goods were transported only to Sławkowa where LHS could not perform its contract obligations (and paying the penalty according to the contractual liability), BestTea must continue the business process using Task Manager to recover by creating a new partial plan and adjusting it to the already executed part of the process.

The invoice-payment phase for purchasing, as well as for all the rest services may be done concurrently according to the corresponding contracts. The phase for delivery note and/or acknowledge receipt may be also incorporated into the business process because it is nothing but a yet another document flow and processing. However, in the current state of development, the system SOA-enT is limited only to query-quotation and order-contract phases. It should be noted that SOA-enT is only an experimental implementation of the proposed new technology consisting of the protocols and the generic service environment representation (i.e. XSD schemata of documents processed by services). From this point of view, IT-SOA serves as an experimental prototype for proving the usefulness of the technology.

5 Social Media in Electronic Markets

Social media and online markets transform the traditional way for doing business and provide the right mix of information technologies to develop e-commerce.

5.1 The Concept of Social Networking

This subsection is based on the work [46] done in the framework of IT-SOA project by the team from the University of Economics in Poznan.

Virtual organization is usually defined as the structure (organization) made up of business entities created dynamically to achieve the business temporary goal. The Virtual Organization Breeding Environment (VOBE) is defined as an association of business entities and institutions supporting them under long-term joint cooperation agreement on the basis of established principles and an infrastructure. The aim of this association is the dynamic creation of virtual organizations to achieve a specific business goal. SOA paradigm can be applied to VOBE (and called SOVOBE) in the manner that the capabilities of business entities are represented as services, and thus described by standards such as WSDL and OWL-S. Consequently, this enables automatic processing of semantic description of services in order to compose them into complex services (i.e. business processes). Due to this application, different virtual organizations created within the framework of SOVOBE can be composed to realize more sophisticated global business goal.

The concept of Social Network (well known in Sociology) was applied to SOVOBE in very interesting and innovative way, that is, it captures the social context which exists between members of SOVOBE and members of virtual organizations. In the case of SOA approach there are also relations between service providers and their customers. Social roles and relations are expressed in the communication protocols between the network members. The social networks augmented in this way are closely related to the social media in the B2B context. Hence, it seems to be possible to incorporate the existing social media into the systems based on SOVOBE framework.

Virtual organizations in the context of small and medium business have been introduced by W. Cellary [47] and [48], where *network virtual organization, is considered as a collection of business units cooperating on the market through a network as if they were a single company.* Also a remarkable claim was presented there: *"The future belongs to small and medium-sized enterprises integrated in virtual network organizations."* Actually, in [48], the concept of electronic market for small and medium enterprises (SMEs) and virtual organization was proposed as *"a virtual platform for cooperation between enterprises."* In [49], also the concept of the so-called *Content Communities on the Internet* that is important for electronic markets was described. Based on the above arguments, the conceptual and architectural similarity of SOVOBE and the platform SOA-enT implemented by the authors of this chapter should be stressed. The similarity concerns the description, publishing, discovering and composing simple services into complex business services. It is an obvious consequence of the fact that both platforms are based on SOA. The social media (introduced to the communication platform SOA-enT for developing electronic markets) are analogous to Social Network in SOVOBE. However, they are much more simple and because of this simplicity more scalable. We appreciate the pioneering approach done by W. Cellary in applying social network concepts in the SOA and virtual organizations. Later, the validity of the approach was confirmed in several publications and reports, e.g. [50, 51], IBM market research [52], and implemented systems, e.g. [53].

5.2 Social Media

Social media are defined as the means of communication for the implementation of social interactions using scalable and widely available information technologies. This term refers primarily to the Web-based (also mobile) technologies, which enable an interactive dialogue between users by means of web browsers.

The social media are often identified with Web 2.0 technologies dedicated mostly for the creation and sharing of web content by users. They are also used by business and electronic markets in consumer-generated media (CGM) where the interactions between customers generate new business value manifested in the increased (or decreased) confidence for business partners. Generally speaking, social media can be defined as a fusion of new communication technologies and social behavior of users.

Rapid, even revolutionary, growth of social media influences the development of electronic markets. Users of social networks (in addition to the social interaction) are increasingly interested in e-commerce, e.g. Groupon. Different kinds of communities created within the portals like Facebook or Google+ provide confidence to the users and ability to manage the trust. Electronic markets are understood here as a meeting place of business partners for the purpose of planning, negotiation, business process arrangement and other activities related to business. Because hundreds of millions of (potential) customers are using social networking websites, they may do business as well. Using social media to create new, more expansive electronic markets has become a reality and it is possible in several ways by:

- incorporating social media into existing technology solutions for electronic markets (e.g. Goldenline, LinkedIn, Grupon);
- developing new technologies, e-business solutions for existing social media, such as f-business, f-commerce (f stands for Facebook);
- developing completely new solutions, i.e. new social media incorporated into new technologies for electronic markets.

Existing social networking sites often adapt the functionalities of a typical e-business, such as engaging in online commerce or providing information services (like CNN International). Other portals are offering Business Directories that stores information about companies, products, services, etc. On the other hand, companies are using social networking to create its brand and corporate image, recruiting new employees. It is usually done by redirecting from social networking sites to the own sites, where further information is available and other dedicated functionalities are offered. Social networks are increasingly used in business for the following reasons:

- They connect people at low cost. Small and medium sized companies often use the services to influence potential customers through advertising campaigns or posting regular announcements.
- They support online business communities (e.g. linkedIn.com).
- Business communities portals work as online meeting places for professionals.
- The business users of virtual communities create their own identity and image, contact directly with each other or within a group that has common interests and similar goals.
- Social networking websites focus huge number of people that are potentially interested in the services, product, employment, etc. They are able to influence people or institutions by creating interactive informational meeting places (Business networking).

The slogan: *B2B social networks* that is referring to the evolution of electronic markets in the direction to social media is becoming increasingly popular. This manifests in social networking that is no longer made from the side of individual users, but from companies that see this as being in their common interest.

An important feature of electronic markets integrated with social media is the ability to manage trust of the customers to services providers (and vice versa) in the C2B relations as well as trust between business partners in B2B relations. This management is based on making recommendations, introducing new companies to the markets, and the assessment activity (evaluation of these companies by customers and business partners). Many companies already have dedicated departments for creating and maintaining a positive image of the company on the main social networking websites.

5.3 Neiberia

The main goal of IT-SOA project was to develop and implement new information technology based on the SOA paradigm for electronic commerce and information society. It is obvious that at the beginning of the project (four years ago) it was not possible to predict all the new trends in the development of electronic markets. Now it is clear, that for successful and effective development of the communications platforms for electronic markets, the growing potential of social media must be used. It has been done by extending the SOA-enT platform with the classic functionalities known from the social media portals as well as introducing new ones. The new extended platform was called Neiberia and is available online as **neiberia.com**. Users of Neiberia can act as clients (regular users) as well as service providers (managers of small and medium sized enterprises). In this sense, communities created within Neiberia correspond to a classical C2B and B2C relations. Hence, Neiberia is not a typical social networking site such as Facebook, or Google+. The fundamental difference consists in the introduction of new structures corresponding to real communities with B2C and B2B relation (e-markets as meeting places) between their members. The communities are essential here; individual users and their business relations are visible only through the communities they belong to.

Neiberia provides business functionalities typical for electronic markets such as offers, commissions, sales, auctions, sharing skills, and recommendations. It also supports planning and realization of joint business projects in the form of quasi-automated business processes, i.e. compositions of simple services. Neiberia provides also traditional forms of communication such as events, chats, forums, galleries, blogs, and commenting. The Neiberia platform was also extended with functionality supporting the business communities (corresponding to Business networking, Business clubs, etc.) for small and medium enterprises (SME). The functionalities related to business allow for:

- Defining business profile of a company.
- Defining the semantic description of the services provided by the company (based on the definitions of service types stored in Dictionary).
- Creating B2B cooperation communities. They can be used by companies of similar type which, e.g. cooperate in order to protect their rights and interests (trade unions), or for the purpose of discovering, planning and execution of business ventures in a group of companies from different branches of the industry.

Similar technologies are emerging, such as http://www.grera.net/ (see [54]) which are defined for small and medium enterprises as an equivalent of Facebook and Google+ for SMEs. So far they are based on the classic (facebook) model and are not so advanced as the Neiberia platform is in applying the business functionalities.

6 Summary

The main goal of the IT SOA project was to develop new information technology based on the SOA paradigm for electronic commerce and information society. In this chapter an example of such technology was presented, i.e. a platform for developing electronic markets of sophisticated business services; the platform is called SOA-enT.

The proposed technology is generic and can be easily adapted for a new application domain by defining new ontology specific to this new domain, i.e., by defining service and document types in the Dictionary and implement Service Managers handling these new types of services. A set of tools is provided that support the process of implementing new instances of Service Mangers including Interface Repository and code libraries handling communication with other system components. In the case of Service Broker and Task Manager no changes are required to adapt them to a new application domain.

Because of rapid growth in popularity of social media, a new extended version (called Neiberia) of the SOA-enT was implemented. It integrates widely known and used (however, so far separately) business and social activities, and proposes a new form of business cooperation mainly for small and medium enterprises. The extended SOA-enT is flexible and allows to create different kinds of electronic communities by the individual as well as business users. The communities may have social, business or mixed character, depending on the configuration of the available tools. The kinds of communities include societies, companies (as communities of employers and employees), cooperation of a group of companies and finally the most important one, i.e. the electronic markets that are built in into these communities. Moreover, communities may be developed around these markets.

References

1. Davenport, Thomas: Process Innovation: Reengineering Work Through Information technology. Harvard Business School Press, Boston (1993)
2. Peltz, C.: Web services orchestration and choreography. Computer **36**(10), 46–52 (2003)
3. Alvarez Napagao, S., Biba, J., Confalonieri, R., Dehn, M., Kollingbaum, M., Jakob, M., Oren, N., Panagiotidi, S., Solanky, M., Vazquez Salceda, J., Willmott. S.: Contract based electronic business systems state of the art. In: IST CONTRACT PROJECT, 10 April 2007
4. Agarwal, V., Chafle, G., Dasgupta, K., Karnik, N., Kumar, A., Mittal, S., Srivastava, B.: Synthy: a system for end to end composition of web services. J. Web Seman. **3**, 311–339 (2005)
5. Kim, J., Gil, Y.: Towards interactive composition of semantic web services. American Association for Artificial Intelligence (2004)
6. McIlraith, S., Cao Son, T.: Adapting golog for composition of semantic web services. Knowledge Representation and Reasoning, pp. 482–493 (2002)
7. Therani, M., Uttamsingh, N.: A declarative approach to composing web services in dynamic environments. Decis. Support Syst. **41**(2), 325–357 (2006)
8. Traverso, P., Pistore, M.: Automated composition of semantic web services into executable processes. Lecture Notes in Computer Science LNCS, vol. 3298, pp. 380–394. Springer, Heidelberg (2004)
9. Coyle, F.: XML, Web Services, and the Data Revolution. Addison-Wesley Information Technology Series (2002)
10. Daconta, M., Obrst, L., Smith, K.: The Semantic Web: A Guide to the Future of XML, Web Services, and Knowledge. Wiley, New York (2003)
11. Kaye, D.: Loosely Coupled: The Missing Pieces of Web Services. RDS Press, Marin County, California (2003)
12. OASIS Web Services Business Process Execution Language (WS-BPEL), May 2006. TC. Web Services Business Process Execution Language v2.0. Committee Draft May 2006
13. Chinnici, R., Moreau, J.-J., Ryman, A., Weerawarana. S.: Web Services Description Language (WSDL) Version 2.0 Part 1: Core Language. W3C Recommendation, 26 June 2007
14. Ambroszkiewicz, S., Bartyna, W., Faderewski, M., Mikułowski, D., Terlikowski, G., Stepniak. M.: A revision of the SOA paradigm from the e-business process perspective. In: Cellary W., Grzech, S. Ambroszkiewicz, W. Brzeziski, K. Zieliski (Eds.), SOA Infrastructure Tools. Concepts and Methods, pp. 419–438. Poznan University of Economics Press, Poznan (2010)

15. Ambroszkiewicz, S., Bartyna, W., Faderewski, M., Mikulowski, D., Terlikowski, G., Stepniak, M.: The SOA paradigm and e-service architecture reconsidered from the e-Business perspective. LNCS **6385**, 256–265 (2010)
16. McGuinness, D., van Harmelen, F.: Owl web ontology language overview. http://www.w3.org/TR/owl-features/. Accessed 10 February 2004
17. Ambroszkiewicz, S.: Entish: agent communication language for service integration. In: Wierzchon, S., Klopotek, M., Trojanowski, K. (eds.) Intelligent Information Processing and Web Mining. Advances in Soft Computing, pp. 49–58. Springer, Berlin (2003)
18. Ambroszkiewicz, S.: Entish: a language for describing data processing in open distributed systems. Fundamenta Informaticae **60**(1–4), 41–66 (2004)
19. Ambroszkiewicz, S., Mikulowski, D., Rozwadowski, L.: Entish: e-lingua for web service integration. In: Abramowicz, W. (Ed.), Proceedings of the 5th International Conference on Business, Information Systems BIS-2002 (2002)
20. Project enTish. http://www.ipipan.waw.pl/mas/
21. Ipi, PAN. IT-SOA platform. http://www.itsoa.ipipan.eu/
22. Doliwa, D., Horzelski, W., Jarocki, M., Niewiadomski, A., Penczek, W., Półrola, A., Skaruz, J.: Harmon riptsize ICS—a tool for composing medical services. Submitted to ZEUS'12 (2012)
23. Doliwa, D., Horzelski, W., Jarocki, M., Niewiadomski, A., Penczek, W., Półrola, A., Szreter, M.: Web services composition—from ontology to plan by query. Control Cybern. **40**(2), 315–336 (2011)
24. Doliwa, D., Horzelski, W., Jarocki, M., Niewiadomski, A., Penczek, W., Półrola, A., Szreter, M., Zbrzezny, A.: Web service composition toolset. In: Proceedings of the International Workshop on Concurrency, Specification and Programming (CS&P'10), volume 237(1) of Informatik-Berichte, pp. 131–141. Humboldt University (2010)
25. Doliwa, D., Horzelski, W., Jarocki, M., Niewiadomski, A., Penczek, W., Półrola, A., Szreter, M., Zbrzezny, A.: Plan riptsize ICS—a web service compositon toolset. Fundamenta Informaticae **112**(1), 47–71 (2011)
26. Jarocki, M., Niewiadomski, A., Penczek, W., Półrola, A., Szreter, M.: A formal approach to composing abstract scenarios of web services. In: Proceedings of the 18th International Conference on Intelligent Information Systems (IIS 2010), pp. 3–22 (2010)
27. Penczek, W., Półrola, A., Zbrzezny, A.: Towards automatic composition of web services: A SAT-based phase. In: Proceedings of the 2nd International Workshop on Abstractions for Petri Nets and Other Models of Concurrency and of the International Workshop on Scalable and Usable Model Checking (APNOC'10 + SUMO'10), pp. 76–96 (2010)
28. Penczek, W., Półrola, A., Zbrzezny. A.: Towards automatic composition of web services: a SAT-based phase. In Proceedings of APNOC'10 + SUMO'10, pp. 76–96 (2010)
29. Ambroszkiewicz, S.: enTish: An Approach to Service Description and Composition. Instytut Podstaw Informatyki Polskiej Akademii Nauk, Warsaw (2003)
30. Lenat, D., Guha, R.V.: Building Large Knowledge-Based Systems: Representation and Inference in the Cyc Project. Addison-Wesley, Reading (1990)
31. OpenCyc.org. The syntax of cycl. http://www.cyc.com/cycdoc/ref/cycl-syntax.html
32. R. Ferrario, N. Guarino, F. Barrera. Towards an ontological foundations for services science: the legal perspective. In G. Sartor, P. Casanovas, M. Biasiotti, M. Fernandez Barrera (eds.), Approaches to Legal Ontologies, Law, Governance and Technology Series, vol. 1, pp. 235–258. Springer, Heidelberg (2011)
33. Rosso, P., Mascardi, V., Cord, V.: A comparison of upper ontologies (technical report disi-tr-06-21). http://www.disi.unige.it/person/MascardiV/Download/DISI-TR-06-21.pdf
34. Schneider, L.: Designing foundational ontologies: the object-centered high-level reference ontology OCHRE as a case study. Lecture Notes in Computer Science, vol. 2813. Springer, Heidelberg (2003)
35. Hoser, B., Hotho, A., Jäschke, R., Schmitz, Ch., Stumme. G.: Semantic network analysis of ontologies. In: Proceedings of the 3rd European Semantic Web Conference, June 2006
36. Herre, H., Heller, B., Burek, P., Hoehndorf, R., Loebe, F., Michalek. H.: General formal ontology (gfo): a foundational ontology integrating objects and processes. http://www.onto-med.de/ontologies/gfo.html

37. Bosak, J., McGrath, T., Holman, G.: Universal business language v2.0. Organization for the Advancement of Structured Information Standards (OASIS), Standard, December 2006
38. Silver Bruce. BPMN Method and Style: A levels-based methodology for BPM process modeling and improvement using BPMN 2.0. Cody-Cassidy Press, 1 June 2009
39. TIBCO Software Inc. TIBCO Business Studio. Process Modeling User's Guide. TIBCO, software release 3.5.3 edition, March 2012
40. Andres, G., Sastoque, A.: Bpmn 2.0 by example, bizagi process modeler. Electronic version, March 2012
41. Ghallab, M., Howe, A., Knoblock, C., McDermott, D., Ram, A. Veloso, M., Wilkins D., Weld. D.: Pddl the planning domain definition language. Technical Report CVC TR98-003/DCS TR-1165, Yale Center for Computational Vision and, Control (1998)
42. Ricky, B., César, M.: An abstract plan preparation language. Technical Memorandum Report L-19280, NASA Langley Research Center Hampton, VA 23681–2199, November 2006
43. Nau, Dana, Tsz-Chiu, Au, William Murdock, J., Wu, D.: Shop2: an htn planning system. J. Artif. Intell. Res. **20**, 379–404 (2003)
44. Risberg, T., Evans, R., Tung, P.: Developing a spring framework mvc application step-by-step
45. Robinson, I., Knight, P.: OASIS Web Services Transaction (WS-TX) TC. Technical report, OASIS (2005)
46. Picard, W., Paszkiewicz, Z., Gabryszak, P., Krzysztofiak, K., Cellary, W.: Breeding virtual organization in a service-oriented architecture environment. In: Ambroszkiewicz, S., Brzeziński, J., Cellary, W., Grzech, A., Zieliński, K. (eds.) SOA Infrastructure Tools: Concepts and Methods, pp. 375–396. University of Economics Press, Poznań (2010)
47. W. Cellary. e-biznes szansą dla małych i średnich przedsiębiorstw. http://www.web.gov.pl/g2/big/2009_10/f0a29fbc31e28564f1fd59e3599c3e88.pdf. Wykład wygłoszony 20 października 2009 w Warszawie podczas Ogólnopolskiego Forum e-Biznesu pod auspicjami Polskiej Agencji Rozwoju
48. Cellary, W.: Elektroniczny biznes w regionach przygranicznych. www.logincee.org/file/6356/library. Accessed 1 March 2004. Referat wygłoszony na Konferencji i warsztatach, Ponad granicami - e-Government w regionach słabo rozwiniętych"
49. Cellary, W.: Content communities on the internet. IEEE Comput. Soc. **41**, 106–108 (2008)
50. Surveys show, Social Business concept gaining traction. http://gillin.com/blog/2011/11/surveys-show-social-business-concept-gaining-traction/
51. Small businesses balance online marketing with offline interactions. http://www.constantcontact.com/about-constant-contact/press/press_2011_1115FallSurvey.jsp. Accessed 15 November 2011
52. Insights from the global chief marketing officer study. http://www-304.ibm.com/businesscenter/cpe/html0/224128.html
53. The community roundtable. http://community-roundtable.com/
54. Creator of Grera J. Alberti. The facebook for SMES: social networking has changed the business world, December 2011

Chapter 4
Application of the Service-Oriented Architecture at the Inter-Organizational Level

Willy Picard, Zbigniew Paszkiewicz, Sergiusz Strykowski, Rafał Wojciechowski and Wojciech Cellary

Abstract In this chapter, flexibility and adaptation of collaborative processes occurring among organizations by applying the Service-Oriented Architecture (SOA) at the inter-organizational level are considered. First, an in-depth rationale for a service-oriented approach to inter-organizational collaboration is presented. The need for SOA at the inter-organizational level is explained by the ubiquity of services and the economically legitimated need of organizations to collaborate to gain and retain competitive advantage. Second, two SOA-based methods are proposed: the CMEAP method supporting flexibility, and a method for adaptation of service protocols supporting process adaptation. The CMEAP method allows administrative procedures to be automatically composed in a flexible manner, based on modeling legislative provisions in a form of elementary processes, decision rules, and domain ontology. The proposed method for adaptation of service protocols allows collaborators to modify the process model ruling their collaboration at run time, taking into account their social relations. Third, two prototype systems, the PEOPA platform and the *ErGo* system, implementing the proposed methods for the needs of the collaborators of the construction sector are detailed. Finally, a case study illustrates how the prototype systems may support the construction processes.

W. Picard (✉) · Z. Paszkiewicz · S. Strykowski · R. Wojciechowski · W. Cellary
Department of Information Technology, Poznań University of Economics,
Mansfelda 4, 60-854 Poznań, Poland
e-mail: picard@kti.ue.poznan.pl

Z. Paszkiewicz
e-mail: zpasz@kti.ue.poznan.pl

S. Strykowski
e-mail: strykow@kti.ue.poznan.pl

R. Wojciechowski
e-mail: rawojc@kti.ue.poznan.pl

W. Cellary
e-mail: cellary@kti.ue.poznan.pl

S. Ambroszkiewicz et al. (eds.), *Advanced SOA Tools and Applications*,
Studies in Computational Intelligence 499, DOI: 10.1007/978-3-642-38957-3_4,

1 Introduction

Although Web services and their associated tools and standards enable an implementation of SOA at infrastructural and technical levels, SOA does not have to be limited to these levels. The OASIS definition of SOA as "a paradigm for organizing and utilizing distributed capabilities that may be under the control of different ownership domains" [39] emphasizes some characteristics of SOA shared with a set of organizations which mutually collaborate to achieve common or compatible goals. Collaboration among organizations enables them to gain and retain their competitive advantages, especially in highly dynamic and competitive business environments. Due to collaboration, organizations may focus on their core competences and offer high quality services to collaborating organizations, which, as a consequence, permits the whole set of collaborating organizations to provide competitive products and services on the market.

Moreover, application of SOA at the inter-organizational level enables collaborating organizations to work in a more *user-centric* manner, i.e., to focus on addressing particular needs of their customers in a personalized way. Services and SOA are appropriate for the implementation of user-centric collaborative organizations: the concept of a service, which is at the fundament of SOA, is defined by OASIS [39] as follows: "a service is the mechanism by which needs and capabilities are brought together". In the context of collaboration among organizations, this definition of a service by OASIS focuses exactly on the matching of the needs of users and the capabilities of an organization or a set of organizations to answer these needs.

Although implementations of SOA at the inter-organizational, infrastructural, and technical levels share many common concerns, such as orchestration of services, reliability issues, and service instance selection, the SOA ecosystem at the inter-organizational level has two specific characteristics. First, the SOA ecosystem at the inter-organizational level is often highly regulated. For example, in the private sector, the pharmaceutical industry is highly regulated, so organizations operating in this industry have to encompass constraints following from the regulations into their collaboration with other organizations. In the public sector, the construction segment is highly regulated: many administrative procedures have to be performed prior, during, and after the construction of a building.

Second, the SOA ecosystem at the inter-organizational level is often a highly competitive and dynamic environment. In a globalized economic environment, organizations are often competing at a global scale. New organizations are created on a daily-basis, however, for instance in the USA, more than 50 % of them do not survive 5 years [62]. Therefore, new potential collaborators appear continuously, while many already collaborating organizations disappear.

In the case of the application of SOA at the inter-organizational level in highly regulated environments, the problem of *flexibility* of collaborative processes arises. In such environments, collaborative processes have to be modeled in a way supporting a large variety of cases, such as testing a new drug for flu on diabetic autistic children or conducting an administrative procedure to issue a building permit for various types of constructions like residential buildings, housing estates, shopping centers, roads, bridges, airports, levees, etc. In the case of the application of SOA at the inter-organizational level in highly competitive and dynamic environments, the problem of *adaptation* of collaborative processes arises. In such environments, collaborative processes have to be modified at run time to deal with unforeseen situations, such as organization withdrawal from collaboration or a natural disaster delaying construction work.

In this chapter, two main problems of the application of SOA at the inter-organizational level are addressed, namely, collaboration flexibility and adaptation. The CMEAP approach—Composable Modeling and Execution of Administrative Procedures—is proposed to tackle flexible administration procedures carried out in an automated manner. Adaptation of service protocols is proposed as an approach to support adaptation of inter-organizational collaboration.

The remainder of this chapter is organized as follows. In Sect. 2, basic notions used in this chapter are defined. Next, the importance and the characteristics of inter-organization collaboration are detailed in Sect. 3. Then, two approaches to computer support for collaborative processes are proposed in Sect. 4, addressing both the flexible automation of administrative procedures and the adaptation of collaboration processes. In Sect. 5, the application of SOA at the inter-organizational level is presented in the context of the construction segment, followed by two case studies detailed in Sect. 6. Finally, Sect. 7 concludes the chapter.

2 Basic Notions

The terms traditionally used in SOA can be used in context of collaborating organizations in the following way. In order to achieve their goals, organizations perform *activities* defined as closed pieces of work [69]. An activity may be a piece of automated work performed by an information system, e.g., a web service for creating invoices, a piece of work performed by a human, e.g., making a decision by a senior executive, or a piece of work performed by an organization, e.g., constructing a residential building. A set of partially ordered activities which realize an objective in a structured manner is called a *process* [69]. A *process instance* is a single enactment of a process. A *process instance state* is a representation of the internal conditions defining the status of a process instance at a particular moment. A *process model* captures the possibility to execute a given activity in a given state. In public administration, the equivalent term for the process is an *administrative procedure* or a *procedure* for short. Formally, the administrative procedure is defined as an ordered set of administrative activities carried out by a public office and other entities which aim at resolving a case through an administrative decision.

Information systems, humans and organizations involved in activities being a part of the process are called *actors*. A *service* is an access to a competence of an actor, called *service provider*, to satisfy a need of another actor, called *service consumer*, where the access is provided via a prescribed *interface* [44]. A service perceived by a service consumer corresponds to an activity performed by a service provider. A *collaboration* arises when two actors alternately and mutually play roles of service consumer and provider. Actors involved in collaboration are called *collaborators*. A process is *collaborative* if some actors involved in it are collaborators.

As a generic organizational structure supporting execution of collaborative processes, the concept of *Collaborative Networked Organizations (CNO)* has been coined by Camarinha-Matos et al. [9]. In this chapter, a Collaborative Networked Organization (CNO) is a network consisting of a variety of actors called *CNO members* that are largely autonomous, geographically distributed, and heterogeneous in terms of their operating environment, culture, social capital and goals, which conduct processes including at least one collaborative process in order to carry out a particular venture due to the demand from CNO clients.

CNOs may also be considered as structures aiming at "organizing and utilizing distributed capabilities under the control of different ownership domains". Following the similarities between CNO and SOA characteristics, concepts underlying SOA may be applied at the coarser level of organizations within the context of CNOs:

- *service provisioning and delivery*: services are provided by actors to other actors in order to be composed in complex processes;
- *service reuse*: a given actor may provide the same service to many actors within the same or different CNOs;
- *service abstraction*: the details of the implementation of services offered by actors are usually hidden from other actors, because the implementation of the core services is associated with the know-how capital that give an actor some advantages over competitive actors;
- *service discoverability*: knowledge concerning services provided by potential actors is publically available, so that those actors may be identified as potential CNO members;
- *service contracting*: services adhere to an agreement, as defined collectively by one or more contracts describing terms of service provision;
- *loose coupling and autonomy*: business logic is encapsulated in services with the intention of promoting reuse; actors have control over the logic encapsulated in services; service loose coupling supports aggregation of services into complex processes with a few well-known dependencies;
- *service composition*: a complex process provided by an actor is a result of composition of services provided by the actor itself or other actors; services can be composed in one of two ways:
  - *service orchestration*: collaboration among actors is controlled by one single actor acting as a coordinator responsible for assignment of activities and supervision of their execution;
  - *service choreography*: actors perform a complex process in collaborative manner, where synchronization of their efforts is based on inter-actor peer-to-peer communication;

- *service monitoring*: services provided by actors are constantly monitored by service providers as well as service consumers in terms of conformance with contract terms; various methods are used to verify service executions including audits, acceptance protocols, key performance indicators, etc.

3 Inter-Organizational Collaboration

Organizations always perform their processes in a particular economic, legal, social, political and technological environment which has impact on their success. Current trends: globalization, development and proliferation of information technology, spread of social media, development of electronic, knowledge-based economy and rising competition, are followed by increased complexity, uncertainty, dynamism, turbulence and diversity of organization processes.

In the SOA approach, the environment mentioned above is referred to as a SOA ecosystem. According to OASIS, a *SOA ecosystem* is defined as "a network of discrete processes and machines that, together with a community of people, creates, uses, and governs specific services as well as external suppliers of resources required by those services" [40].

3.1 SOA Ecosystem at the Inter-Organizational Level

In this section, the concept and characteristics of the SOA ecosystem at the inter-organizational level are presented. A special emphasis is put on economic justification of inter-organizational collaboration and organizational forms supporting it.

3.1.1 Rationale for Collaboration Among Organizations

The main reason for collaboration among organizations is the need for *competitive advantage*. The first theoretical framework permitting to understand the fundamental need for collaboration among organizations has been proposed by David Ricardo [55] in his book *Principles of Political Economy and Taxation*. David Ricardo indicated the strategy of *specialization* as a way to boost efficiency of organization operation. Specialization means concentration of an organization on operations where it has comparative advantage and taking benefits from exchange of goods and services with other specialized organizations. Ricardo has explained that this approach is effective even if an organization is able to produce all the goods and services more efficiently than the other organizations. Ricardo legitimates the collaboration of countries and/or organizations, as a means to improve global efficiency.

The concept of comparative advantage has been extended by Porter [54] with the concept of a value chain. *Competitive advantage* is defined as an advantage over competitors gained by offering consumers greater value. A *value chain* is a set of activities related to production processes, marketing, supply, client support, etc. which all together lead to service provision or product delivery that has a value for a final customer. Basing on the value chain concept, two organization strategies leading to competitive advantage were proposed: cost advantage and differentiation. An organization develops a *cost advantage* by reconfiguring its value chain to reduce costs of as many stages of the chain as possible. Reconfiguration means making structural changes, such as adding new production processes, changing distribution channels, or trying a different sales approach. *Differentiation* stems from uniqueness and perceived value. An organization focusing on the activities it does best and creating innovative and unique products and services, naturally rises above its competitors. An organization can achieve a differentiation advantage by either changing individual value chain activities to increase uniqueness in the final product or by reconfiguring the entire value chain.

Barney [6] has proposed that "a firm is said to have a competitive advantage when it is implementing a value creating strategy not simultaneously being implemented by any current or potential player". Focusing on the resources and attributes which provide the competitive advantage to an organization has a deep impact on its performance outcomes, and therefore, should be a fundamental aspect in every business strategy. As a consequence, organizations should specialize to gain a competitive advantage.

Strategy of focusing on the organization operation areas which provide the competitive advantage stressed by Porter and Barney should be a fundamental aspect of every business strategy. This approach leads to narrowing the areas of organization expertise and operation. Meanwhile, in current economy the production and service provision require a large set of skills and resources that a given organization is usually not able to handle efficiently. Thus, modern value chains cover not one but a number of specialized organizations, integrated with each other to perform activities defined within the value chain. Such collaboration creates an opportunity for efficient, cost effective differentiation at each phase of the value chain. Therefore, organizations should not only specialize, they should also integrate with other organizations by providing services to their clients, consuming services of the others allowing exchange of digital and material products.

3.1.2 Pervasive Services

In modern approaches to management of inter-organizational collaboration, the SOA paradigm plays a key role as a way to integrate heterogeneous information systems coming from different organizations [13]. Service orientation is emerging at multiple organizational levels in response to growing needs for greater integration, flexibility, and agility. The world economy is currently in an advanced stage of transformation from a goods-based economy to a services-based economy in which value creation, employment, and economic wealth depend more and more on the service sector [57]. Service-orientation is one of the largest applied paradigms with relevance to accounting, finance, supply chain management and operations, as well as strategy and marketing. The trend is especially significant, visible and economically justified in the sector of small and medium-sized organizations. Integrated, service-based collaboration permits such organizations to leverage the strategies of specialization, differentiation and cost advantage and to compete on the global market with large multinational organizations. The fact of strong service orientation of organizations is confirmed by the statistic data:

- services sector accounts for 80 % of United States gross domestic product [17];
- services play a similarly important role in all the OECD countries [16];
- although, in 2009 in European Union due to financial crisis, services turnover fell by 8.5 % compared with the year before, it rebounded in 2010 increasing by 5.0 % [20] almost reaching the level from before the crisis;
- global spending on software as a service (SaaS) will rise 17.9 % in 2013 to $14.5 billion, according to Gartner [30]; SaaS market growth will remain strong through 2015, when spending on the software is expected to reach $22.1 billion; North America is the most mature and largest SaaS market, expected to generate $9.1 billion in revenue this year, compared to $7.8 billion last year [30];
- due to increasing spending on IT services in the healthcare, retail, and transportation sectors, the IT service market is forecast to reach an estimated US $1,147 billion in 2017 with a Compound Annual Growth Rate of more than 5 % during 2012–2017 [35];
- companies that implement a service-oriented architecture are able to reduce costs for the integration of projects and maintenance by at least 30% [66];
- in 2010, 32 % of individuals aged 16 to 74 from 27 member states of European Union have used the Internet in the last 3 months, for interaction with public authorities, i.e., using the Internet for one or more of the following activities: obtaining information from web sites of public authorities, downloading forms, submitting filled in forms [22];
- in 27 member states of European Union in 2010, about 85 % of the 20 basic e-government services were fully available online, i.e., it was possible to carry out full electronic case handling [21];
- according to the OECD data, in the year 2010, 82 % of businesses from 25 member countries and 42% of citizens from 26 member countries of OECD have used the Internet to interact with public authorities [43].

3.1.3 Supporting Organizational Structures

Efficient collaboration among autonomous organizations based on services is difficult mainly due to differences among them including geographic, legislative, and cultural differences, diverse markets of products and services, constant changeability of customers and suppliers, as well as changeability of law, technology and methods of work. This raises the question of the organizational structures supporting collaboration [11, 12]. In this context, the concept of CNO presented in Sect. 2 has been divided into a number of subtypes of organizational structures described in the literature and applied in practice, such as middle-age guilds, partners clubs, and clusters. The following organizational structures supporting collaboration among organizations have been intensively scrutinized:

- *communities of practice*, defined as "a set of relations among persons, activity and world, over time and in relation with other tangential and overlapping communities of practice" [33],
- *virtual teams*, defined as "a group of people who interact through interdependent tasks guided by common purpose" that "works across space, time, and organizational boundaries with links strengthened by webs of communication technologies" [34],
- *virtual enterprises*, defined as "a temporary consortium of autonomous, diverse and possibly geographically dispersed organizations that pool their resources to meet short-term objectives and exploit fast-changing market trends" [15],
- *virtual organizations*, defined as "a geographically distributed organization whose members are bound by a long-term common interest or goal, and who communicate and coordinate their work through information technology" [4].

The services available in the SOA ecosystem are contributing to business processes on markets of services having various forms and levels of formality. For instance, the service market may take the form of catalogs of organizations [18], service auctions [24], IT service parks [49], dynamic business networks [8], Web service ecosystem [7], B2B e-marketplaces [1] or public administration service platforms [27, 38]. Within these markets, service providers offer multiple services that customers can dynamically and on-demand bind into their business processes. The evolution in the area of service publication, discovery and usage recently has shifted from tight coupling of intra-organizational systems to inter-organizational coupling of partners (value chains) using well defined service level agreements to open market of services with new models of licensing comprising abstract processes and dynamic service selection and instantiation [45]. In this context, a concept of *Virtual Organization Breeding Environment* (VOBE, sometimes abbreviated to VBE in the literature) has been proposed to support dynamic partner and service selection and instantiation of CNOs. A VOBE is "an association of organizations with the main goal of increasing preparedness of its members towards collaboration in potential virtual organizations" [10]. VOBE allows potential collaborators to prepare their future collaboration with other VOBE members before a business opportunity occurs by publication of provided services and provision of other data useful for identification of collaboration chances.

While the concept of VOBE is currently widely accepted in the CNO research community, they vary in terms of the architecture and implementation. Existing VOBEs have been created in an ad hoc manner and have an infrastructure allowing limited support for efficient integration of VOBE members on business and technical levels. In this context, *Service-Oriented Virtual Organization Breeding Environments* (SOVOBEs) have been proposed as VOBEs organized systematically on both technical and organizational level around the concept of a service allowing SOVOBE members to collaborate better [52].

3.2 Complexity of Collaboration in the SOA Ecosystem

In this section, the heterogeneity and dynamism of SOA ecosystem at the inter-organizational level are described. Then, the concepts of dynamism, flexibility and adaptation are introduced as an approach to addressing the characteristics of SOA ecosystem by organizations.

3.2.1 Heterogeneity and Dynamism of the SOA Ecosystem

In the global economy, the SOA ecosystem is neither homogeneous nor static. Following Porter's Five Forces theory [53], SOA ecosystem heterogeneity comes from supplier power, barriers to entry, rivalry among organizations, threat of substitute services and buyer power which are different in different business domains. Also, organizations operating in the SOA ecosystem are characterized by different readiness to personalize provided services, level of computer support in service provision, business models, organizational culture, level of formalization of collaboration and internal processes, geographical localization, ability to adapt to SOA ecosystem changes, etc. Dynamics of SOA ecosystem follow from constant changes in the set of organizations, individuals and information systems operating in the SOA ecosystem, changes in their operations and their results having impact on the whole ecosystem.

Complexity of collaborative processes directly follows from the heterogeneity and dynamic nature of the SOA ecosystem. Collaborative process actors constantly gain knowledge through the analysis of information concerning the process context. Moreover, actors learn from each other both explicit and tacit rules governing the execution of a collaborative process. As a consequence of the instantly gained knowledge, actors change the way they perceive the process, activities, semantics of the decisions being made, and the way these decisions have been made. Due to the usually long-lasting character of the collaborative processes, the set of collaborating actors and their roles change, so the set of actors having the holistic vision and understanding of the whole process may be small. Finally, similar instances of a collaborative process—e.g. having a similar goal, involving a similar set of actors, performed at the same time—may be interrelated, which means that the course of execution of one process instance and its result may influence the course of execution of another instance [44].

3.2.2 Run Time Modification of Collaborative Processes

Due to collaborative processes characteristics as well as dynamism and heterogeneity of the SOA ecosystem, it is hard to create a single correct executable model of a particular collaborative process that describes its execution from the start to the very end. Correct execution of collaborative processes require methods for easy and efficient modification of process models ruling collaboration or particular collaboration process instances, where the modification is potentially performed at run time. Such modifications aim to address the actual needs of actors involved in the process and actual context of collaborative process execution.

To precisely define problems related to process modifications, the concepts of process dynamism, adaptation and flexibility have been proposed [56].

Process Dynamism. Process dynamism means changing a process model. The change can be tiny, resulting from the need for corrections or improvements, or drastic, significantly altering the shape of the model. There are four main sources of dynamism: new technology, new management methods, new policies of the organization, and new legislation. Process dynamism is currently well supported by workflow management systems and business process execution engines that allow several versions of a given process, and their instances, to execute simultaneously.

Process Adaptation. Process adaptation means adjusting a process model for an active instance to exceptional circumstances that may or may not be predicted before starting the instance. Adaptation refers to one or possibly a few instances only and does not involve a permanent change of the process model. Process adaptation raises the question: if exceptional circumstances are predictable, why not include them in a process model? An answer is that including all exceptional circumstances in a single model would result in significant complexity and thus make the model difficult to understand and then to execute. For practical reasons, it is better to have a simpler model containing only the main paths, initiate all instances according to this model and, in the event of exceptional circumstances in a given instance, adapt the model for this instance only.

Process Flexibility. Process flexibility means the lack of the full specification of a process model, and constructing such a model uniquely and separately for each active instance at run time. The model is built from pre-defined single activities or sets of activities (fragments of a process), which are selected on an ongoing basis in accordance with the current circumstances occurred in the instance.

As the possibility of modification of collaborative process instances at run time plays a crucial role in the execution of collaborative processes in a SOA ecosystem, in this chapter a special attention is put on methods supporting process flexibility and adaptation. Collaborative processes requiring either flexibility or adaptation may be found both in the private and public sector. The level of regulations concerning a given process distinguishes processes that should be flexible from those that should be adapted. If such regulations are extensive and detailed, then constraints are strict, and the set of possible activities is finite, even if potentially large. Then, flexibility of processes is more appropriate. On the contrary, if the regulations concerning a process are vague and leave room for interpretation, which means that constraints set on a process are weak, then adaptation of processes is more appropriate than flexibility.

An example of a process to which flexibility is well-suited is a healthcare treatment process, highly regulated, independently whether a hospital is private or public. For each patient, the treatment process must be constructed step-by-step, based on current patient's health state, and results of the previous treatment activities, while the further treatment activities can be selected from a set of available medical procedures following from existing medical knowledge, technology, and pharmaceuticals.

An example of a process to which adaptation is well-suited is a rescue process provided by emergency management centers which have developed procedures of rescue actions for different types of events. Specific activities, in particular the selection of the type and number of rescue units and the organization of collaboration between them, are always adapted to the circumstances of an event which differs from one another.

In this chapter we focus on processes arising in the construction segment, both in the public and private sectors. The legislation governing this segment is very detailed, complex, and spread among various legal acts. These acts can be divided into two main groups. The first group consists of general acts; for example, Administrative Procedure Code which specifies general rules, common for all administrative procedures regardless of the segment they apply to. These general acts are superior ones to the second group of acts, which have much more detailed character and specify rules of a specific class of procedures. For example, Building Code regulates details of a procedure for granting a building permit. Moreover, the course of each procedure can be influenced by acts which are not directly related to administrative procedures at all. For example, Civil Code regulates the issue of granting a power of attorney. If an investor applying for a building permit has appointed an attorney, in such case the administrative procedure must include activities arising out of the Civil Code. Therefore, flexibility is required to manage procedures (processes) arising in the public sector (public administration) for the construction segment. On the contrary, in the private sector of the construction segment, managing processes requires adaptation. Given a construction project, a general process model is predetermined, because it follows from the investment type, technologies planned to be used, legal regulations, etc. A detailed model of the process instance is, however, adapted on an ongoing basis to emerging exceptional circumstances arising during construction. Examples of such circumstances are: too low temperature preventing from pouring concrete or the discovery of an unexploded bomb on construction plot.

4 Approaches to Computer Support for Collaborative Processes

Two approaches to support flexibility and adaptation of collaborative processes are presented in this section. The Composable Modeling and Execution of Administrative Procedures (CMEAP) approach deals with flexibility of processes performed in public administration, while service protocols address the problem of adaptation.

4.1 Flexible Automation of Administrative Procedures

The concept of administrative procedure automation is based on two main assumptions:

- Information systems take over the execution of all routine activities which are known in advance how to perform them. In administrative procedures, the routine activities are primarily associated with processing information contained in documents and registers, which is perfectly suited to be taken over by information systems.
- Humans (clerks) released from the routine activities can entirely dedicate their time to making decisions based on information prepared and provided by the systems.

Defined as above, the automation of administrative procedures requires creating very detailed models of these procedures; such models are necessary to direct and supervise the work of information systems. The models must take into account all possible variants of procedure realization, and for each variant they must include all possible operational activities. This section includes the review of traditional approaches to process modeling and executing and outlines problems of applying those approaches to automation of administrative procedures. Then the CMEAP approach based on the concept of process flexibility is presented. As mentioned in Sect. 2, in the public administration area, the term *administrative procedure* (or *procedure* for short) is used instead of the term *process*.

4.1.1 Traditional Modeling and Execution of Administrative Procedures

One of the most common purposes of administrative procedure modeling is work standardization. Public agencies hire analysts to develop the optimal models of administrative procedures and then seek to ensure that their employees (clerks) always carry out the procedures according to those models. The purpose of the standardization is to guarantee the constant high quality of work regardless of the individual competence of a given employee. Currently, the most often used notations for creating work-standardizing models are: (1) EPC (Event-driven Process Chain) standard of ARIS (Architecture of Integrated Information Systems) methodology [5] and (2) BPMN (Business Process Modeling Notation) standard [42] developed by Object Management Group (OMG).

Nowadays, in most of public agencies the execution of administrative procedures is supported by workflow management systems [69]. Such systems store administrative procedure models (often referred as administrative procedure definitions), enable creating and managing their instances, and controlling their interactions with participants; e.g., clerks and information systems (e.g., public registers).

Workflow management systems can be divided into two categories: ad hoc and production [31].

The *ad hoc workflow management systems* do not make direct use of administrative procedure models. In these systems, work and its flow are performed and supervised by clerks who are responsible for activities ordering and making coordination decisions during the execution [23]. The ad hoc workflow management systems are capable of notifying about new work waiting to be done and when it is completed, of registering what was done, by whom, and when. A clerk must manually, outside the system, consult with the administrative procedure model how to do the work and to whom forward its results. The level of administrative procedure automation offered by the ad hoc workflow management systems is very limited.

A significant step towards the automation of administrative procedures was made with the advent of the *production workflow management systems*. These systems can be fed with models of administrative procedures and then they are capable of using these models to automate the execution of those procedures [23]. However, the production workflow management systems require the models to be created according to the monolithic approach. The *monolithic approach* means that a model has a form of a single end-to-end flow. The shape of the flow is defined and embedded in the model at design time and does not change during the execution. Even if the administrative procedures are modeled as a set of models and submodels, the submodels are invoked during execution according to the design.

Therefore, the monolithic approach to modeling makes it virtually impossible to create comprehensive administrative procedure models taking into account all possible execution variants of these procedures, and within each specific variant, all detailed operational activities which should be performed to guarantee that a procedure course complies with all legal circumstances applicable in the variant. This problem follows from the fact that the course of administrative procedure is regulated by the legislation. Two issues relating to the nature of the legislation significantly influence the form of administrative procedure models:

- *Complexity of administrative legislation*: If the models of administrative procedures were supposed to include all the details reflecting all possible variants following from the provisions of legislation, they would have to be very complex.
- The *many-to-many relationship between models and legislation*: one specific act has an impact on many administrative procedures, and thus on their models, while one administrative procedure depends on the provisions of many acts. As a consequence, any change in the text of any act entails the need to update models of all the procedures to which this act applies if the monolithic approach to administrative procedure modeling is applied.

Due to these problems, currently the models of administrative procedures are usually created at a high level of generality. Such models do not include detailed actions, and therefore they are easy to create, rather small in size, and slightly susceptible to updates reflecting changes in legislation. For example, at the beginning of most administrative procedure models there is an activity named "Check the completeness and correctness of an application and all required attachments". A model containing such an activity does not specify the conditions under which the specific attachments are required, or what conditions should be met by these attachments. Also, another typical activity occurring in the middle of administrative procedure models is "Inform the relevant authorities of the action taken" or "Turn to the appropriate authorities for review and agreement". The models containing the activities defined at a high level of generality remain valid even in the case of multiple changes in legislation.

Thus, the models created in accordance with the monolithic approach can be used to automate the administrative procedures using workflow systems [63], but such automation is very limited. Clerks are supported by workflow systems in monitoring the course of the procedures. The system can also indicate which activities to undertake and registers completed ones. However, the execution of the activities remains a responsibility of a clerk, who has to analyze and interpret the general specifications of activities contained in the models, and decide how to translate these specifications into low-level operational activities for different cases. In this way, a detailed model of the administrative procedure performed for a specific case is created ad hoc by clerks at run time. Therefore, this type of automation should be classified as the human-driven automation.

The main problem of the workflow systems is subjectivity. The course of a specific procedure depends on the individual competence and experience of a clerk who conducts it. There is a high probability that a less competent clerk will misinterpret a general activity specified in the model and as the result conduct the whole procedure in a wrong way. Similarly, courses of two discrete cases with the same conditions may significantly differ from one another just because they were conducted by two different clerks, although they should be identical.

As a response to the limitations imposed by the monolithic approach, Adams in [3] has proposed a concept of *worklets*. Based on the activity theory, he has assumed that each activity in a process can be performed in several different ways. The ways have been named worklets. The single worklet consists of operations needed to complete the activity it refers to. A worklet selection is performed at run time and is based on examining contextual information related to a given process instance. However, in the worklet approach it is assumed that there is a top-level process model ruling the selection of appropriate ways to accomplish activities. The approach introduces a relaxation of the rigid monolithic modeling, but does not eliminate its main drawback which is the necessity to create a model for a whole process in the end-to-end form.

4.1.2 CMEAP Approach to Flexible Automation of Administrative Procedures

In this section, a novel approach to modeling and executing of administrative procedures, called Composable Modeling and Execution of Administrative Procedures (CMEAP), is presented [59]. The approach enables generating models of administrative procedures consisting of all operational activities that must be performed for different cases due to the applicable legal circumstances. At the same time, the level of details in activity mapping is high enough that there is no need to interpret them by clerks, and thus it is possible to automate their execution within information systems.

In the CMEAP approach, modeling of administrative procedures based on the end-to-end monolithic approach is entirely abandoned. Instead of associating a single end-to-end model or a set of firmly connected models and submodels with the course of each procedure, the CMEAP approach is based on modeling legislative provisions in a form of elementary processes, decision rules, and domain ontology. The course of the administrative procedure for a specific case is dynamically composed from elementary processes based on current legal circumstances occurring during execution of that procedure. The legal circumstances are represented by instances of ontology concepts and recognized by decision rules. In this way, every execution of an administrative procedure has its own unique model, closely corresponding to the specifics of the case the execution applies to. Therefore, in the CMEAP approach, modeling of administrative procedures consists of creating elementary processes, modeling ontology, and defining decision rules.

An elementary process is a sequence of activities determined by legislation to be executed in a course of an administrative procedure and constituting an operational entity. Activities of the elementary process may be performed by humans, IT systems or both; i.e., humans assisted by IT systems. An example of an elementary process is checking whether the person in a photo intended to be placed in an ID card wears dark glasses and if so, verifying a disability certificate prescribing continuous use of such glasses.

The ontology is a conceptual model of a specific domain describing concepts of this domain [64]. The instances of ontology concepts are called *facts* and represent real or abstract objects from the domain which can be involved in the execution of administrative procedures. An example of a real object is an applicant named John Brown represented by the fact being an instance of the domain concept *Applicant*. An example of an abstract object is the legal capacity of the applicant John Brown represented by the fact being an instance of the domain concept *LegalCapacity*. The concepts in the ontology may have attributes describing their characteristics. The *Applicant* concept attributes are *firstName*, *lastName* and *SSN*, which for the applicant John Brown take actual values, e.g.: *John*, *Brown* and *518-84-4887*, respectively. The *LegalCapacity* concept has an attribute *type*, which for John Brown's legal capacity takes value *full* while other possible values are *partial* and *none* [58].

The decision rules specify legal circumstances which when occur during the course of an administrative procedure make necessary to execute a specific elementary process. The legal circumstances in the administrative procedure execution are represented by facts. A single decision rule is a structure consisting of two sections, the *when* section (often also called the *if* section) and the *then* section. The *when* section contains conditional expressions relating to the existence (or absence) of certain facts and to the attributes of these facts. The *then* section contains a set of actions. The conditional expression in the *when* section is considered to be true if there is a fact which attribute values are consistent with the values specified in the condition of the *when* section. The rules having the *when* section evaluated to the true value are selected to be fired. Firing a rule involves executing actions included in the *then* section. As the rules are independent of each other, it is possible to select more than one rule to be fired.

The CMEAP approach consists of the following logical components [59]:

- Knowledge Base;
- Composition Server;
- Process Server.

The interrelation of the components together with their behavior is presented in Fig. 1.

Knowledge Base. The Knowledge base is a central repository storing knowledge artifacts of administrative procedures; i.e., elementary processes, ontology, and decision rules. The knowledge is divided into terminological and assertional one. The terminological knowledge includes elementary process models, ontology model, and decision rule definitions. The assertional knowledge includes instances of elementary processes, facts (instances of ontology concepts), and instances of decision rules.

Composition Server. The Composition server is the main component of the system, supervising and steering all other components. The server is responsible for dynamic composition of administrative procedure courses. The server operates according to the following general algorithm.

The first phase of the algorithm is the analysis of facts which emerge at a given moment of administrative procedure execution. Based on the analysis results, there is made a choice of decision rules to be fired. Firing a rule involves executing actions

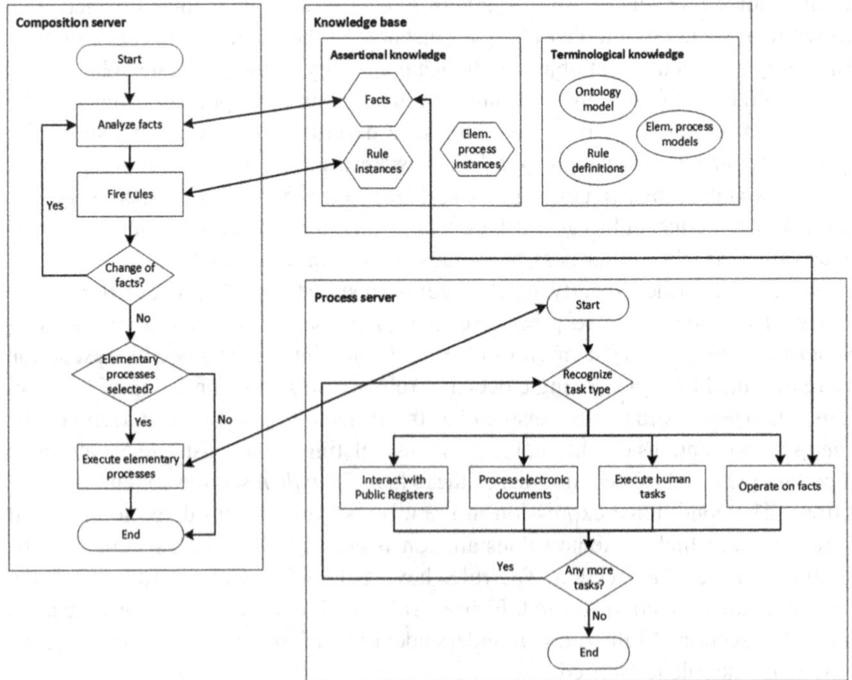

Fig. 1 Components of the CMEAP approach

included in its *then* section. The result of executing a single action can be asserting, updating or retracting a fact (a change of facts in general), or selecting the elementary processes which should be executed at this stage of the administrative procedure execution.

If a change of facts has occurred as the result of firing a rule, the Composition server goes back to the fact analysis phase and checks if this change has evaluated any rules to be fired. The loop is repeated until the facts and rules reach the stable state; i.e., there are no rules evaluated to be fired.

Process Server. If some elementary processes are selected as the result of firing rules, the Composition server triggers the Process server to execute them. The tasks included in the elementary processes can be divided into four types:

- Interactions with public registers;
- Processing electronic documents;
- Executing human tasks;
- Operating on facts.

Interactions with public registers include accessing and processing information collected in these registers. A public register is a collection of information on citizens, businesses, things, or entitlements. Examples of information collected in the public registers are: date of birth, marital status, date of death, names of persons entitled to represent a company, lists of decisions issued by public authorities. Examples of the decision lists are the following: a list of building permits issued, a list of professional licenses authorizing their holders to deliver certain types of services, details of a local spatial development plan. According to legal circumstances that occur in the course of an administrative procedure, interaction with public registers include updating existing data, inserting new data, or reading data for further use in the course of the administrative procedure. In the latter case, data retrieved from the public register are inserted into the Knowledge base as assertional knowledge.

Processing electronic documents includes an analysis of their content. In order to make such analysis feasible, electronic documents must have a structured form; i.e., the semantics of each piece of information included in them must be precisely specified. The appropriate technology here is XML Schema [65]. Information retrieved from electronic documents can be fed into public registers and/or can be inserted into the Knowledge base as facts, because this information is an essential source of legal circumstances relevant in the execution of the administrative procedure and therefore it can affect its further course.

Human tasks represent activities to be performed by humans (clerks). In an ideal electronic government based on the administrative procedure automation, human tasks involve only non-routine activities, i.e., those that cannot be implemented as a software package executed by a computer system [68]. The principle activity of this type is decision-making. Currently, however, human tasks, also often include routine activities which, due to technical problems, cannot be implemented as software. These activities include processing of paper documents or unstructured electronic ones, and processing data in public registers which do not have interoperable interfaces for external IT systems. As the process of replacing paper documents with electronic ones is progressing and the number of public registers with interoperability capabilities increases, human tasks will involve a declining number of routine activities.

Operating on facts includes tasks of technical nature related to reading, inserting, updating, and deleting facts from the assertional part of the Knowledge base.

If as the result of executing an elementary process, a change in facts occurs, the Composition server proceeds to the first phase of the algorithm. The composition of the whole administrative procedure model completes, and so the execution of the procedure, when the whole system reaches the stable state; i.e., when there are no more elementary processes selected for execution as a result of firing rules, and when there are no more changes in facts as a result of executing elementary processes.

4.2 Adaptation of Collaborative Processes

Efficient computer support for CNOs involving private organizations has to take into account the characteristics of collaboration introduced in Sect. 3. Methods effectively supporting collaboration in the private sector must exhibit the following characteristics:

- *support for adaptation*: computer support for collaboration should allow actors to adapt to new situations. Adaptation should enable modifications of the set of actors, the set of activities, the set of services, and their ordering. The adaptation method should provide means to restrict possible modifications that may be applied during an adaptation process;
- *a prescriptive approach*: computer support for agile collaboration should rely on prescriptive models. Such models define constraints on the sets of actors, activities, services, their relationships, and potential changes of these sets during adaptation;
- *a computer supported approach*: according to the computer supported approach, information system provides tools for human actors to make decisions, instead of making decisions on behalf of them, in opposition to the automated approach. The decision making should remain under responsibility of human actors, due to the influence of social norms and tacit knowledge on decisions, which are elements that may hardly be taken into account by information systems;
- *a collaborative approach*: adaptation of collaborative process models itself should be conducted in a collaborative manner.

The above characteristics constitute a foundation for definition of requirements concerning methods supporting three main phases of collaborative process execution: (1) specification of a collaborative process model [50], (2) instantiation of a collaborative process model [46] and (3) adaptation of collaborative processes instances [50].

Below, specific requirements concerning methods supporting each of the above three phases are defined. Then an approach based on the concept of service protocols is presented.

4.2.1 Requirements for Computer Support for Collaborative Processes

Specification of a Collaborative Process Model. Two aspects of collaborative processes in CNOs have to be addressed to tackle unpredictability and emergence that arise during CNO operations. The unpredictable aspect of CNO collaborative processes refers to the difficulty to plan in advance a partially ordered set of activities to reach an expected goal. The emergence aspect of CNO collaborative processes refers to the influence of the process instance execution on the process instance itself and/or the process model. To address these two aspects, the following main requirements for computer support for methods supporting modeling collaboration are defined:

- *separation of activities implementation from the process model*: a collaborative process model should include potential interactions among collaborators, however, the interactions should be decoupled from implementation of the activities performed by collaborators. As a consequence, activities of a given collaborative process model may be implemented in different ways, using different technologies, or different locations/hosts;
- *modeling shared responsibility*: in a collaborative process, responsibility for execution of activities is shared by different actors. Collaborative process model should express the responsibility of service consumers for invocation of services as well as the responsibility of service providers for the execution of services;
- *constraints on actors*: collaborative process model must support the definition of constraints on actors, both service providers and consumers. Constraints on actors can be then used as means to define the obligations that an actor has to fulfil to participate in the collaborative process. Constraints on actors should concern different aspects of actors such as their competences to perform a given activity;
- *relational constraints*: due to the importance of social aspects in CNO operation, collaborative process models have to support the definition of relational constraints between actors. Relational constraints concern activities that may be performed only by actors with appropriate relations with other actors. Computer support for collaborative processes should treat relational constraints as an integral part of the model of collaboration processes;
- *reusability*: a collaborative process model should be reusable in a similar way as a class models a set of objects in object-oriented programming: different instances of a collaborative process may rule different CNOs.

Instantiation of a Collaborative Process. During collaborative process instantiation, a set of actors and services is selected. Proper selection of actors and services determines to large extent the success of CNO operation [47]. The specific requirements for a method supporting instantiation of a collaborative process are the following:

- *continuous instantiation*: the selection of actors and services should not refer to the whole collaborative process at once at the beginning of CNO operation. Instead, in order to address unpredictability and emergence of collaborative processes, instantiation should be performed throughout the whole CNO operation;
- *multi-stage approach*: as the selection is not performed for the entire collaborative process at once, the selection process itself should be divided into stages, where each stage concerns the selection of a subset of actors and services;
- *multi-criteria analysis*: a method supporting collaborative process instantiation should encompass the constraints concerning actors, services and relations among them that are captured in the collaborative process model;
- *analysis of correlations among different instantiation processes*: the results of selection performed at previous stages should be considered during the next stages of the selection process. For instance, if a selection is performed for an activity "cleaning of construction site" it might be useful to consider who was selected to perform preceding activity "building demolition", as particular techniques used to demolish a building may force special cleaning techniques;
- *collaborative approach*: the selection of actors and services requires the involvement of people potentially belonging to different organizations, having different experience and knowledge, who are able to analyze and decide about the pertinence of the choice of given actors and services;
- *dynamic set of selection process actors*: the set of collaborators who are selecting actors and services, referred to as the selection process participants, may change over time.

Adaptation of Collaborative Processes and Their Instances. When a situation faced by collaborators imposes changes on the model ruling their collaboration, adaptation has to be performed. Specific requirements for methods supporting the adaptation of collaborative processes and their instances are the following:

- *adaptation as a collaborative process*: adaptation of a collaborative process is a collaborative process itself. The CNO collaborative process, referred to as the adapted collaborative process, is modified by another collaborative process, referred to as the adapting collaborative process;
- *dynamic set of adapting process actors*: a set of actors adapting a collaborative process, referred to as adapting collaborators, may change over time. This set consists of freely defined set of actors including actors executing the adapted collaborative process, referred to as adapted collaborators;
- *weak coupling between adapting and adapted processes*: the coupling between the adapting and adapted collaborative processes should be weak, i.e., no constraint should be set on the overlapping of the sets of adapting and adapted collaborators;
- *structured collaboration*: the collaboration within an adapting collaboration process should obey rules that limit the activities that may be performed by a given adapting collaborator at a given moment. Rules may originate from social norms, collaboration techniques, or collaboration culture;

- *modeling multi-faceted changes*: the model of adapting collaborative process should allow adapting collaborators to define the changes they propose to tackle the situation being a reason for adaptation;
- *support for contextual change propositions*: adaptation of collaborative processes should encompass the contextual character of change propositions. Contextual information about a given change proposition improves the understanding of the reason for the submission of the change proposition;
- *reusable models of collaboration during adaptation*: a given model concerning collaboration during adaptation should be generic enough to be applied in various adaptation situations.

4.2.2 Support for Collaborative Processes in Existing Approaches

Different approaches to modeling and management of processes has been proposed in the literature. Existing approaches do not satisfy the requirements imposed on methods supporting collaborative processes [50] due to the following features:

- *static set of process actors*: in the existing approaches, experts in process modeling and experts in a given knowledge domain collaborate at design-time to define process models. Next, these models are instantiated and performed by the employees of the organizations in which the process models have been deployed. Employees participating in the process instances are knowledge domain experts but are rarely involved in the definition of the process model, each situation requiring a modification of the process model is tackled by a set of experts in process modeling that occasionally consult knowledge domain experts. Such approach imposes limits on efficient adaptation mechanism as changes in the process cannot be performed only by actors actually performing the process;
- *singular service consumer*: in the existing approaches, a single service consumer and multiple service providers are assumed. For instance, the execution of a BPEL process consists in the invocation of service provided by various service providers but the BPEL engine executing the BPEL process is the only service consumer [41];
- *limited constraints*: in the existing approaches, process models focus on the set of activities and their partial ordering. The concept of role is used to limit the execution of a given activity to actors with appropriate rights. The role definition is usually limited to a label associated with a set of activities that may be performed;
- *unsupported social aspects*: although in the existing approaches the importance of social aspects in collaborative processes has been largely studied, existing methods of process modeling still lack support for relational constraints. Most process modeling languages and notations do not provide designers of process models with means to explicitly capture social requirements concerning actors in process models;
- *one-time instantiation*: in the existing approaches, the instantiation of a process consisting of the assignment of actors, activities, and services is done at once for the whole process and this assignment cannot be modified at run time.

4.2.3 Service Protocols

While CMEAP is proposed as an approach to support flexibility in administrative procedures, service protocols are proposed as an approach to support adaptation of collaborative processes in a dynamic environment.

As a preliminary remark, service protocols are different than communication protocols: although communication protocols define message formats and rules for the exchange of messages among computer systems, service protocols focus on three aspects of human-to-human interactions: processes, service-orientation, and social aspects.

A formal model of service protocols and their adaptation may be found in [50].

Specification of a Service protocol. A Service protocol consists of three elements: a process model, a service-oriented summary of a process model, and a service network schema.

As mentioned in Sect. 2, a *process model* defines a set of partially ordered activities to be performed during process execution.

A *service-oriented summary of a process model* provides a representation of the activities of the associated process model in SOA terms, independently of the process modeling language, e.g., BPEL or BPMN. In a service-oriented summary, each activity of the process is associated with a service represented by a *service description*. A service description is a triplet defining the "who" (the service consumer), "what" (the service interface), and "whose" (the service provider) part of the activity.

Information about *service entities*, i.e., service providers, service interfaces, and service consumers, are captured in a *service network*. A service network is a set of linked service entities, i.e., service providers, service interfaces, and service consumers. Service network aims at capturing properties and relations among service entities. In the proposed approach, a service network is the source of service implementation used to instantiate service protocol.

A *service network schema* is a graph that restricts the set of potential service entities that may participate in a service protocol by defining constraints on nodes and arcs [51], as predicates. Constraints imposed on nodes, i.e., isolated actors and service interfaces, are referred to as *classes of service entities*. Constraints imposed on arcs are called *social requirements*. The constraints should be taken into account when selecting service entities, i.e., actors and service interfaces, during instantiation of the collaborative process model. A service entity is an instance of a class of service entities iff it satisfies all the constraints defined by the class of service entities.

The concept of service protocol may be applied at four levels that differ mainly with regard to the availability of information concerning the chosen service consumers, providers, and interfaces:

- at the *abstract level*, a service-oriented summary provides a service-oriented representation of a collaborative process model, a service network schema provides constraints on service entities and social requirements, and both the service oriented summary and the service network schema are linked to associated service descriptions (from the service-oriented summary) with classes of service entities (from the service network schema);

- at the *prototype level*, service entities of a service network are associated with both service elements of the service-oriented summary and the classes of service entities of the service network schema. At the prototype level, the service network provides only a partial implementation of an abstract service protocol, as some elements of the service-oriented summary and some classes of service entities of the schema may not be associated with any service entity of the service network;
- at the *executable level*, the service network associated with both the service-oriented summary and the service network schema provides a complete implementation of an abstract service protocol: all the service elements of the service-oriented summary and all the classes of service entities of the schema are associated with service entities of the service network;
- at the *instance level*, an executable service protocol is enacted. At the instance level, service entities defined at the executable level consume and provide services modifying the state of the process model.

Service protocols address the requirements concerning methods supporting specification of collaborative processes. Service protocols provide a prescriptive model of collaboration: specification of the allowed partially ordered set of activities is provided by the process model, with their representation as service descriptions in the associated service-oriented summary, while the choice of the service elements is restricted by the service network schema. Four levels of abstraction cope with the requirements of reusability and separation of activity implementation from the process model. Service network schema encompassing classes of service entities and social requirements satisfies requirements concerning constraints of actors and relations as an immanent part of the process model. Finally, by distinguishing the roles of service consumer and service provider in service description associated with activities, service protocols tackle the problem of modeling responsibility for service invocations and execution.

Instantiation of a Service Protocol. A service protocol is instantiated as a result of a selection process of service entities from a service network according to the constraints defined in a service network schema. The selection of appropriate service entities is done from among all the service entities existing in the service network, where selected entities must be instances of appropriate classes of service entities and satisfy relational constraints defined in the service network schema.

The selection process consists of a set of stages, possibly overlapping in time, where each stage aims at identifying service entities being instances of a subset of classes of service entities defined in a service network schema.

The execution of the selection process encompasses:

- the definition of stages, i.e., classes of service entities to be instantiated in each stage,
- the execution of all defined stages,
- if applicable, as selection process proceeds, i.e., subsequent stages are executed, redefinition of the set of stages.

Every stage of selection process consists of four phases:

1. *constraints modification*: if needed, possible modification of constraints defined in the service network schema;
2. *contractor analysis prior negotiations*: a set of acceptable potential collaborators and their services is created for each class of service entities;
3. *contractor analysis during negotiations*: a set of potential collaborators created in phase 2 is updated on a basis of negotiations conducted with them;
4. *stage conclusion*: the best service entity for each class of service entities is selected and, if applicable, final collaboration conditions are set.

In each phase, collaboration is performed. The collaboration takes form of negotiations, i.e., an exchange of offers concerning possible assignment of a particular service entity to the class of service entities. Negotiations are restricted to selected CNO members. Actors involved in the selection process are CNO members, actors supporting selection process, and organizations being potential new service customers and service providers in CNO collaborative process. Supporting institutions may include domain experts, public administration unit representatives, legal and technical advisors, non-government organizations. Collaborators selecting service entities for the service protocol are generally different from collaborators executing the service protocol being instantiated. Some collaborators take part only in the selection process. At the same time, some CNO members are excluded from the selection process. Finally, some of collaborators takes part in both. The set of collaborators may change depending on selecting process stage and a set of classes of service entities being instantiated.

The collaboration among actors takes place on two levels:

- among CNO members and actors supporting selection process—appears in all phases;
- among CNO members and organizations being candidates for service consumers and service providers—takes place in phases 2–4.

In the proposed approach, a service network contains information about service entities and their relations, including for instance organization competences, legal status or formerly performed projects. In case of service interfaces, information may include service average execution time or service cost.

The submission of offers by selection process participants in phases 2 and 3 are supported by an appropriate decision supporting tool encompassing the selection process participants' needs specified as:

- *individual preferences*: optimal values, reject values and weights referring to classes of service entities and social requirements used in selection process stage;
- *individual fitness functions*: functions taking into account potential collaborator characteristics and selection process participant preferences used for ranking service entities and their sets according to required satisfaction levels.

The definition of preferences and fitness functions allows the service entities best suited for a particular class of service entities to be automatically suggested. Social requirements are also taken into account for the suggestion of a set of service entities matching a set of classes of service entities. Different tools may be used for this purpose. For instance, a genetic algorithm may be used for generation of the best acceptable service entity sets ranked according to a fitness function.

The approach proposed above addresses the requirements defined for CNO instantiation. The concept of stages encompassing instantiation of a subset of classes of service entities addresses the requirement of continuous instantiation and multi-stage approach. Requirement of multi-criteria analysis is satisfied by analysis of classes of service entities and relation constraints supported by decision support mechanism that does not make decision on behalf of the selecting collaborators, but generates suggestions, improving the efficiency of selection. Analysis of relations existing in a social network allows correlations among different instantiation processes stages to be analyzed. Finally, the above approach to service protocol instantiation tackle the problem of collaborative processes: the selection process is performed by a number of participants being CNO members or not, where the group of selection process participants is dynamic.

Adaptation of a Service Protocol. The model of adaptation of service protocols proposed in this section is based on the concept of a meta-protocol [50]. A meta-protocol is a compound of two service protocols: adapting and adapted ones. An adapting service protocol rules the collaboration between adapting collaborators aiming at defining the changes to be applied to an associated adapted service protocol.

To address the requirements of a dynamic set of adapting collaborators and weak coupling between adapting and adapted processes, analogously to the selection process the set of adapting collaborators is in general different from the set of adapted collaborators. Some collaborators are adapted, but not adapting ones. Other collaborators are adapting, but not adapted ones, while yet other collaborators are both adapted and adapting ones. During adaptation of service protocols, adapting collaborators confer about changes that may be applied to the adapted service protocol to tackle the situation faced by the adapted collaborators.

The adapting collaborators collaborate according to the adapting service protocol. An important type of activities in adapting service protocols are activities applying *change propositions* to the adapted service protocol, referred to as *adapting actions*.

Adapting statements are activities performed by the adapting collaborators during the execution of the adapting service protocol instance, aiming at the definition of an adapting action. The execution of adapting statements may be considered as a negotiation concerning potentially many change propositions.

A change proposition differs from an adapting statement. A change proposition defines a potential change to be applied to an adapted service protocol instance. An adapting statement defines the position of an adapting collaborator about an adapting proposition: in an adapting statement, an adapting collaborator may submit a change proposition for discussion, another adapting collaborator may reject this change proposition.

The collaborator performing an adapting statement may express his/her opinion concerning the change proposition that may take different forms such as: commenting, acceptance, rejection, supporting, and abhorring. The set of available forms of expressing opinions is limited by the adapting service protocol. The possibility to express a given opinion is controlled by the adapting service protocol. The possibility to express acceptance of a given change proposition may be limited to a few adapting collaborators having appropriate power of decision.

The opinion of a given adapting collaborator concerning a change proposition may be justified within an adapting statement. The explicit justification for the opinion of his/her author with regard to a given change proposition improves the collaboration because it reduces the misunderstanding of other adapting collaborators with regard to the adapting statement. Although the proposed model of adaptation of service protocols does not define a set of adapting statement types, each type of adapting statement in an adaptive service protocol should provide support for expressing justification. The proposed model of adaptation of service protocols does not assume a given structure for adapting collaborators' justifications. In some applications, a justification may be an unstructured plain text document. In other applications, it may be structured, and contain multimedia such as illustrative video or audio recordings.

An adapting statement may contain a list of links to appropriate information for a good understanding of the adapting statement by collaborators.

Adaptation of service protocols exhibits the characteristics of methods effectively supporting collaboration in the private sector mentioned in Sect. 4.2. First, adaptation of service protocols allows collaborators to adapt to new situations by applying changes to the adapted service protocol instance, which is the first feature, i.e., support for adaptation. Second, meta-protocols constraint the interactions among adapting collaborators by enforcing the adapting service protocol to be respected, which is related to the requirement of a prescriptive approach. In the proposed approach, meta-protocols support the adaptation process, but do not make decision on behalf of the adapting collaborators, addressing the requirement of a computer supported approach. Finally, the proposed method is collaborative, conforming to the fourth required feature of methods effectively supporting collaboration in the private sector.

5 SOA in the Construction Sector

The PEOPA and *ErGo* systems are technical implementations of theoretical concepts presented above. PEOPA is a system supporting execution of administrative procedures, while *ErGo* is designed to support collaboration among private organizations. Both systems have been developed for application in the construction sector.

In this section, first characteristics of the construction sector justifying usage of the SOA approach and other concepts presented in previous sections are introduced. Second, real-estate construction investment lifecycle is described. Then, a nature of interactions among major players in the investment process and the nature of administrative procedures performed in this sector are presented. Finally, functionality of PEOPA and *ErGo* systems is described in detail.

5.1 Characteristics of the Construction Sector

In this section, selected characteristics of the construction sector, especially important for real-estate developers, are presented, starting with the complexity of the construction investment lifecycle and ending with the heterogeneity of real-estate developers. In the presented characteristic of the construction sector, the term "real-estate developer" is used as an orchestrator of the development process, i.e., an organization that makes a land or a building suitable for commercial, industrial, or residential purposes, delegating the construction activities to subcontractors.

5.1.1 Complexity of the Construction Investment Lifecycle

A development process concerns the construction of new buildings on a purchased plot of land. The development process is complex and consists of many phases. It differs from one construction investment to another, some phases being optional. However, a common structure can be established: a preparation phase is followed by a construction phase, an operation and maintenance phase, and an estate selling phase.

Preparation Phase. Each development process starts with the preparation phase. The ultimate goal of the preparation is to obtain a construction permit for the construction phase. In the preparation phase, a real-estate developer identifies and acquires a plot of land, potentially supported by a financial institution. The acquisition of the plot of land involves the former owners of the plot and the real-estate developer as well as public agencies, e.g., a cadastral agency. The acquisition is often based on topographic surveying services, aiming at gathering data about the contours of the terrain, as well as the land features. Geotechnical surveys are also often required to evaluate the quality of the ground, which influences the price of the plot of land.

When the plot of land has been bought, a construction project has to be developed. Architects are usually responsible for the design of the blueprints of the buildings. However, the preparation of a construction project usually involves many other organizations, mainly public agencies that have to delivery certificates, opinions, and permits. An example is the involvement of the Department of Transport and Roads that have to approve the part of the project concerning the connection of the developed property to the existing road network. The city of Vancouver mentions the following set of departments, work groups and staff consulted when necessary for advice concerning construction permits: Area Planner, Heritage Planner, Industrial Lands Planner, City Drug Policy Coordinator, Social Planning, Housing Centre, Building Processing Centre, Fire and Rescue Services, Environmental Protection, Licenses and Inspections, Legal Services, Police, Vancouver Coastal Health, Park Board [14].

When the construction permit is delivered by the appropriate local public agency, the preparation phase is over and the development process evolves to the construction phase.

Construction Phase. The goal of the construction phase is to build buildings according to the construction permit delivered during the preparation phase.

In the construction phase, real-estate developers do not build themselves. They delegate all the construction-related activities to subcontractors. The role of real-estate developers is to coordinate construction activities performed by subcontractors. Real-estate developers are therefore considered as service consumers, while their subcontractors as service providers.

During the construction phase, a large variety of specialists with various competences are needed to build or renovate the buildings: electricians, plumbers, carpenters, masons, excavators, drywall hangers and painters. These specialists are usually contracted by real-estate developers with short-term contracts. Even in middle-size projects, real-estate developers often sign a dozen contracts with a given subcontractor. As a consequence, the management of all the contracts signed during the construction phase is often a complex task.

The construction phase ends when the building is ready for occupancy, i.e., after the delivery of the certificate of occupancy by the appropriate authority. The delivery of the certificate of occupancy is usually preceded by a series of inspections, e.g., fire alarm and elevator inspections.

Operation and Maintenance Phase. The goal of the operation and maintenance phase is to assure that the built facility performs the functions for which a property was designed and constructed. The operation and maintenance phase is optional.

In the operation and maintenance phase, real-estate developers manage day-to-day activities needed for proper functioning of the building. Operation and maintenance activities include janitorial, cleaning, lift maintenance, and waste removal activities.

Activities are usually outsourced to organizations specializing in a given type of activities. As a consequence of the wide spectrum of activities to be performed during the operation and maintenance phase, the real-estate developer usually has to contract activities from a large number of organizations providing different services. Besides day-to-day activities, a set of activities related to the rental of commercial space are performed during the operation and maintenance phase, especially for commercial investments, e.g., malls and warehouses. The management of rental contracts as well as tracking the issues reported by the renting organizations is the core aspect of the operation and maintenance phase. In the case of large malls in which the number of renting organizations exceeds one hundred, the management of rental contracts is a complex task. Similarly, capturing, updating, and resolving reported customer issues is also a crucial and demanding activity.

Selling Phase. The goal of the selling phase is to transfer the ownership of the built estate property from the real-estate developer to new organizations, as in the case of commercial properties, or individuals, as in the case of residential properties. The selling phase is optional.

In the operation and maintenance phase, the real-estate developer aims at providing potential buyers with information regarding the estate property. Potential buyers usually investigate the estate property prior to signing the property transfer contract. The investigation process is usually referred to as "due diligence". Real-estate developers are involved in the preparation of property condition assessment (PCA) to assess the value of the estate property. Ten areas are covered in PCAs: the building site, the building envelope, structural elements of the building, interior elements, roofing systems, HVAC (Heating, Ventilation, and Air Conditioning), plumbing, electrical systems, vertical transportation systems, life safety systems. An important part of PCAs concerns the costs of immediate and necessary future repairs. PCA reports provide precise information about the state of the estate property. Therefore, PCAs are often a basis for price negotiations and financial planning.

Besides the assessment of the value of the estate property, assessment of the financial results of the operation and maintenance phase for a given estate property is the key information for potential buyers that need to estimate the profitability of the estate property. As a consequence, appropriate documents concerning financial aspects have to be prepared, such as annual audited and unaudited financial statements, budgets, financial forecasts, external and internal auditors' reports, schedule of prepaid expenses, schedule of guarantees, last annual cash flow statement. Due diligence in a commercial property transaction often "include securing a title insurance policy regarding the ownership of the property and the encumbrances to which it is subject, and requiring the owner to secure an attornment from each tenant establishing agreement as to lease terms currently in force, and to research the zoning laws applicable to the property, building code compliance of the premises, the existence of any special assessments of property taxes applicable to the property, and the sales price history of the property" [67].

In the selling phase, the real-estate developer has to address a variety of aspects related with the estate property to be transferred. As a consequence, the real-estate developer has to collaborate with many individuals and organizations providing a wide spectrum of competences. Managing the documents generated during the due diligent process is a complex task, often supported by IT systems, such as virtual data rooms [36].

5.1.2 Heterogeneity of Real-Estate Developers

The construction investment market, in which real-estate developers play a central role, is highly heterogeneous. Its manifestation is the heterogeneity of real-estate developers themselves, both in terms of volume and use of IT systems to support the construction phase.

During the realization of the ITSOA project [25], the use of information systems supporting the realization of investment processes in the construction sector has been evaluated in a survey [32]. The evaluation focuses on real-estate development organizations operating in the Greater Poland (in Polish: "Wielkopolska"). The evaluation clearly confirms the heterogeneity of the construction investment market.

The survey was conducted electronically between May 1st and June 30th, 2011 via two communication channels: on appropriate professional Web sites and on the telephone. Among 37 filled survey forms, 9 survey forms (24 %) were filled online, and 28 by phone (76 %). Among 78 contacted organizations, 28 organizations (36 %) completed the survey.

The target audiences of the survey were subcontractors, builders, construction suppliers and other organizations related to building construction sector (renovation, contracting, construction, building, and home building industry). The survey had been prepared for managers working on all levels of organizational structure that had a direct or indirect impact on the shape of the investment process, e.g., scope of activities to be performed, set of subcontractors to be selected, supervision of the process execution.

Most of the polled persons are construction engineers, the remaining ones being estate agents, assistant managers, chairman of the board, project managers, main directors, secretaries, directors of sales, supervisors, financial analysts, accountants, work managers and specialists.

The number of employees in the polled organizations ranges from 3 to 280, so the survey was conducted mainly among small and medium real-estate development organizations.

In the years 2009–2011, the number of investment processes performed by the polled organizations ranged from 1 to 600 depending on the organization size. Most of the polled persons declares that the main type of investment processes concerns family houses and housing estates but do not exclude other types of investments.

The average number of subcontractors varies depending on the type of investment processes: from 1 (especially for single houses) to 50 (especially for whole housing estates). The number of permanent subcontractors is similar among polled organizations, usually up to 5.

The heterogeneity of the construction investment market concerns also the information systems used to support the construction phase of development processes. As a result of the evaluation, two groups of organizations are distinguished with regard to the information systems they use to support the construction phase of development processes:

- a group consisting of 29 out of 37 (78 %) organizations that declare that they use only basic office software packages, e.g., Microsoft Excel, Word and Project;
- a group consisting of 8 out of 37 (22 %) organizations that declare that they use information systems tailored to the management of investment processes.

The first group consists of small and medium size development organizations which mainly use Microsoft Excel, Word and Project for project schedule, verification and supervision of works, contracts and orders preparation. On a scale from 1 to 4, a rate of 4 meaning the higher level of suitability and usability, these tools have been rated 3 or 4.

The second group consists of rather medium development organizations (from 30 to 60 employees) which use various tailored IT solutions for the five operational areas taken into account in the survey: investment process planning, subcontractor selection, investment process execution management, communication with subcontractors, and contract management. The financial capacity, the number of realized investments (above 20), the size and complexity of faced situations of the organizations of the second group explain a wider adoption of IT tools by them. The list of information tools mentioned by the polled organizations is varied as follows:

- Primavera: 1 of 37 (2 %) organizations, a Project Portfolio Management system (PPM) by Oracle,
- Cobra: 1 of 37 (2 %) organizations, a program for scheduling by Deltek,
- Korab: 1 of 37 (2 %) organizations, an in-house developed program to evaluate subcontractors,
- Symfonia: 1 of 37 (2 %) organizations, a program for subcontractors maintenance, Customer Relationships Management systems (CRM): 2 of 37 (6%) organizations, a complex platform for all the partners of the investment process,
- Autocad: 5 of 37 (14 %) organizations, Archicad: 1 of 37 (2 %) organizations, Comos: 1 of 37 (2%) organizations, Nemeczek: 1 of 37 (2%) organizations, Emerwin: 1 of 37 (2 %) organizations, CAD programs for project design and drafting,
- Forte: 2 of 37 (6 %) organizations and Norma Pro : 3 of 37 (8 %) organizations, programs for costing, analyze and preparation of cost estimates,
- CalcDesc: 1 of 37 (2 %) organizations, an in-house developed program for investment process management.

Both groups use the Internet to seek subcontractors. They communicate with subcontractors via email, telephones, Skype or face-to-face. Most of the subcontractors are selected on the basis of former collaboration or recommendations by partners of the real-estate developer.

5.2 Characteristics of Collaboration in the Construction Sector

Four characteristics of collaboration during the construction phase have to be taken into account to support agile construction for real-estate developers: long-term dynamic collaboration, continuous process instantiation, document-based process execution, and the important role of social aspects.

5.2.1 Long-Term Dynamic Processes

In the construction phase of development processes, collaboration processes last long time and are characterized by high dynamics. The construction phase of a development process usually lasts months, often years. During the construction phase, various collaboration processes among the real-estate developer and its subcontractors usually take place, often in parallel. A good example is the construction of a skyscraper. Subcontractors responsible for electrical systems are working on lower storeys of the skyscraper, while subcontractors responsible for roofing systems are working on the top of the skyscraper. Moreover, the duration of collaboration processes varies. The installation of electrical systems starts when the foundations are built, to settle the main cable infrastructure, and ends at the end of the construction process when the final customers provide information about the type of outlets they have chosen for their property. Collaboration processes are also highly dynamic: subcontractors often work simultaneously on various construction sites to increase their profitability. As a consequence, it is quite frequent that a subcontractor has to be replaced by another one because of scheduling incompatibility with the real-estate developer construction schedule. Moreover, exogenous factors, e.g., weather conditions or law modification, also increase the dynamics of collaboration by enforcing changes to tackle new situations. In the construction phase of development processes, adaptation is not an exceptional case but a normal procedure.

5.2.2 Continuous Process Instantiation

As a consequence of the long-term and dynamic character of collaboration during the construction phase of development processes, the selection of subcontractors is performed in a continuous manner during the whole construction phase. The selection of subcontractors concerns only activities to be performed on a short-term horizon, usually during the next three months. When adaptation is the normal case and interactions potentially last years, it is not appropriate to select all the subcontractors for all the activities at the beginning of the construction phase. The selection of subcontractors in a continuous manner permits to limit the risk of inadequacy of the selected subcontractors to new situations. Otherwise, a subcontractor selected at the beginning to the construction phase may be inadequate to execute its activities due to the difference between the planned situation and the current one.

5.2.3 Document-Based Process Execution

In the traditional approach to project management, the scheduling of projects is usually supported by Gantt or PERT charts. A Gantt chart captures the activities to be performed, their start and end dates, and their potential dependencies on other activities. A PERT chart captures as a labeled graph activities to be performed, the time required to perform each activity, and the dependencies between the activities. A classical use of PERT chart is the reduction of critical paths to shorten the total duration of the project.

In the construction phase of development projects, most real-estate developers do not use either Gantt or PERT charts. More generally, no formal model of the process is established. The main reason for the lack of formal process modeling is the shared ubiquity of implicit knowledge: real-estate developers and subcontractors know the activities to be performed to build a given edifice, from their experience, practice, and know-how.

Instead of Gantt and PERT charts, the activities to be performed by a subcontractor for a real-estate developer, as well of the terms of the collaboration are defined in contracts. Contracts may therefore be considered as a basic document modeling the execution of the construction phase of development processes. Contracts are potentially modified bilaterally by appendices that modify certain terms of the collaboration.

Finally, progress payment claims are a common means to check the correct partial execution of a given activity.

Therefore, the execution of the construct phase of development processes is mainly driven by three types of documents: contracts, appendices, and progress payment claims.

5.2.4 Importance of Social Aspects

The importance of social aspects is the fourth characteristic of interactions during the construction phase to be taken into account to support agile construction for real-estate developers. First, social aspects influence the choice of subcontractors for the execution of activities during the construction phase. As an example, a trusted or recommended subcontractor is usually preferred over another subcontractor that is less trusted or that has not been recommended.

Second, social aspects also affect the interactions among the real-estate developer and the subcontractors during the execution of the activities. As an example, a subcontractor that would be an authority on a particular construction technique would probably be more superficially inspected for the progress payment claims than a subcontractor that has already been in conflict with other subcontractors.

5.3 *Administrative Procedures for the Construction Sector*

Realization of vast majority of construction projects requires a number of administrative procedures to be carried out by public agencies. The main legal acts regulating the course of administrative procedures in the construction sector is Construction Law, Master Planning Law, and Administrative Law regulating the general course of all administrative procedures. The number, type and scope of these procedures depend on the nature and extent of investments to be erected. The most often conducted administrative procedures related to the construction sector are presented in this section.

The interaction between a real-estate developer and public administration takes place mainly in the preparation and construction phases, and relatively rarely in the operational and maintenance phases. In the preparation phase, the most important administrative procedure is the issuance of a decision on a building permit. Another procedures that occur frequently in this phase are for obtaining a planning decision and a decision on environmental conditions. The most important administrative procedure in the construction phase is aimed at issuing an occupancy certificate.

5.3.1 Building Permit

For the majority of construction projects, construction works can start only on the basis of a final decision on a building permit. The decision on the building permit is issued by a competent public authority at the request of the investor, and after carrying out the relevant administrative procedure. An investor who fulfills the conditions for obtaining a building permit, may request the issue of a separate decision on approval of the construction project, prior to obtaining of a building permit. The decision on the construction design approval is valid for a specified period of time but not longer than one year.

The decision on a building permit expires if the construction work does not commence within 3 years from the date when the decision became final, or if the construction work is interrupted for more than 3 years.

In the Construction Law, there are certain types of investments specified that can be realized without a building permit. Some of these investments require that competent public authorities have to be notified about the intention of their commencement. The construction work can be started if within 30 days after the notice, the authority does not raise any objections, and no later than 2 years after the date of the work commencement declared in the notice.

Before submitting an application for the issuance of a building permit, the investor is often required to obtain additional permits, opinions and arrangements required by specific regulations. Obtaining these documents may require carrying out administrative procedures by appropriate public authorities. As a result of execution of each of these procedures an appropriate administrative decision is issued which establishes the specific circumstances of these permits, opinions and arrangements.

If the area of an investment is contained in the local master plan then the building design must be in strict compliance with the terms and conditions specified in the plan and the building permit should be issued in accordance with the plan. A local master plan should describe the function of the land, requirements for general public areas and planning conditions including all restrictions imposed by law.

If no master plan is adopted for the area where the investment is planned, a planning decision must be obtained as a result of an administrative procedure performed by an appropriate public authority. In this case, a final building permit must be issued according to the planning decision. A planning decision specifies the investment type and detailed planning conditions.

5.3.2 Planning Decision

A planning decision for a given area may be issued only if all of the following requirements are fulfilled:

- at least one neighboring plot which can be accessed from the same public road is developed in a way allowing to specify requirements for new buildings,
- the area has access to a public road,

- the existing or planned infrastructure is sufficient for the planned investment,
- the area does not require consent to change of purpose of agricultural or forest lands to non-agricultural or non-forest,
- the decision is compliant with other specific regulations, such as the Environmental Law, the Monument Protection Law.

A planning decision may concurrently be issued to several applicants. When one of the applicants obtains a building permit based on the planning decision, then the planning decisions issued to the remaining applicants become invalid. A planning decision may also become invalid when a new local master plan is adopted for the area covered in the decision and the plan imposes requirements different from those imposed by the decision.

An administrative procedure for issuing a planning decision may by suspended for no longer than 9 months from the date of submitting the application. The authority resumes the procedure and issues a planning decision when:

- within 2 months from the date of the suspension, the municipal council does not adopt a resolution to proceed with the preparation of the local master plan, or
- during the suspension of the procedure no local master plan is adopted.

However, if the application for a planning decision concerns an area for which there is a duty to adopt a local master plan, the administrative procedure is suspended until the adoption of the plan.

5.3.3 Occupancy Certificate

After the completion of the construction work for an investment for which a building permit is required a competent authority must be notified. The investment can be used if within 21 days after the notice, the authority does not raise any objections. In some circumstances specified in the Construction Law, before an investment can be occupied a final decision on a certificate of occupancy is required to be issued if: the investment required a building permit and it is categorized as an object that requires such certificate, or the use of a building is to take place before all construction work is finished. The competent authority shall issue a decision on the occupancy certificate after a mandatory inspection. The purpose of the inspection is validation of the construction in accordance with the provisions and conditions of the building permit. The validation includes:

- compliance with the design of a building plot or land;
- compliance with the design of a building;
- construction products particularly important for structural and fire safety;
- in the event of the imposition in the building permit of a requirement for demolition of existing buildings not intended for further use or temporary buildings, the fulfillment of this requirement, if the demolition deadline specified in the permit has already passed;
- cleaning up the construction site.

An investment, for which an occupancy certificate is mandatory, may also require a notification of other public authorities, such as State Sanitary Inspectorate or State Fire Service. The lack of a position of the authorities within 14 days from the date of receipt of the notice is considered as they do not have any objections or comments.

5.3.4 Decision on Environmental Conditions

Before a planning decision or a building permit is applied to, it may be necessary to assess the environmental impact of an investment. The assessment is required for the investments which may have significant impact on the environment or a protected area of Natura 2000 [19]. The assessment of the environmental impact is performed through an administrative procedure for issuing a decision on environmental conditions. If a competent authority finds the need to conduct the environmental impact assessment, an investor is obliged to prepare a report on the planned investment's impact on the environment. The decision also is issued if the authority does not find it necessary to assess the impact on the environment.

The environmental impact assessment for an investment can be carried out twice:

- in the context of an administrative procedure for issuing a decision on environmental conditions, and
- in the context of administrative procedures for a planning decision and a building permit; reassessment of the environmental impact is required in three cases:
  - if the requirement for the reassessment is specified in the decision on environmental conditions;
  - if the investor has applied for the reassessment if changes in conditions for the investment implementation are necessary;
  - if the authority carrying out the procedure has found that changes have occurred in the planned investment design relative to the findings of the decision on environmental conditions.

The environmental impact assessment is required for the investments which:

- could always have a significant impact on the environment;
- could potentially have a significant impact on the environment, but only if the obligation to assess the environmental impact has been determined by the authority competent for issuing decisions on environmental conditions.

The environmental impact assessment is also required for the investments that are not classified as having a significant impact on the environment, but may have a significant impact on a Natura 2000 site and are not directly related to the protection of the area and not the result of this protection.

5.3.5 Demolition Permit

If there is an existing building in the investment area the investor must obtain a permit for its demolition. If the building is entered in the register of historical monuments, a demolition permit may be issued after obtaining a decision of the Chief Conservator of Monuments, acting on behalf of the minister competent for culture and protection of national heritage. In relation to buildings not listed in the register of historical monuments, but included in the communal records of historical monuments, a demolition permit is issued by the competent authority in consultation with the regional conservator of monuments.

5.4 PEOPA Platform for Flexible Administrative Procedure Automation

In this section, the PEOPA platform is presented. The platform implements the CMEAP approach for flexible administrative procedure automation depicted in Sect. 4.1.

The general architecture of the PEOPA platform is presented in Fig. 2. The PEOPA platform consists of the following main components:

- Modeling environment;
- Knowledge base of administrative legislation;
- Administrative procedure execution server;
- Administrative case portal.

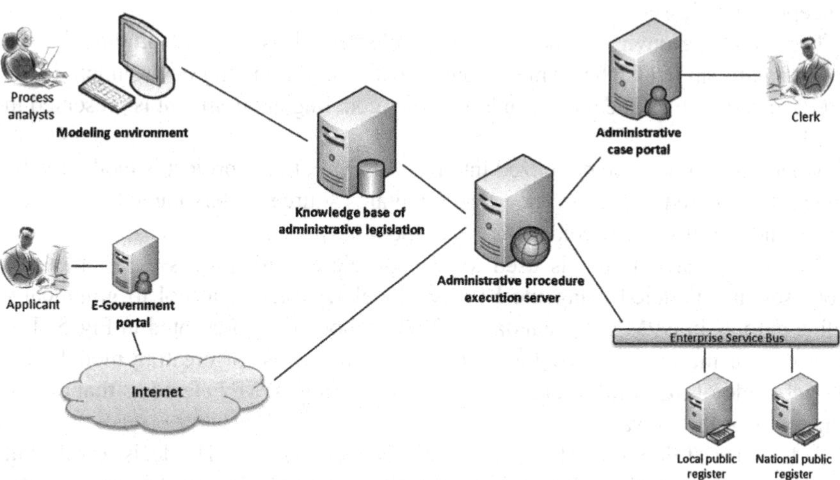

Fig. 2 Architecture of the PEOPA platform

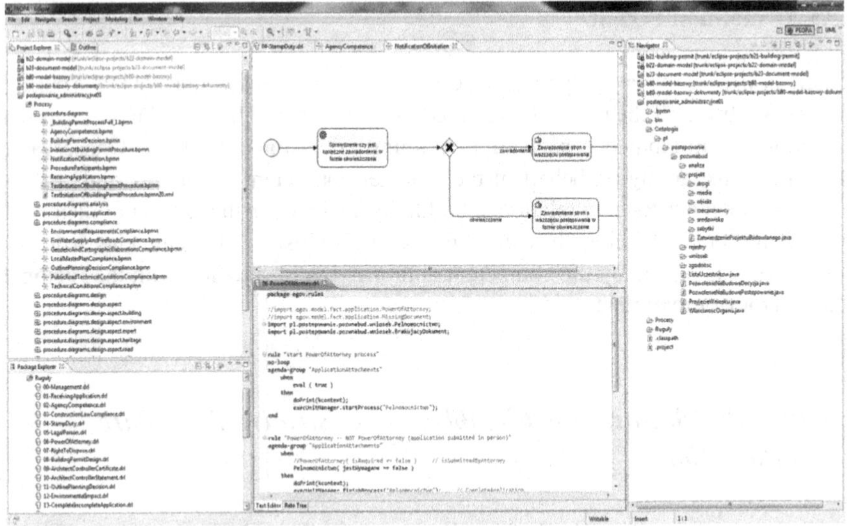

Fig. 3 The main window of the modeling environment

5.4.1 PEOPA Modeling Environment

The PEOPA modeling environment is a graphical tool used by process analysts to model administrative procedures. The models are developed based on the legislation analysis and consultation with lawyers and employees of public offices. According to the assumptions of the CMEAP approach, the models are represented in the form of the following types of artifacts: elementary processes, decision rules, and ontology concepts (fact classes).

The modeling environment has been implemented as a set of plugins for the Eclipse platform [61]. The primary component of the modeling environment is the PEOPA perspective. The main window of the modeling environment is presented in Fig. 3.

Modeling artifacts are organized into projects. Creating a project is facilitated by a wizard presented in Fig. 4. The project contains source folders named Processes, Rules, and Ontology, one folder for each type of artifacts.

Processes source folder is used to manage elementary processes. Elementary processes are modeled using the BPMN Modeler, which is a tool to graphically define models in BPMN 2.0 standard [42]. The modeler is presented in Fig. 5. The main part of the modeler window is filled with a canvas for creating models. On the right side of the editor there is a palette containing BPMN elements that can be dragged onto the canvas.

Rules source folder is used to manage units of decision rules. The decision rule unit consists of a set of decision rules associated with a particular legal issue; for example, individual's legal capacity. The editor with sample definitions of decision logic for

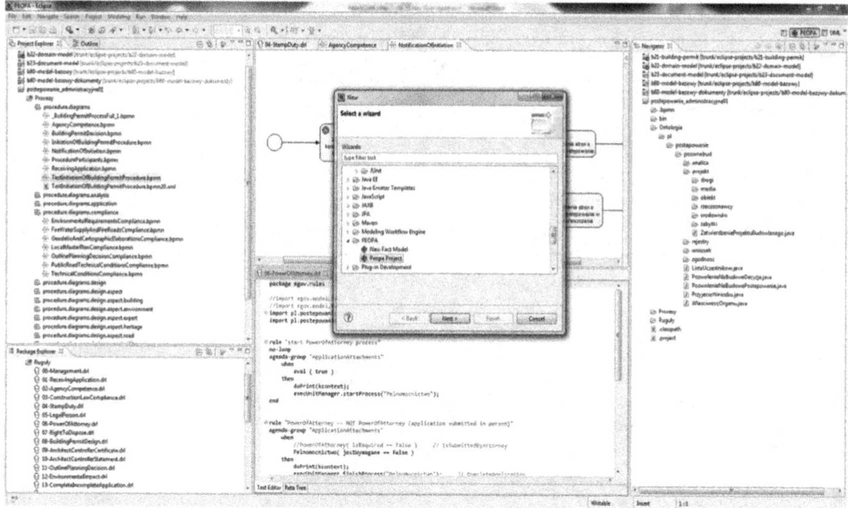

Fig. 4 Example definition of a fact model

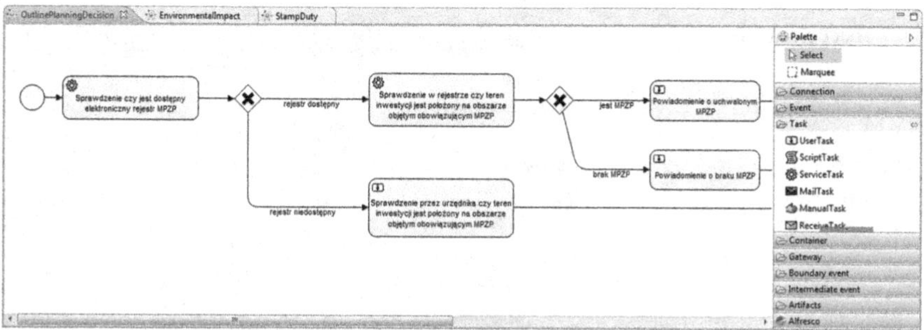

Fig. 5 Elementary process modeler

determining an individual's legal capacity is presented in Fig. 6. The decision rule units are defined using a rule editor for the Drools rule engine [28]. The definitions of the rules may be specified in the general Drools language or in a language specific to a given domain.

Ontology source folder is used to manage ontology concepts, i.e., fact classes. The fact class is defined using a wizard. The wizard used for defining a fact class concerning legal capacity of an individual is presented in Fig. 7.

A fact class consists of attribute definitions. Each attribute definition specifies a name and a type of an attribute. An attribute can be of a simple type or a reference type that refers to another fact class. In the example, the fact class LegalCapacity is composed of three attributes:

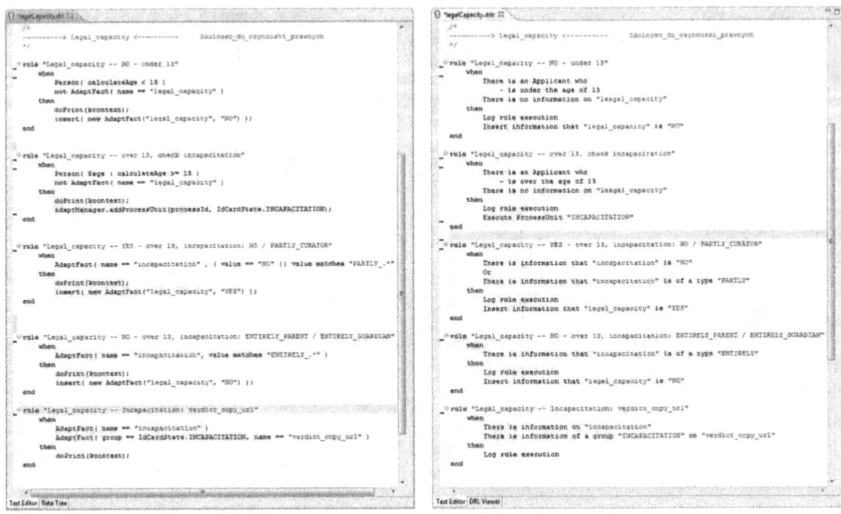

Fig. 6 Definition of an example decision rule unit in the general Drools language (*left part*) and in a domain specific language (*right part*)

Fig. 7 Fact class wizard

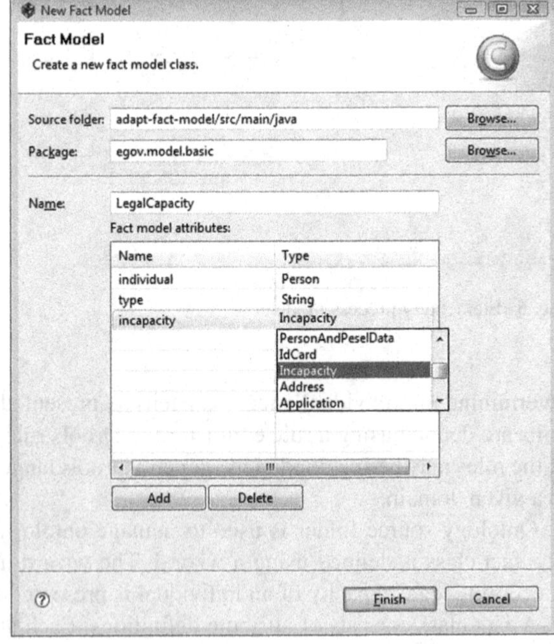

- individual: an attribute of a reference type that refers to Person fact class; denotes an individual which the legal capacity concerns,
- type: an attribute of the String type; denotes the type of legal capacity; can be full, restricted, or null which means the entire lack of legal capacity,
- incapacity: an attribute of a reference type related to Incapacity fact class; denotes an incapacitation status.

As a result of using the fact class wizard, the definition of a fact class in the form of a JavaBean component is automatically generated.

5.4.2 PEOPA Knowledge Base of Administrative Legislation

The PEOPA Knowledge base of administrative legislation is the implementation of the CMEAP approach Knowledge base.

The functions of the PEOPA Knowledge base are delivered to other components of the PEOPA platform through the service interface. The interface is implemented in Web services technology. Service interface allows remote access to the Knowledge base functions regardless of the IT infrastructure used to implement this access.

The service interface of the PEOPA Knowledge base consists of the following web services:

- KnowledgeBaseService web service;
- KnBaseAccountService web service;
- KnBaseAddressService web service;
- KnBaseAgencyService web service;
- KnBaseDocumentService web service;
- KnBaseEntityService web service;
- KnBaseProcedureService web service.

The PEOPA Knowledge base has been implemented in Java 7 using the following frameworks:

- Axis 2 framework for web service interface [60],
- Hibernate 4.1.4 library for object-relational mapping [29],
- SQL Server database for permanent data storage [37].

The KnowledgeBaseService web service delivers operations for managing decision rule units and definitions of elementary processes.

The KnBaseAccountService web service delivers operations for managing user accounts and privileges.

The KnBaseAddressService web service delivers operations for managing addresses and contact data.

The KnBaseAgencyService web service delivers operations for managing public administration agencies, clerks, and positions.

The KnBaseDocumentService web service delivers operations for managing administrative documents, digital images of administrative documents, and metadata describing administrative documents.

The KnBaseEntityService web service delivers operations for managing entities of administrative procedures.

The KnBaseProcedureService web service delivers operations for managing administrative procedure hierarchy and instances of administrative procedures.

5.4.3 PEOPA Administrative Procedure Execution Server

The PEOPA Administrative procedure execution server is the implementation of Composition and Process Servers of the CMEAP approach presented in Sect. 4.1.

As in the case of the PEOPA Knowledge base, functions of the PEOPA Execution server are delivered through the service interface implemented in Web service technology.

The service interface of the PEOPA Execution server consists of the following web services:

- AuthorizationService web service;
- CaseInfoService web service;
- HumanTaskService web service;
- ProcedureRuntimeService web service;
- DocumentService web service;
- CommentService web service.

The PEOPA Knowledge base has been implemented in Java 7 using the following frameworks and components:

- Axis 2 framework for web service interface [60];
- JBoss Drools decision rule engine [28];
- Activiti process execution engine [2].

The AuthorizationService web service delivers operations for authorizing clerks within the system.

The CaseInfoService web service delivers operations for retrieving detailed information about cases, retrieving a list of cases on the basis of certain criteria, retrieving information about documents and comments attached to a case, retrieving information about tasks performed within a case.

The HumanTaskService web service delivers operations for managing tasks intended to be performed by a clerk.

The ProcedureRuntimeService web service delivers operations for managing applications for conducting administrative procedures and managing instances of administrative procedures.

The DocumentService web service delivers operations for creating, retrieving, updating and deleting administrative documents, their digital images, assigning documents to tasks and instances of administrative procedures.

The CommentService web service delivers operations for creating, retrieving, updating and deleting comments and assigning them to tasks and instances of administrative procedures.

In the course of administrative procedures, the PEOPA Execution server may need to interact with the local and national public registers. Such an interaction is conducted by means of enterprise service buses available on the intranet of public administration.

5.4.4 PEOPA Administrative Case Portal

The PEOPA Administrative case portal is a web tool intended for public clerks to manage execution of administrative procedures. The portal delivers functions for performing the following main activities:

- registration of an application,
- attribution of an application,
- starting a new case by the launch of a new administrative procedure,
- executing an administrative procedure in order to settle the case,
- browsing completed administrative procedures.

Within execution of administrative procedures, the portal delivers the following operational functions:

- claiming and carrying out human activities (tasks),
- delegating activities to other employees,
- reading notifications about the status of administrative procedures and results of executing non-human activities by the system,
- displaying XML documents in a human-friendly manner,
- editing XML documents in electronic forms,
- entering and displaying comments for administrative procedures and individual activities,
- creating documents and uploading their digital images.

All clerks using the portal must have their own password-protected accounts. Each account is linked to a set of privileges specifying which functions the account owner is granted to perform.

The following sections present details of portal functions.

Registration of an Application. Launching an administrative procedure begins with registration of an application submitted by an applicant—an individual or a representative of a company. Figure 8 presents a form for registering application (*Rejestracja wniosku–Application Registration*). Registration usually takes place in the Filling Office. Using the form, the office employee enters metadata describing the application being submitted and indicates the department the application will be forwarded to for further processing. The form includes a box for attaching a file containing the application and its attachments written in the XML format. This option is used when the application is submitted electronically through an e-government web portal and due to technical limitations there is no direct connection between the portal and the PEOPA platform.

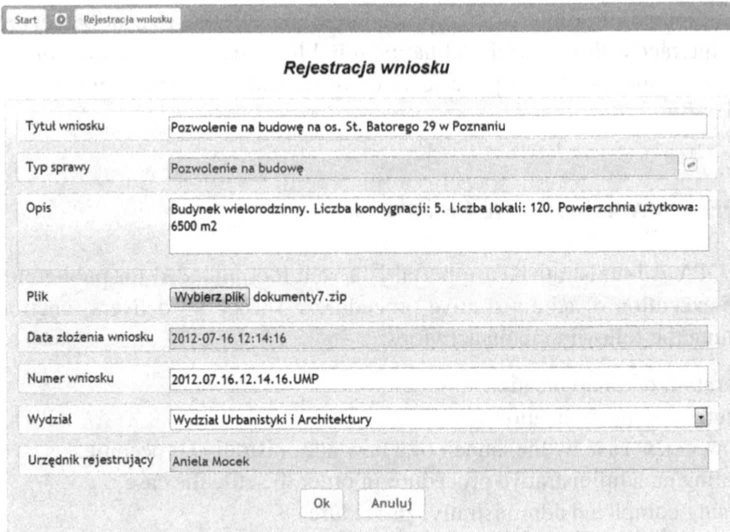

Fig. 8 The form for registering an application

Once completed and approved, the data entered in the form are stored in the system.

Attribution of an Application. Once registered, the application is forwarded to the department indicated during the registration. It appears on the list of applications for attributing. This list is visible for an employee of the department who is granted the privilege for attribution. Such an employee is usually a person occupying the position of a department manager. The form for application attribution is presented in Fig. 9 (*Dekretuj wniosek–Attribute Application*).

There are two types of attribution: to a department and to selected employees. The choice depends on the method used to organize work in a department. Attribution to a department means that all employees receive the possibility to pick up the attributed application. Attribution to selected employees means the arbitrary limitation of the number of employees who receive such a possibility. A special case of the latter attribution type is selecting one employee only. In this case, the manager deprives the employees of self-management capabilities and arbitrarily takes decisions on workloads.

Starting a New Case. After being attributed, an application is removed from the list of pending for attribution and appears on the list of pending to initiate a case. The case can be initiated by a clerk with appropriate privilege only. Figure 10 presents the main screen of such a clerk (*Podjęcie nowej sprawy–Starting a new case*; *Sprawy w toku–Ongoing cases*; *Zadania do wykonania–Tasks to be performed*; *Sprawy zakończone–Complete cases*). In addition to starting a new case, the clerk may also perform other

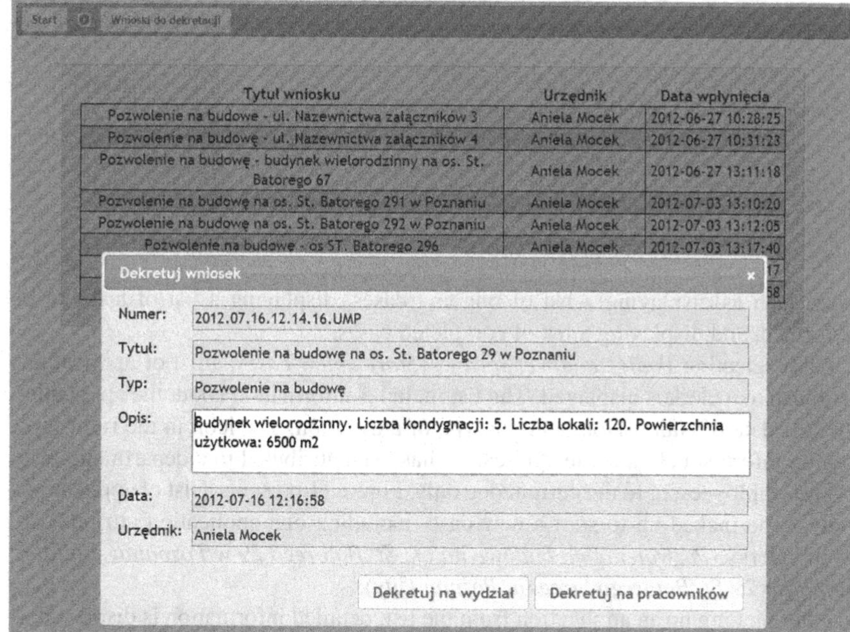

Fig. 9 The form for application attribution

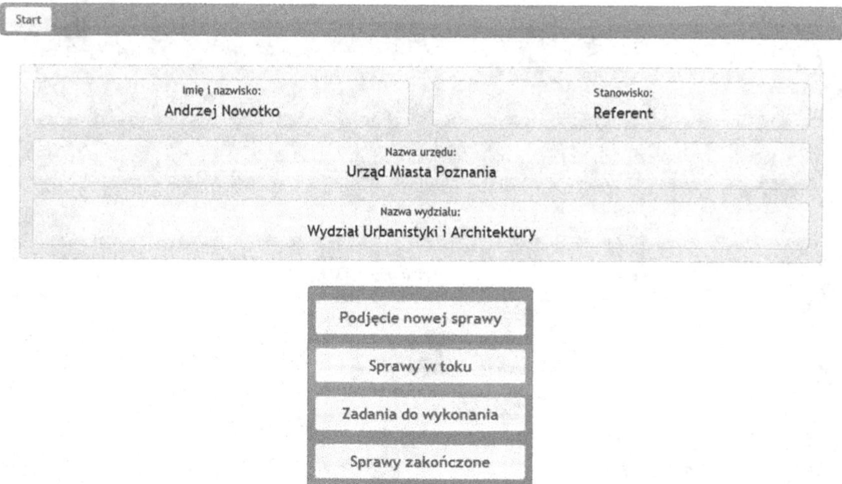

Fig. 10 The main screen of the portal displaying for a clerk with executing administrative procedure privilege

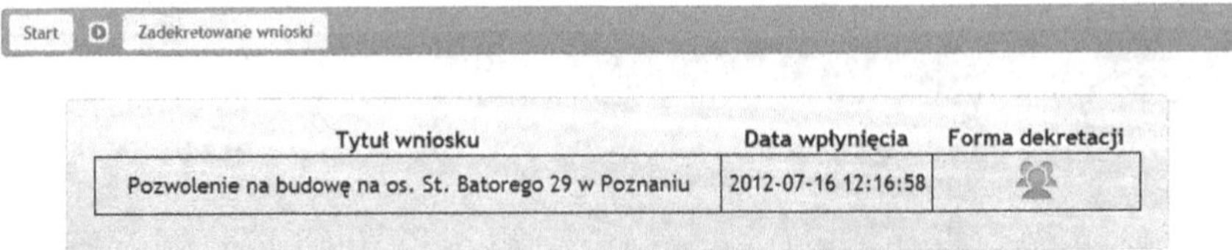

Fig. 11 A list of applications waiting to be picked up by a clerk to start a new case

tasks, such as displaying a list of ongoing cases, displaying a list of tasks to be performed, and displaying a list of completed cases.

After selecting *Podjęcie nowej sprawy (Start a new case)*, a list of applications attributed to a clerk is displayed. The list includes information about the application title, the date of submission, and the type of attribution. The icon in the right-most column informs a clerk if the application has been attributed to a department, or to several employees, or to the current one only. Figure 11 presents a list of applications ready to be picked up to start a new case, including one application attributed to several clerks (*Pozwolenie na budowę na os. St. Batorego 29 w Poznaniu–Building permit for 29 St. Batorego Street in Poznan City*).

After picking up an application from the list, detailed information is displayed as well as a button to start the case (Fig. 12, *Podejmij sprawę–Start the case*).

Start
Zadekretowane wnioski
Tytuł wniosku
Data wpłynięcia
Forma dekretacji
Pozwolenie na budowę na os. St. Batorego 29 w Poznaniu
2012-07-16 12:16:58
Szczegóły sprawy
Numer: 2012.07.16.12.14.16.UMP
Tytuł: Pozwolenie na budowę na os. St. Batorego 29 w Poznaniu
Typ: Pozwolenie na budowę
Opis: Budynek wielorodzinny. Liczba kondygnacji: 5. Liczba lokali: 120. Powierzchnia użytkowa: 6500 m2
Data: 2012-07-16 12:16:58
Urzędnik: Aniela Mocek
Przydzielono do: Andrzej Nowotko
Julia Krajna
Podejmij sprawę
Anuluj

Fig. 12 The details of an attributed application

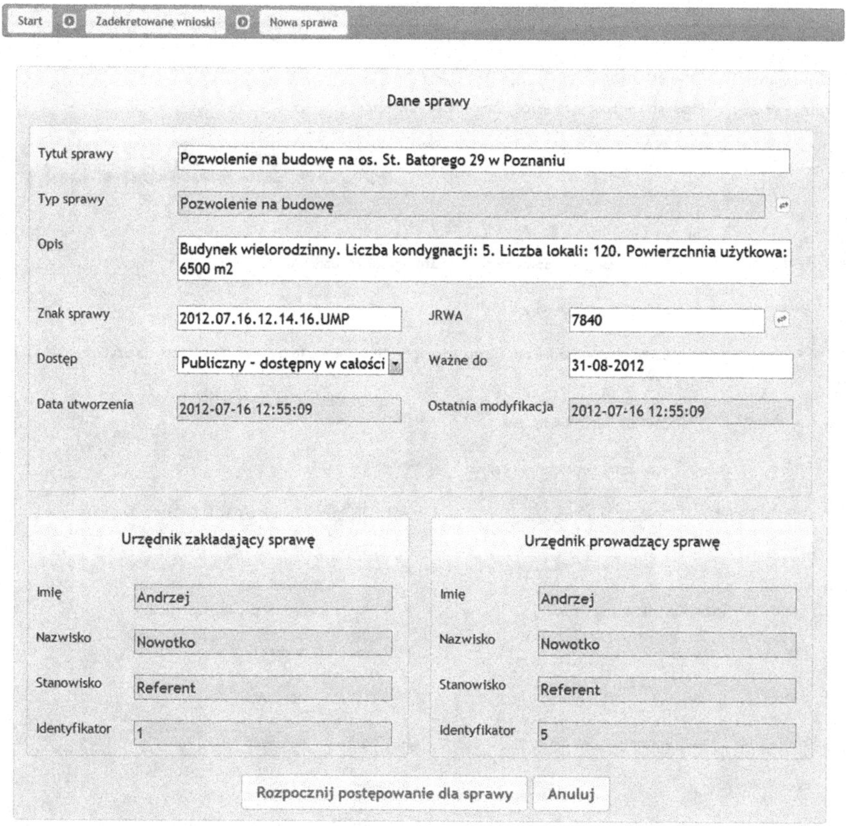

Fig. 13 The form to enter data for creating a new administrative procedure

Pressing the *Podejmij sprawę (Start the case)* button gets displayed a form to enter data for creating a new administrative procedure for the case. Such a form is presented in Fig. 13 (*Dane sprawy–Case description*).

After creating a new administrative procedure, a screen with four tabs gets displayed. The *Szczegóły sprawy (Case description)* tab contains description of the case the administrative procedure is executed for; the *Przebieg (Course)* tab—a list of tasks to be performed by a clerk within the procedure; the *Teczka (Case folder)* tab—folder for documents attached to the case; the *Komentarze (Comments)* tab—comments submitted by clerks on the case. The screen is presented in Fig. 14.

Executing an Administrative Procedure. The execution of an administrative procedure is performed using the *Przebieg (Course)* tab presented in Fig. 16. The tab contains a list of tasks. There are two types of tasks: human tasks assigned by the PEOPA platform to be performed by a clerk in the specific moment of procedure execution and notifications of actions taken by the system, in particular the automatic execution of activities.

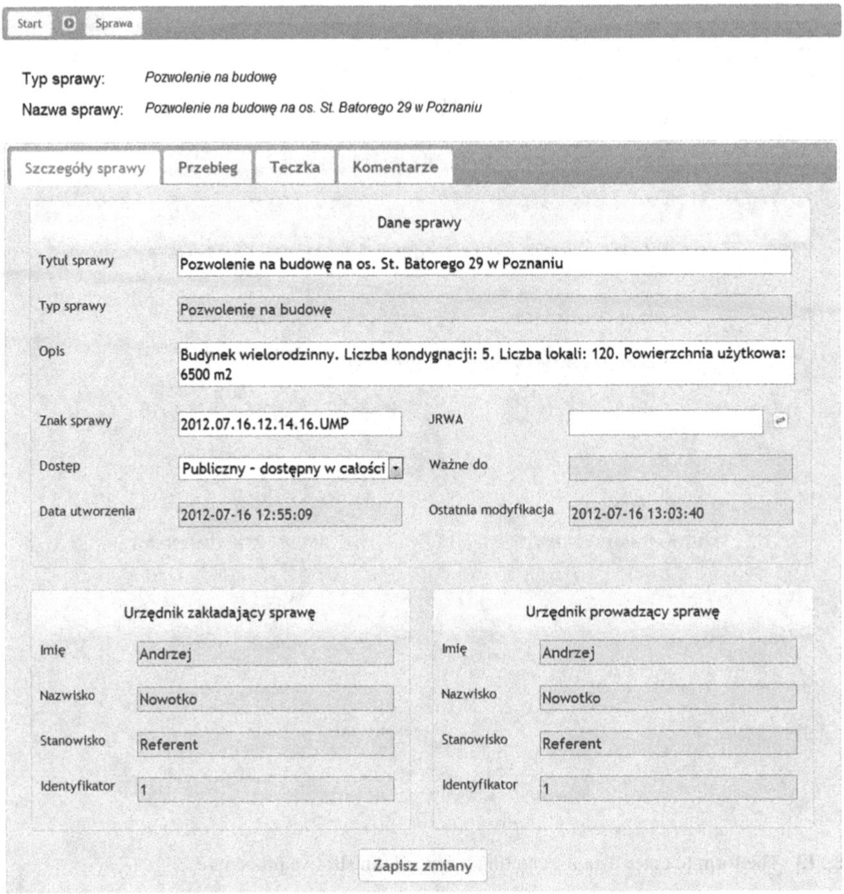

Start Sprawa

Typ sprawy: *Pozwolenie na budowę*

Nazwa sprawy: *Pozwolenie na budowę na os. St. Batorego 29 w Poznaniu*

Szczegóły sprawy | Przebieg | Teczka | Komentarze

Dane sprawy

Tytuł sprawy: Pozwolenie na budowę na os. St. Batorego 29 w Poznaniu

Typ sprawy: Pozwolenie na budowę

Opis: Budynek wielorodzinny. Liczba kondygnacji: 5. Liczba lokali: 120. Powierzchnia użytkowa: 6500 m2

Znak sprawy: 2012.07.16.12.14.16.UMP — JRWA:

Dostęp: Publiczny - dostępny w całości — Ważne do:

Data utworzenia: 2012-07-16 12:55:09 — Ostatnia modyfikacja: 2012-07-16 13:03:40

Urzędnik zakładający sprawę: Imię: Andrzej; Nazwisko: Nowotko; Stanowisko: Referent; Identyfikator: 1

Urzędnik prowadzący sprawę: Imię: Andrzej; Nazwisko: Nowotko; Stanowisko: Referent; Identyfikator: 1

Zapisz zmiany

Fig. 14 The screen for managing an administrative procedure

Figure 15 presents a task *Wyświetlenie formularza z wnioskiem o wydanie pozwolenia na budowę i zatwierdzenie projektu budowlanego (Display a form for an application for a building permit and approval of a construction project)*. For this task, there are three tabs: *Szczegóły zadania (Task details)*, *Załączniki (Attachments)*, and *Komentarze (Comments)*. In the *Szczegóły zadania (Task details)* tab the content of the task is presented. The *Załączniki (Attachments)* tab includes a list of documents and their digital images attached to the task, and the *Komentarze (Comment)* tab includes a list of comments on the task. In the task discussed here, in the *Szczegóły zadania (Task details)* tab there is a form displaying the submitted application. The form could be unfilled; thus a clerk's task would be to fill it with data from a paper application. In the discussed task, the form is completed with data retrieved from the XML document included while registering the application.

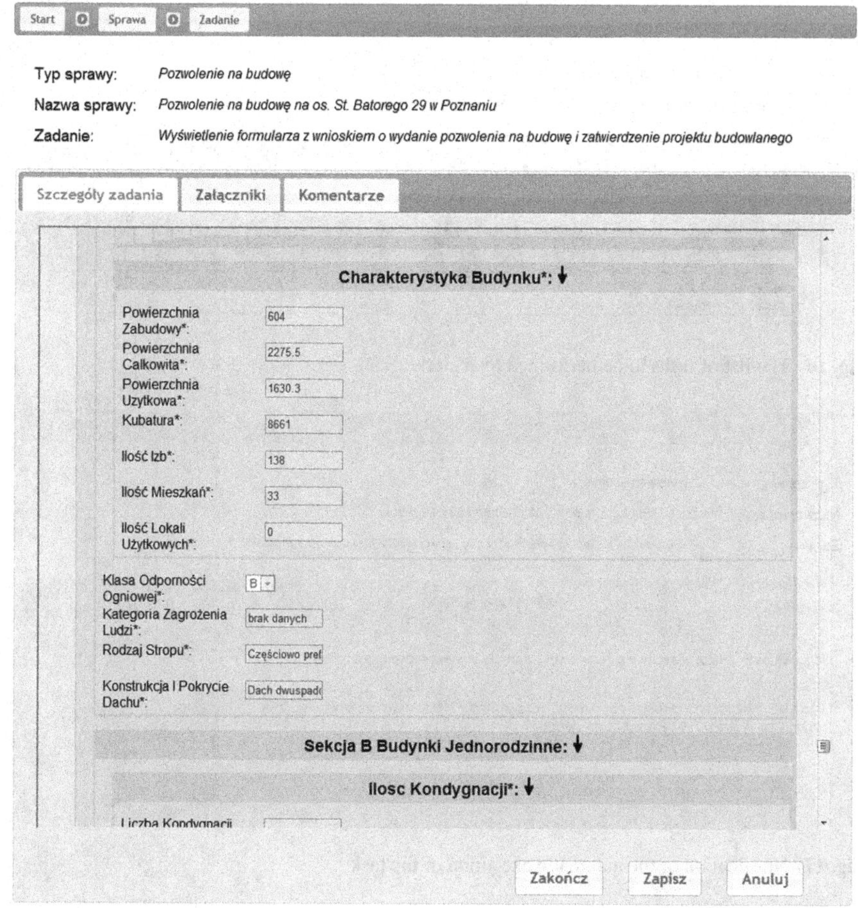

Fig. 15 The task “Display a form for an application for a building permit and approval of a construction project”

After completing the task *Wyświetlenie formularza z wnioskiem...(Display a form for an application...)*, the system updates the task list. The updated list is presented in Fig. 16.

After selecting the first item from the list presented in Fig. 16, a notification of the automatic execution of tasks by the system is displayed. This notification is shown in Fig. 17.

After selecting the second item from the list presented in Fig. 16, a task *Sprawdzenie poprawności merytorycznej wniosku z terminologią ustawy Prawo budowlane, a także czy wniosek dotyczy inwestycji liniowej (Verify if the application is aligned with the terminology of the Building Code and if it refers to the line investment)* is displayed. The task is presented in Fig. 18.

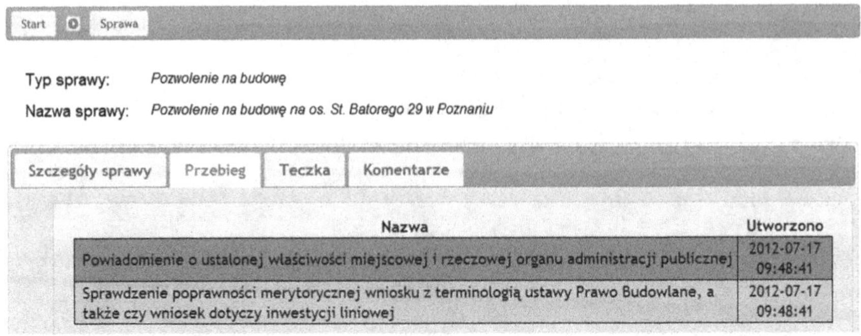

Fig. 16 The list of tasks to be performed by a clerk

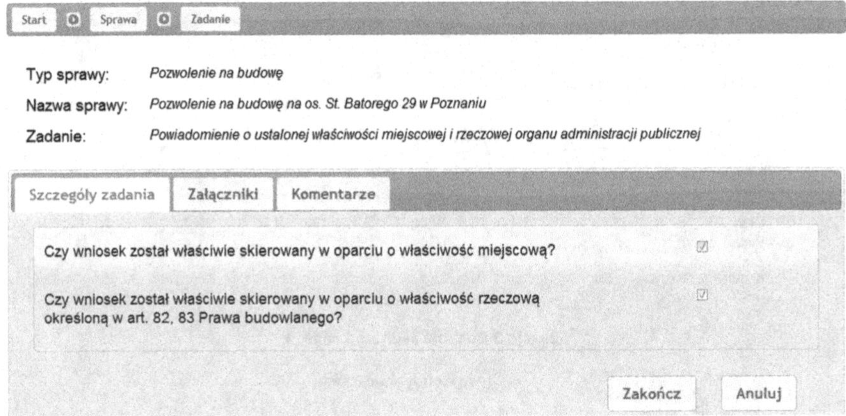

Fig. 17 Notification of the automatic execution of the task

Depending on the answers provided by the clerk within the task *Sprawdzenie poprawności merytorycznej wniosku ...(Verify if the application is correct)*, the system determines the further course of the procedure until submitting the final decision which indicates the end of the procedure.

5.4.5 Summary

The PEOPA platform, presented in this section, uses the CMEAP approach to compose an administrative procedure model in a flexible manner. The resulting model is closely tailored to the legal circumstances of the case the procedure is conducted for and provides a detailed map of all administrative tasks that need to be performed within this procedure. This makes it possible to automate the procedure in two dimensions: selection of activities and execution of activities. Automation of selection means that the PEOPA platform using terminological knowledge stored in

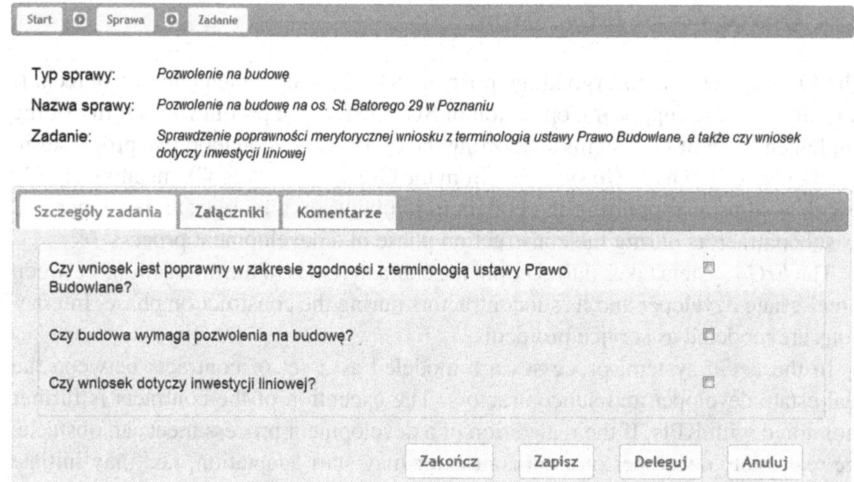

Fig. 18 The task "Verify if the application is aligned with the terminology of the Building Code and if it refers to the line investment"

the Knowledge base replaces a clerk in interpreting the law and making decisions on activities to perform on a given stage of the procedure. Automation of execution means that the PEOPA platform replaces a clerk in the operational execution of activities. Automation of selection is an inherent feature of the CMEAP approach and thus the PEOPA platform, automation of execution is largely derived from the technical and legal circumstances: the access to public registers in online manner, ability and willingness to use the electronic documents, etc.

The use of the PEOPA platform also significantly reduces the complexity of the administrative procedure models compared to the traditional monolithic approach. Thus, the process of updating models in order to reflect the changes in legislation is much easier; i.e., in the case of a change in a legal act, instead of modifying a number of administrative procedure models, merely a single or few elementary processes have to be altered. In this way, the updating process becomes less time consuming, less expensive, and less prone to errors. Furthermore, the replacement of the monolithic models of administrative procedures with the elementary process models corresponding to specific legal aspects contributes to elimination of inconsistencies in representations of the same aspects in various monolithic procedure models.

5.5 The ErGo System for Support of Collaborative Processes

In this section, the *ErGo* prototype which implements service protocols and adaptation tools for real-estate developers is presented.

5.5.1 Overview of the *ErGo* System

The *ErGo* system (http://ergo.kti.ue.poznan.pl/) [32] is an implementation of technical infrastructure supporting operation of SOVOBE and a proof of feasibility of the implementation of IT systems supporting the execution of collaborative processes in an adaptive way. The *ErGo* system—from the Greek word ("ergo"), meaning "task", "work"—aims at supporting the collaboration between a real-estate developer and its subcontractors during the construction phase of a development process.

The *ErGo* system takes into account the characteristics of the interactions between a real-estate developer and its subcontractors during the construction phase. Interactions are modeled as service protocols.

In the *ErGo* system, processes are modeled as a set of contracts between the real-estate developer and subcontractors. The execution of the contracts is further monitored with KPIs. If the realization of a development process meets an obstacle, the real-estate developer or a subcontractor may start adaptation, i.e., may initiate a discussion concerning the modification to be applied to the set of subcontractors and/or the contracts related with the investment.

5.5.2 *ErGo* Applications

The *ErGo* system consists of six applications tackling various complementary aspects of the construction phase of development processes. The first application—*ErGo* Organizations—is responsible for the management of the description of organizations and their competences. The second application—*ErGo* Services—is responsible for the management of the description of business services provided by organizations. A real-estate developer manages all the contracts with subcontractors with the third application—*ErGo* Investments. The subcontractor contracting process is supported by the fourth application—*ErGo* MatchMaker. Templates for various types of development processes are defined in the fifth application—*ErGo* Investment Types. Finally, the sixth application is m-*ErGo*, a mobile application tailored to the need of construction managers working on the construction site.

***ErGo*Organizations.** The *ErGo* Organizations application allows users to manage the description of organizations, especially their competences. Choosing the right organization to perform a given activity is not an easy undertaking. All the competence requirements have to be satisfied by the chosen subcontractor to avoid cost increases, missing deadlines or even breach of contract. A precise description of competences of organizations is an important element for the management of development processes.

The functionality of the *ErGo* Organizations application is organized as follows: first, a group of functions provides users with means for registering new organizations and updating the data concerning already registered organizations. Second, another group of functions allows users to manage the competences of organizations, based on the competence model proposed in [48]. Third, another group of functions gives

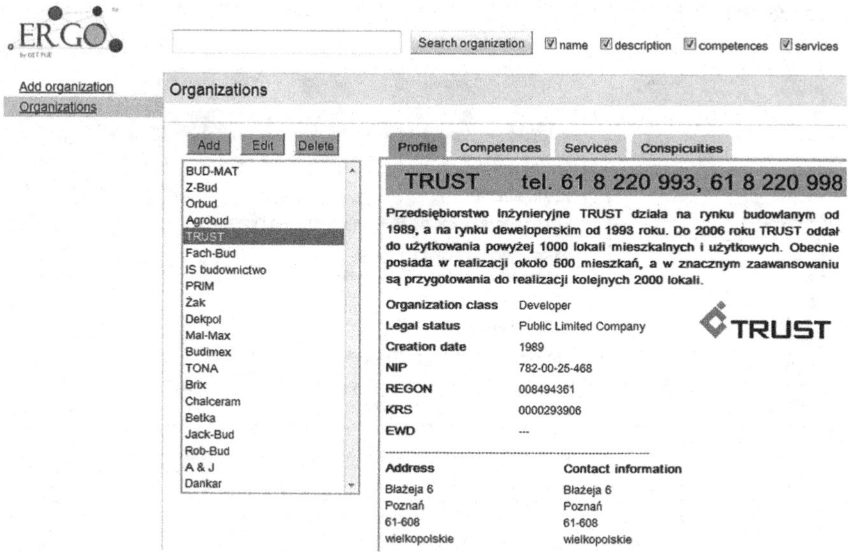

Fig. 19 Main panel of the *ErGo* Organizations application

users means for retrieving organizations satisfying a set of requirements concerning either their profile or their competences. Finally, the *ErGo* Organizations application provides functions for searching and filtering registered organizations.

A screen capture of the *ErGo* Organizations application is presented in Fig. 19. Users select an organization from the list of organizations on the left side. Detailed information about the selected organization is then displayed in the central panel. Competences of the organization are available on the associated tabbed panel. Organizations can be filtered out with the Search field on the top.

***ErGo* Services.** The *ErGo* Services application allows users to manage the description of the services provided by organizations. The list of services that organizations provide is an important element of the description of organizations. The choice of an organization as a subcontractor depends on the services the organization provides. The goal of the *ErGo* Services application is to support the management and search of services provided by organizations.

The functionality of the *ErGo* Services application is organized as follows: first, a group of functions provides users with means for registering new types of business services and updating the already registered ones. Second, another group of functions provides users with means for managing the business services provided by organizations registered in the *ErGo* Organizations application. Third, another group of functions allows users to retrieve organizations satisfying a set of requirements concerning the business service they provide. Finally, functions for searching and filtering business services registered in the *ErGo* system are provided by the *ErGo* Services application.

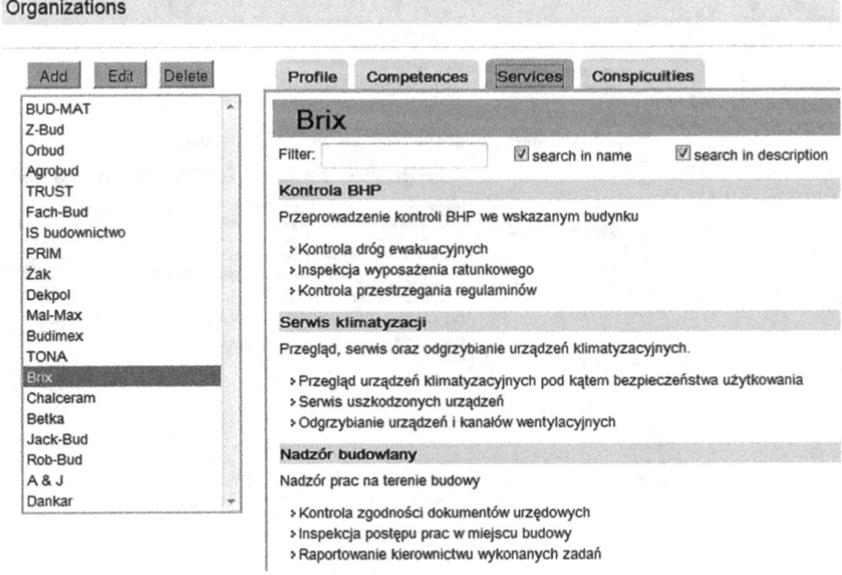

Fig. 20 Main panel of the *ErGo* Services application

A screen capture of the *ErGo* Services application is presented in Fig. 20. The graphical user interface of the *ErGo* Services application is part of the *ErGo* Organizations application. For a given organization, details concerning the business services provided by the organization are available on the "Services" tabbed panel. Services can be filtered out by name and/or description.

***ErGo* Investments.** The *ErGo* Investments application allows users to manage contracts for the construction phase of development processes. In the context of service protocols, the *ErGo* Investments application is based on an implementation of prototype service protocols and instances of service protocols modeling development processes in their construction phase. During the construction phase of a development process, the sequence of activities to be performed is usually not explicitly specified in advance. Moreover, a development process is usually not entirely determined when its construction phase starts. Activities to be performed during the construction phase of a development process are usually not planned ahead for more than three months.

The *ErGo* Investments application allows users to control the realization of the construction phase of a development process in a seamless manner, by providing means for continuous preparation of contracts, appendices and progress payment claims.

The *ErGo* Investments application enables the adaptation of development processes. In the *ErGo* system, and in the *ErGo* Investments application in particular, adaptation is a collaborative task: potential modifications of the execution of the development process are discussed by stakeholders of the development process. The discussion consists of an exchange of adapting statements among stakeholders. An adapting statement is related to a contract, a progress payment claim, an appendix, a group of contracts, and an investment.

Adapting collaborators express various types of positions in their adapting statements: a change proposition is rated, commented or extended. Rating and commenting positions are useful to ease the final decision about the execution of a change proposition: a high rating of a given change proposition may influence the decision maker responsible for the execution of change propositions.

The functionality of the *ErGo* Investments application is organized as follows: first, a group of functions provide users with means for registering new contracts, appendices, and progress payment claims, and updating the already registered ones. Second, another group of functions provides users with means for managing KPIs. Third, another group of functions provides users with means for adaptation. Finally, the *ErGo* Investments application provides functions for searching and filtering contracts, appendices, and progress payment claims registered in it. A screen capture of the main panel of the *ErGo* Investments application is presented in Fig. 21. In the middle of the panel, a table presents the data concerning development processes (e.g., "Budynek mieszkalny–Głogowska 199", *ang. "Residential building–Głogowska 199"*), groups of contracts (e.g., "Dach–pokrycie, obróbki, detale", *ang. "Roof–covering, treatment, details"*), contracts (e.g., "Umowa UM/88", *ang. "Contract*

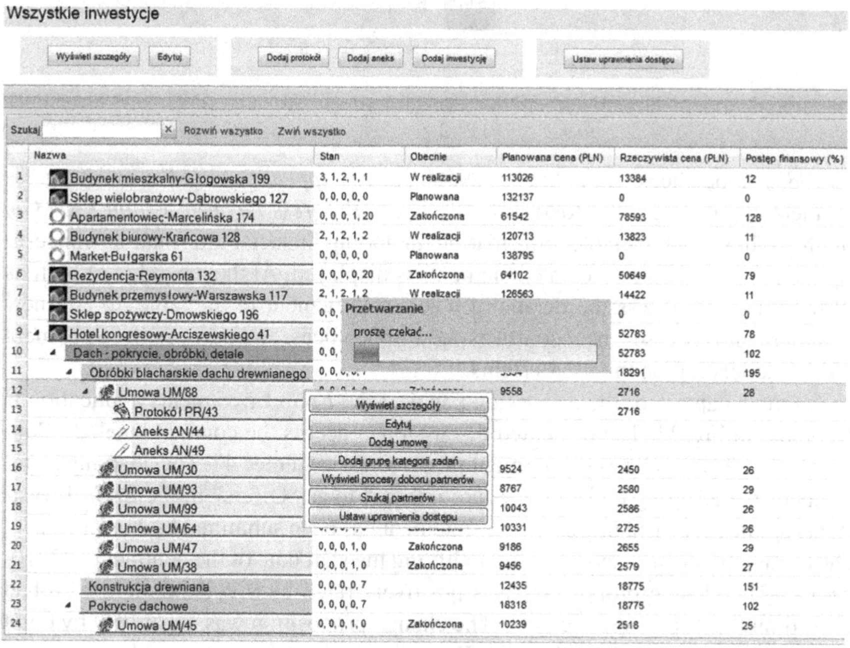

Fig. 21 Main panel of the *ErGo* Investments application

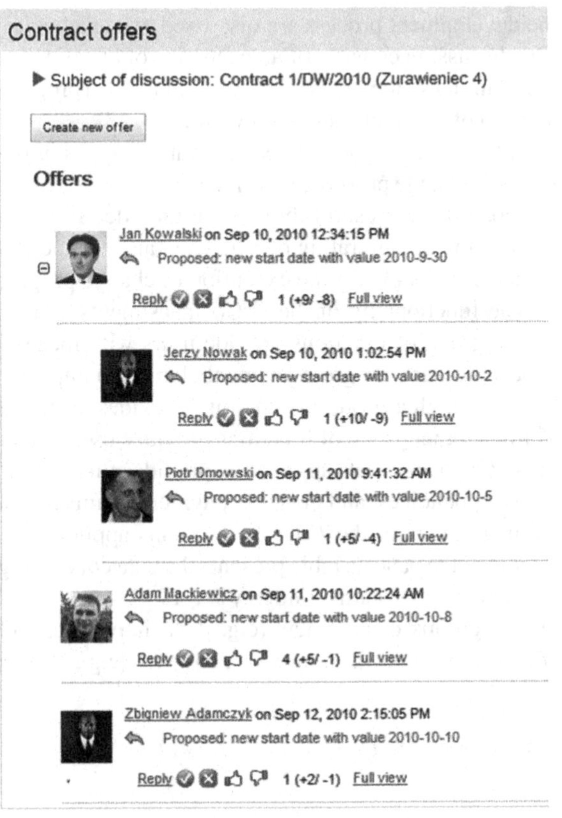

Fig. 22 Discussion about a contract in the *ErGo* Investments application

UM/88"), appendices (e.g., "Aneks AN/44", *ang. "Appendix AN/44"*), and progress payment claims (e.g., "Protokół PR/43", *ang. "Progress Payment Claim PR/43"*). On the right side of the table, information, including financial information, concerning the progress of the associated elements is displayed. At the top, a list of buttons allows a user to access the details of a selected element in the table, to add new development processes, new contracts, new appendices, and new progress payment claims, depending on the user's rights.

A screen capture of the discussion panel of the *ErGo* Investments application is presented in Fig. 22. The presented discussion concerns the contract labeled "Contract 1/DW/2010" of the development process "Żurawieniec 4". The adapting statements of the adapting collaborators are referred to as "Offers" in the *ErGo* Investments application. The first adapting statement has been submitted by Jan Kowalski. The change proposition consists in modifying the start date of the contract to "2010-09-30". The change proposition was positively rated by 9 collaborators who had clicked on the "Approved" icon. The change proposition was negatively by eight collaborators who had clicked on the "Disapproved" icon. A user with appropriate rights accepts the change proposition by clicking on the "Accept" icon ✔, or rejects it by clicking on the "Reject" icon ✖.

***ErGo* MatchMaker.** The *ErGo* MatchMaker application allows users to select subcontractors during the construction phase of development processes. In the context of service protocols, the *ErGo* MatchMaker application provides means to select appropriate actors to instantiate service protocols modeling development processes in their construction phase.

The *ErGo* MatchMaker application supports continuous selection of subcontractors, where the selection is performed in a collaborative way by persons directly responsible for the realization of the investment process. A user selects organizations that can perform required activities, then aggregates the chosen organizations. Both single organizations and groups of organizations are potentially discussed with other collaborators. The discussion concerning the chosen organizations and groups of organizations is structured in an identical manner to the *ErGo* Investments application.

Internal discussions concern the potential subcontractors and the terms of the contracts to be negotiated with the chosen organizations. Internal discussions are followed by external discussions in which the representatives of chosen organizations are involved. The goal of the external discussion is the negotiation of the final terms of the contract. When the external discussion ends with a satisfying compromise, the contract is signed and its representation is added to the *ErGo* Investments application.

In the *ErGo* MatchMaker application, the selection process of subcontractors is based on a set of requirements concerning a group of organizations to be contracted. The requirements are defined in templates of development processes, referred to as "investment types". An investment type is an abstract service protocol. Requirements used by the *ErGo* MatchMaker application are based on the service-oriented summary, the service network schema, and the mapping functions of the abstract service protocol. As an example, an investment type for the construction of residential buildings defines requirements concerning the architect and electricians, and their relations. These requirements are captured in the service network schema. A screen capture of the *ErGo* MatchMaker application is presented in Fig. 23. The presented panel allows users to discuss a group of potential subcontractors. On the top of the panel, a description of the group of activities to be contracted and the associated development process is presented. In the middle of the panel, a table presents the activities to be contracted and the organizations that have been proposed as potential subcontractors. In Fig. 23, only one organization, i.e., "Dekoratornia", has been proposed as a potential subcontractor for the building of a reinforced concrete slab. Negotiations are ongoing with the chosen organizations. At the bottom of the panel, another proposition has been submitted by Jakub Flotyński.

***ErGo* Investment Types.** The *ErGo* Investment Types application allows users to manage investment types. In the context of service protocols, the *ErGo* Investment Types application is based on an implementation of abstract service protocols modeling development processes in their construction phase. An investment type is an abstract service protocol. The service-oriented summary, the service network schema

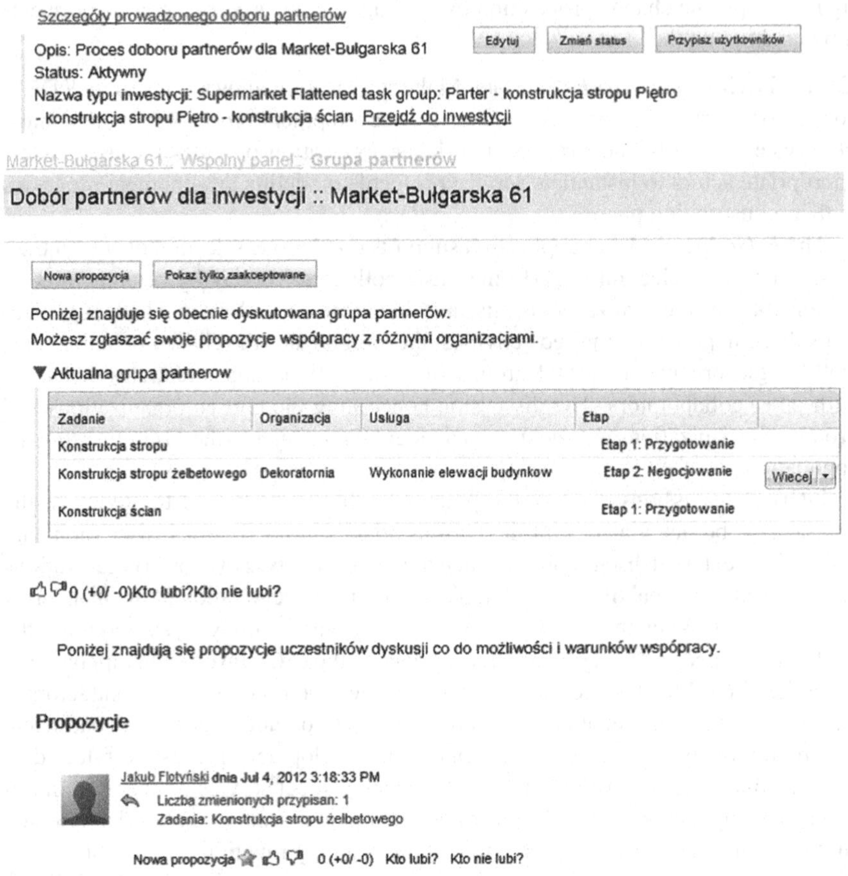

Szczegóły prowadzonego doboru partnerów

Opis: Proces doboru partnerów dla Market-Bułgarska 61 Edytuj Zmień status Przypisz użytkowników

Status: Aktywny

Nazwa typu inwestycji: Supermarket Flattened task group: Parter - konstrukcja stropu Piętro - konstrukcja stropu Piętro - konstrukcja ścian Przejdź do inwestycji

Market-Bułgarska 61 :: Wspolny panel :: Grupa partnerów

Dobór partnerów dla inwestycji :: Market-Bułgarska 61

Nowa propozycja Pokaż tylko zaakceptowane

Poniżej znajduje się obecnie dyskutowana grupa partnerów.
Możesz zgłaszać swoje propozycje współpracy z różnymi organizacjami.

▼ Aktualna grupa partnerow

Zadanie	Organizacja	Usluga	Etap	
Konstrukcja stropu			Etap 1: Przygotowanie	
Konstrukcja stropu żelbetowego	Dekoratornia	Wykonanie elewacji budynkow	Etap 2: Negocjowanie	Wiecej
Konstrukcja ścian			Etap 1: Przygotowanie	

0 (+0/ -0)Kto lubi?Kto nie lubi?

Poniżej znajdują się propozycje uczestników dyskusji co do możliwości i warunków współpracy.

Propozycje

Jakub Flotyński dnia Jul 4, 2012 3:18:33 PM
Liczba zmienionych przypisan: 1
Zadania: Konstrukcja stropu żelbetowego

Nowa propozycja 0 (+0/ -0) Kto lubi? Kto nie lubi?

Fig. 23 Partner selection in the *ErGo* MatchMaker application

and the mapping functions of an abstract service protocol are further used by the *ErGo* MatchMaker application as a set of requirements for the selection of subcontractors.

An investment type contains also groups of contracts, referred to as "category groups" in the *ErGo* system. Groups of contracts usually cover a set of activities to be performed at the same stage of the construction phase, e.g., roofing related contracts. The subcontractor selection process supported by the *ErGo* MatchMaker application usually concerns a group of activities to be contracted within a given group of contracts.

The functionality of the *ErGo* Investment Types application is organized as follows: first, a group of functions provides users with means for registering new investment types and managing already registered ones, including managing groups of contracts, requirements, and contract templates. Second, another group of functions

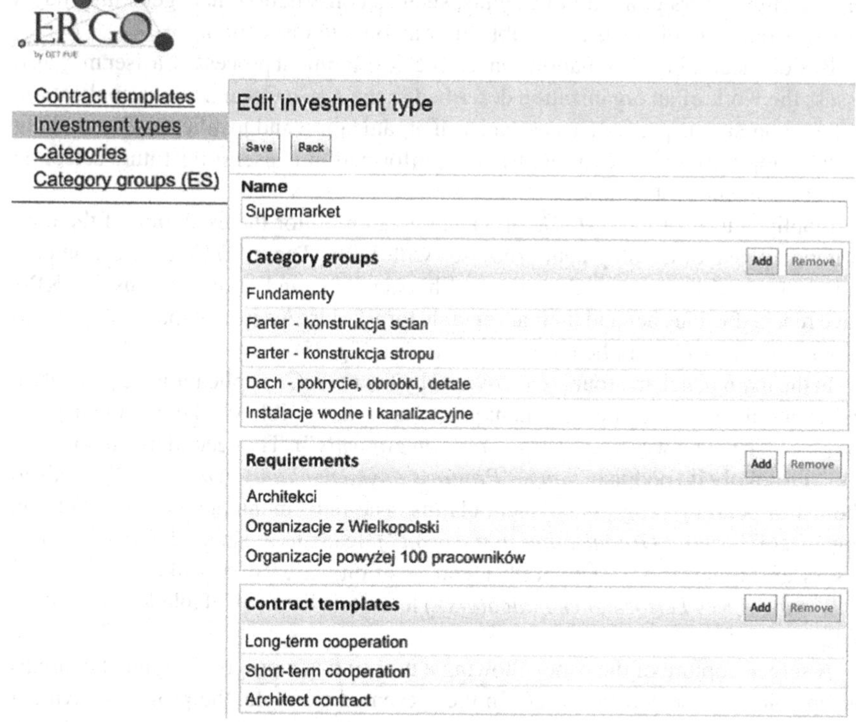

Fig. 24 Edition panel of the *ErGo* Investment Types application

provides other *ErGo* modules and applications, e.g., *ErGo* MatchMaker and *ErGo* Investments, with access to investment types.

A screen capture of the *ErGo* Investment Types application is presented in Fig. 24. In the presented panel, an investment type concerning the construction of a residential building is presented. The investment type for residential building contains five groups of contracts (e.g., "Fundamenty", *ang. "Foundations"*), three requirements (e.g., the first requirement concerns the architects), and three contract templates that can be used to contract activities from subcontractors.

m-*ErGo*. The m-*ErGo* application is devoted to mobile devices and is tailored to the need of construction managers working on the construction site.

Most activities associated with the construction phase of a development process are performed at the construction site. The m-*ErGo* application provides access to a wide range of functions of the *ErGo* system that are useful on the construction site, from accessing information about a particular contract, to signing a progress payment claim issued by a subcontractor.

After successful login, user accesses information concerning development processes, e.g., the start and end dates of a given contract, the list of progress payment claims. Users granted more rights, such as construction managers, may flag a progress payment claim as acceptable after an on-site inspection.

Besides accessing information concerning development processes, a user may also assess the work of an organization described in the *ErGo* Organizations application. After an on-site inspection, a user ranks an organization and justifies his/her ranking with an appropriate justification. Ranking information is useful for future selection processes of subcontractors.

Finally, a user of the m-*ErGo* application can monitor the execution of the construction phase of development processes with KPIs. The m-*ErGo* application provides a user with access to the values of his/her KPIs and notifies the user if KPIs have reached values beyond their acceptable thresholds. A screen capture of the main panel of the m-*ErGo* application is presented in Fig. 25.

In the main panel, the four areas covered by the m-*ErGo* application are presented: information concerning development processes is accessible via the first menu position, i.e., "Twoje inwestycje" (*ang. "Your investments"*). The second menu position, i.e., "Protokoły do podpisu" (*ang. "Progress payment claims to be signed"*) leads to the list of pending progress payment claims. The third menu position, i.e., "Ocena organizacji" (*ang. "Organization assessment"*), leads to a panel allowing a user to assess an organization. The fourth menu position, i.e., "Wskaźniki efektywności KPI" (*ang. "Key Performance Indicators"*) leads to a list of available KPIs and their values.

A screen capture of the panel allowing a user to flag a progress payment claim as acceptable is presented in Fig. 26. In the presented example, the progress payment claim "PR/37" is flagged as acceptable (by clicking on the "Accept" button) or flagged as non-acceptable (by clicking on the "Reject" button).

5.5.3 Summary

The *ErGo* system, presented in this section, enables agile construction support for real-estate developers. Supporting the construction phase of development processes requires a different approach to project management than the one implemented in most project management information systems based on Gantt or PERT charts. Gantt and PERT charts are unnecessary with regard to the ubiquity of shared tacit knowledge concerning the sequence of constructing activities to be performed to build a property. In the *ErGo* system, contracts, progress payment claims, and appendices are at the core of the modeling of interactions between a real-estate developer and its subcontractors. Such an approach is fully aligned with the culture and the customs of the construction sector. As a consequence, the *ErGo* system is more comprehensible to real-estate developers and subcontractors than traditional project management information systems.

The application of service protocols to the construction phase of development processes allows real-estate developer to orchestrate larger development processes. Without an appropriate support for the interactions among the real-estate developer

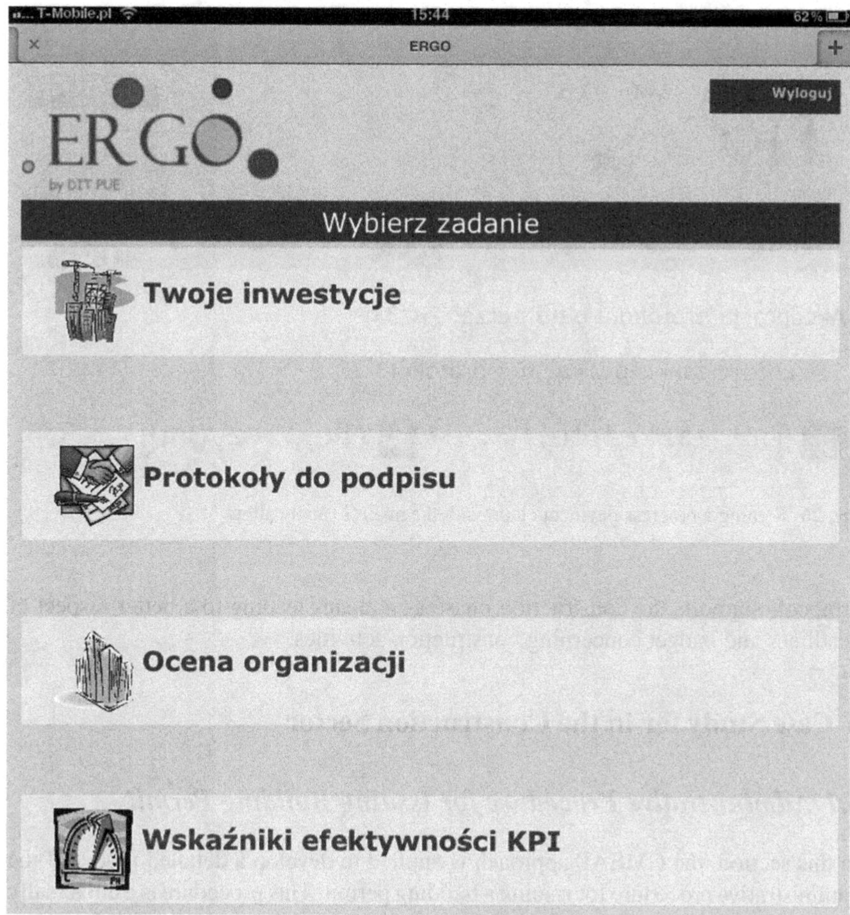

Fig. 25 Main panel of the m-*ErGo* application

and its subcontractors, the real-estate developer, especially small and medium size one, cannot efficiently orchestrate large development processes which requires the management of too many contracts. In the *ErGo* system, abstract service protocols and modeling investment types define and group contracts to be negotiated, signed, and executed. Appropriate KPIs support the management of a large number of contracts by monitoring the state of all the contracts of a given development process.

Adaptation of service protocols addresses directly the needs for agile interactions between the real-estate developer and its subcontractors. In the construction sector, changes of the interactions between the real-estate developer and its subcontractors are frequent and usual. Adaptation of service protocols enables the real-estate developer to rapidly react to new situations, e.g., bad weather conditions, an excavator breakdown, and an unavailable partner. As a consequence, adaptation of service

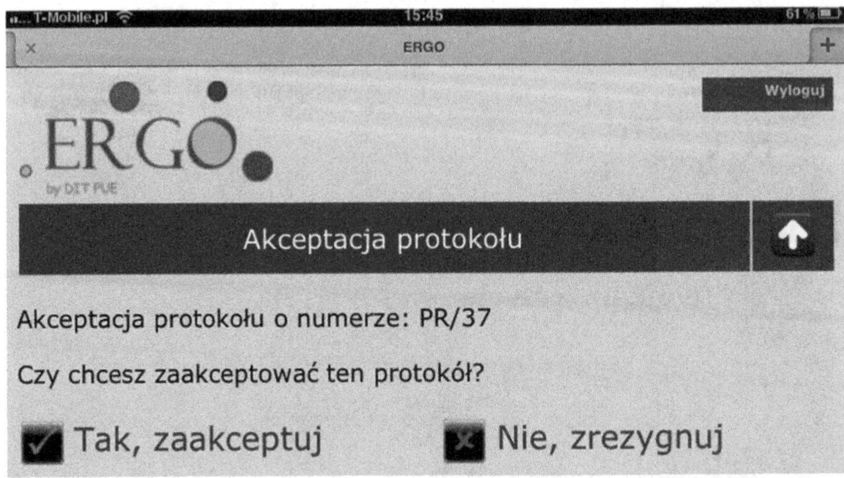

Fig. 26 Signing a progress payment claim with the m-*ErGo* application

protocols supports the construction phase as a means leading to a better respect of deadlines and budget concerning construction activities.

6 Case Study for in the Construction Sector

6.1 Administrative Procedure for Issuing Building Permit

In this section, the CMEAP approach is applied to develop a detailed model of the administrative procedure for issuing a building permit. This procedure is a universally understandable example for readers from different countries, because it is one of the most widely spread administrative procedures for citizens and businesses. The model of this procedure was created under the Polish legislation, but most of the laws governing the course of this procedure are similar to the laws in other countries [26]. The CMEAP approach is general and, therefore, can be applied to any administrative procedures governed by the laws of any country.

6.1.1 Procedure Specification

The building permit should cover an entire construction project. In the case of construction project involving more than one object, the building permit may, at the request of the investor, apply to a selected object or a group of objects. In this case, the investor is obliged to submit the plot or land development plans for the entire construction project.

An application for a building permit should be submitted along with the following attachments:

- four copies of the building design, including the opinions, approvals, permits and other documents required by specific regulations;
- the statement confirming the investor's right to dispose of the real estate for building purposes;
- the planning decision, if required under the provisions on planning and spatial development.

The building design should meet the requirements specified in the planning decision, if such a decision is required. The scope and content of a building design should be tailored to the specifics and nature of the project and the complexity of the construction work. The building design should contain the following documents:

- the plot or land development design, drawn on an up-to-date map, including: specification of the boundaries of the plot or land, location, contours and layout of existing and planned building objects, technical infrastructure network, sewage disposal or treatment, communication system, and arrangement of green areas, and indicating the characteristic elements, dimensions, ordinates and distances between the building objects, in relation to the existing and planned buildings in the surrounding areas;
- the architectural and construction design, specifying the function, form and structure of a building object, the energy and environmental characteristics and proposed the required technical and material solutions for demonstrating the principle of adopting the building object in relation to existing buildings;
- if necessary, statements of appropriate organizational units on the supply of energy, water, heat, gas, sewage collection and the conditions for connecting the object to the water, sewage, heat, gas, electricity, telecommunications, and road networks,
- the statement of a competent road management unit on connection capabilities of the plot with a public road in accordance with the regulations on public roads;
- if necessary, the results of geological engineering surveys and geotechnical conditions for the foundation of building objects.

Prior to issuing a decision on a building permit or a separate decision on approval of the building design, the competent authority should verify:

- compliance of the building design with the provisions of a local master plan or a planning decision, if there is no local master plan, and with the requirements for environmental protection, in particular those specified in the environmental decision;
- compliance of the plot or land development design with regulations, including the technical and building provisions;
- completeness of the building design and the required opinions, approvals, permits and checks, and the information on safety and health protection, as well as other documents submitted along with the application;
- preparation of the building design by a qualified person.

If any infringement in the documents attached to the application is found, the competent authority imposes an obligation to remove the indicated deficiencies, with a deadline for their removal. After its expiry, the authority issues a decision to refuse approval of the building design and issuing a building permit.

The competent authority issues a decision declining a building permit and approval of the building design if at the plot or land designated for the development, there is a building object under a demolition order.

If the application and all attached documents are complete, correct and complying with the provisions, the competent authority is obliged to issue a decision on a building permit and approval of the building design.

A detailed model of the administrative procedure for issuing a building permit is composed of 60 elementary process models, 30 decision rule units, and 50 fact classes. To illustrate the CMEAP approach, selected parts of the administrative procedure model focused on specific legal aspects are presented in the remainder of this section.

6.1.2 Case Study

In this case study, an investor submits an application for a building permit and approval of the building design of a residential building to a competent authority. The investment is characterized by the following circumstances:

- The area of investment is not included in a local master plan.
- The investment area covers an area entered in the communal records of historical monuments and is not included in the register of historical monuments.
- The investment is located in the vicinity of a Natura 2000 site.
- There are exemptions from technical and building regulations in the building design.

In the presented case, the competent authority in the course of an administrative procedure has to cooperate with other authorities of the public administration, as shown in Fig. 27.

Lack of Local Master Plan. If the investment is located in an area not covered by a local master plan, a planning decision has to be obtained by an investor and attached to the application for a building permit. In this case, a final building permit must be issued in strict compliance with the terms and conditions specified in the planning decision. Therefore, before making a decision on a building permit the competent authority has to verify compliance of the building design with the requirements specified in the planning decision.

The elementary processes related to the lack of a master plan are the following:

- Checking if the investment area is covered by a local master plan and verifying if a planning decision is attached, if required.
- Verifying the compliance of the building design with the provisions of a planning decision.

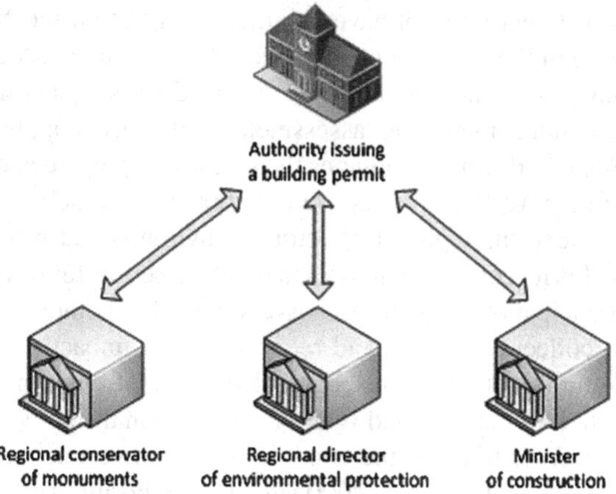

Fig. 27 Collaboration of public authorities within an administrative procedure for issuing a building permit

Investment Area in Communal Records of Historical Monuments. Carrying out construction works related to a building object or an area entered in the register of historical monuments requires obtaining a permit to carry out these works, issued by the competent regional conservator of monuments. The conservator's permit has to be obtained by an investor and attached to the application for a building permit.

In relation to the building objects not listed in the register of historical monuments, but included in the communal records of historical monuments, a building permit or the permit to demolish a building object should be issued by the competent authority in consultation with the regional conservator of monuments. The conservator of monuments is obliged to issue the opinion regarding the application for a permit to build or demolish a building object within 30 days from the day when it was received. Failure to express an opinion within the time limit is considered as lack of objections to the design solutions presented in the application.

The elementary processes related to the fact that the investment area covers an area entered in the communal records of historical monuments and is not included in the register of historical monuments are the following:

- Checking if a building object or area is included in the communal records of historical monuments.
- Applying to a competent regional conservator of monuments for approval of the building design.

Impact on a Natura 2000 Site. If the authority competent to issue a building permit finds that the investment could potentially have a significant impact on a Natura 2000 site, the investor is obliged to submit the application for issuing a building permit along with the documents required for the impact assessment to the regional director of environmental protection for agreement.

The regional director of environmental protection, after receiving the documents examine whether the investment project may have significant impact on the Natura 2000 site. If the regional director of environmental protection recognizes that the investment will not have significant impact on the Natura 2000 site, it waives the need of the assessment. However, if the director recognizes that the investment may have significant impact on the Natura 2000 site, it issues a decision on the obligation to conduct the impact assessment. In this decision, the regional director of environmental protection obligates the investor to prepare and submit a report on the impact of the investment on the Natura 2000 site and defines the scope of the report.

Next, the regional director of environmental protection makes a request to the authority competent to issue a building permit for ensuring an opportunity for public participation in the impact assessment. The authority carries out public consultation to collect comments and requests on the impact of the planned investment on the Natura 2000 site. The regional director of environmental protection analyzes the collected material and issues a decision on the agreed conditions for the investment or refuses to agree the conditions if the investment can have significantly negative impact on the Natura 2000 site. The authority examines the decision and based on the analysis of the collected evidence issues a decision on a building permit.

The elementary processes related to the fact that the investment is located in the vicinity of a Natura 2000 site are the following:

- Checking if an investment area is located in the vicinity of a Natura 2000 site and application to a competent regional director of environmental protection to examine whether the investment may have significant impact on the site.
- Carrying out public consultation to collect comments and requests from on the public on the impact of the planned investment on the Natura 2000 site.
- Verifying the compliance of the building design with the decision of the regional director of environmental protection on the investment's impact on the Natura 2000 site.

Exemptions from Technical and Building Regulations. In particularly justified cases, an investor can realize an investment according to a building design with exemptions from technical and building regulations. The specific needs of the investor, unusual shape of the plot, or other circumstances may make it impossible to realize an investment, for example, with maintaining appropriate distances from neighboring plots. To solve such problems the investor may obtain approval for the requested exemptions from regulations from the authority competent to issue a building permit.

To obtain the approval the investor has to justify the need for the exemptions and provide the proposals of substitutive solutions. The approval of the exemptions requires receiving an authorization from the minister, who set forth the technical and building regulations. The application for an authorization is submitted to the minister by the authority responsible for issuing a building permit, and the minister's authorization is addressed to this authority. If the exemptions apply to regulations set forth by various ministers, an authorization must be obtained from each of them. The minister may grant an authorization upon the fulfillment of additional requirements.

After receiving an authorization from the minister, the competent authority shall either refuse or grant the approval for the exemptions from the regulations. The authority is not bound by the authorization granted by the minister, because the

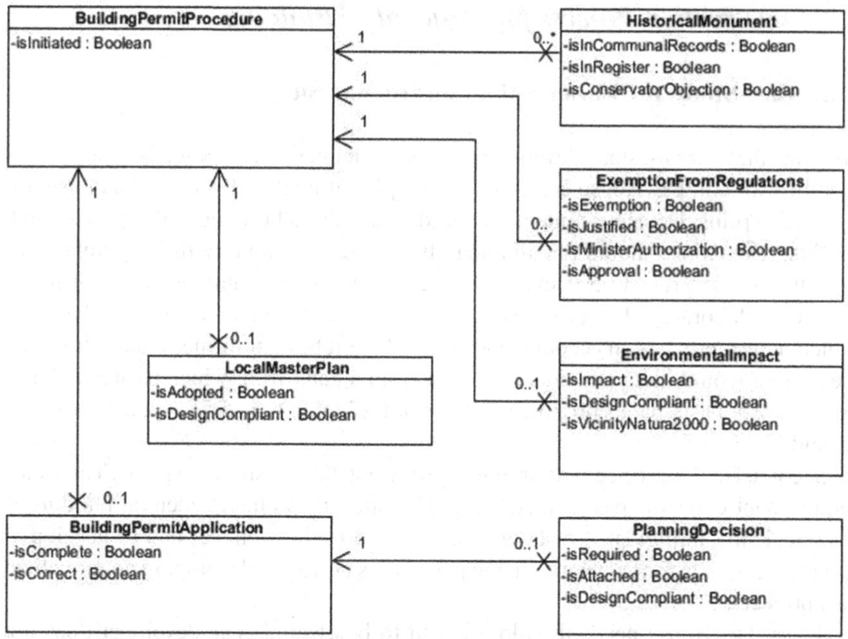

Fig. 28 Ontology of concepts related to the selected legal aspects of the administrative procedure for issuing a building permit

authority is obliged to investigate and comprehensively consider each particular case.

The elementary processes related to the fact that the building design contains exemptions from technical and building regulations:

- Checking if there are any exemptions from technical and building regulations in the building design.
- Applying to the competent minister for authorization to accept the exemptions in making a decision refusing or granting the approval for the exemptions in the building design.

6.1.3 Ontology

Based on analysis of the legal acts related to the selected aspects, the ontology of concepts has been designed, as presented in Fig. 28.

In the course of the administrative procedure for issuing a building permit the instances of the ontology concepts (facts) are created. The facts represent legal circumstances occurring during execution of that procedure for a specific case. The selection of elementary processes to be executed is carried out by the decision rules that are activated based on the analysis of the facts.

6.2 Construction Process for Residential Building

6.2.1 Case Study 1: Clearance of a Construction Site

Assume that a real-estate developer company, named DevHouse, is planning its next investment. DevHouse currently owns a plot of land with an old abandoned oil refinery in ruins. DevHouse intends to build a residential building in this plot of land.

First, DevHouse should obtain a loan from a bank to finance the investment. To ease the whole procedure, DevHouse wants to focus on banks with which it has already collaborated. Therefore, DevHouse plans to negotiate a loan with banks at which it currently has an account, and from which it has got former loans. However, DevHouse would like to avoid a next loan from a bank from which it already has a current loan. Also, banks proposing interest rates higher than 5.5 % for a 3-year loan should be rejected.

Second, DevHouse needs an architect to develop the construction plans. The major requirement concerns past and current collaboration. As the residential building is a strategically important investment for DevHouse, the architect had to develop at least five projects for DevHouse in the past, but is currently developing no more than two projects.

Third, DevHouse needs the plot of land to be cleared. The developer does not want to be in charge of the clearance process. The architect should be responsible for identifying a site preparation company, which will supervise the preparation activities, i.e., demolition and rubble removal. The two companies that will perform the preparation activities, i.e., a demolition company and a debris hauling company, should be known and trusted by the architect. Moreover, they should be able to collaborate efficiently. The remaining activities required to build a residential building, e.g., foundation construction, masonry, carpentry, are not taken into account in this example.

The *ErGo* system supports DevHouse in definition of tasks to be performed within the construction process and specification of the described requirements. DevHouse uses *ErGo* Investments application to define a set of predicted activities to be performed. Using *ErGo* Investment Types functionality DevHouse can also assign particular requirements that a particular organization, i.e., a bank, must satisfy. Requirements defined with the use of *ErGo* Investment Types concern also services of organizations. In this case requirements concerning interest rate can be included. Finally, social requirements can be defined among classes of actors, i.e., acquaintance among architect and demolition company. Information concerning the number of actors envisioned for a particular activity is also captured in investment type, i.e., two companies performing the preparation activities. In this way abstract service protocol is developed. Then DevHouse creates new construction process in *ErGo* Investments application. Among information provided as a description of investment, DevHouse must indicate previously developed investment type as a basis for the construction process. Activities that do not have any requirements assigned may be defined both in *ErGo* Investment Types or *ErGo* Investments application. The advanced functions supporting partner selection available in the *ErGo* MatchMaker application are available only for activities with defined requirements concerning actors.

Although a classical case in the construction sector is described in the this case study, a set of issues related to this case are still to be addressed.

Multiple Service Consumers. In Case Study 1, a process consists of not only various service providers but also various service consumers. In Case Study 1, the first service consumer is DevHouse that is seeking for financing from banks. The second service consumer is the architect who needs a site preparation company to supervise the cleaning of the plot of land. Finally, the site preparation company itself is a service consumer when it consumes the demolition and rubble removal services provided by a demolition company and a debris hauling company.

The *ErGo* system allows various organizations to be assigned as service customers. The assignment is performed manually, i.e., organizations are selected from *ErGo* Organizations application database by users or are indicated by system on a basis of requirements defined in investment type. Analysis of organizations in terms of requirements satisfaction is implemented in the *ErGo* MatchMaker application. Information concerning service consumer and service provider assigned to the activity is stored in *ErGo* Investments application in a form of contract regulating conditions of service provision and signed by both actors.

Constraints on Actors. In Case Study 1, a process model may define that each individual playing the Real-estate Developer role may perform the "negotiate a loan" activity. A real estate developer may negotiate a loan if financing is needed. However, if the investment may be fully financed by the real-estate developer from its own resources, then a loan (and the associated negotiations to obtain it) may be avoided.

In Case Study 1, a constraint on an actor being a real-estate developer may state that its number of investments has to be greater than ten. Any real-estate developer with a lesser number of investments should not be allowed to participate in the formerly presented process. Similarly, banks proposing interest rates higher than 5.5 % for a 3-year loan should be rejected.

In the *ErGo* system, the definition of a constraint concerning actors and their associations with activities is captured in *ErGo* Investment Types application.

Relational Constraints. In Case Study 1, the choice of an architect is limited by relational constraints: the architect should have at least five former projects performed for DevHouse, but currently performing less than two projects. Similarly, the choice of a bank is limited by relational constraints: DevHouse is interested only in banks at which it has a bank account, had already obtained a loan from those banks, and currently has no loan from them.

In the *ErGo* system, the definition of relational constraints is captured in *ErGo* Investment Types application. Information concerning constraint is used in *ErGo* MatchMaker application during evaluation of possible partner groups. The information concerning constraint is available to users throughout the whole collaborative process of partner selection.

6.2.2 Case Study 2: Removal of Dangerous Chemicals

Assume that the real-estate developer company DevHouse has obtained the financial resources from MoniBank. The construction of the residential building can start with the clearance process of the plot of land with the old oil refinery.

The site preparation company PrepSite supervises the preparation activities, i.e., demolition and rubble removal. The demolition company Demolisher is responsible for the demolition activity, while the debris hauling company CleanIt is responsible for the rubble removal. These three companies are known and trusted by the architect in charge of the project, Archibald Tex.

During the first three weeks, the clearance process proceeds without any major perturbation. During the fourth week, CleanIt identifies presence of many barrels of hydrofluoric acid in the ruins. Hydrofluoric acid is a dangerous corrosive and a toxic chemical substance, involved in many explosions and industrial accidents. The removal of hydrofluoric acid is a specialized activity that CleanIt is not able to perform without putting the security of the construction site and its vicinity in jeopardy.

The new situation requires a new plan for the removal of hydrofluoric acid. A special working group, consisting of DevHouse, PrepSite, CleanIt, and Archibald Tex, is formed to work it out. The working group invites a safety inspector of the Occupational Safety and Health Administration to provide guidance concerning regulations about chemical removal. Moreover, following Archibald Tex's recommendation, the working group asks an expert in industrial chemicals for a security audit with regard to environmental pollution.

In the security audit, the risk of major irreversible damages for the environment is clearly demonstrated. As a consequence, the working group decides that a new collaborator specializing in dangerous chemical removal has to be contracted by DevHouse. The architect requires the chosen company to be a collaborator of PrepSite, CleanIt, and Demolisher. CleanIt suggests securing the barrels and continuing the site clearance. The safety inspector informs that under existing regulations the demolition activity should be momentarily suspended until the hydrofluoric acid is removed from the construction site. Moreover, a certificate delivered by the Occupational Safety and Health Administration testifying that a company has the appropriate competences concerning dangerous chemicals is required to manipulate hydrofluoric acid.

The ChemOver company, a certified specialist company in dangerous chemical removal, is identified as satisfying the requirement of the architect, i.e., ChemOver is a former collaborator of PrepSite, CleanIt, and Demolisher. Moreover, after a round of negotiations, DevHouse accepts to contract the dangerous chemical removal service provided by ChemOver, as the terms, including the price, offered by ChemOver were aligned with the expectations of DevHouse.

Following the safety inspector's legal advices, the demolition activity is suspended. Then, after signing the contract with DevHouse, ChemOver removes the dangerous chemicals. With the construction site finally clean from hydrofluoric acid,

CleanIt and Demolisher have come back to finish the clearance process of the construction site.

In the *ErGo* system, adaptation of the service protocol, i.e., modification of a set of activities envisioned in a construction process, can be performed in the *ErGo* Investments application. The adaptation process takes the form of discussion (cf. Fig. 22). The functionality of the application allows the participants of the construction process to communicate and propose changes in the construction process. Once new activities are created, it is possible to launch partner selection process for these activities that later can be conducted in the *ErGo* MatchMaker application. The *ErGo* MatchMaker functionality encompasses negotiations among process participants. At every stage new participants can be assigned to negotiation and can contribute to the ideas. Possible organization and their services that are considered to be assigned activities in the construction process are identified from among organizations and services stored in *ErGo* organization and *ErGo* services application respectively. If needed organizations can be assigned to activity directly in the *ErGo* Investments application, no decision support is provided then to users. Assignment of organization to activity results in creation of contract.

Adapted and Adapting Collaborators. In Case Study 2, the set of adapted collaborators consists of DevHouse, Archibald Tex, MoniBank, PrepSite, CleanIt, and Demolisher. These collaborators are all engaged in the construction of the residential building. The set of adapting collaborators consists of DevHouse, Archibald Tex, PrepSite, CleanIt, the safety inspector, and the expert in industrial chemicals.

Some collaborators are adapted, but not adapting collaborators: MoniBank and Demolisher. These collaborators are not involved in the adaptation of the service protocol in which they are participating. MoniBank is involved in the construction process as a financial institution but not in the adaptation process concerning the hydrofluoric acid removal.

Other collaborators are adapting, but not adapted collaborators: the safety inspector and the expert in industrial chemicals. These collaborators are not participating in the service protocol they are adapting. The safety inspector is involved in the process of defining a solution for the hydrofluoric acid removal, but he is not involved in the construction process.

Other collaborators are both adapted and adapting collaborators: DevHouse, Archibald Tex, PrepSite, and CleanIt. These collaborators are involved in the adaptation of the service protocol in which they are participating. DevHouse is involved in the construction process, as the real-estate developer, and in the adaptation process concerning the hydrofluoric acid removal.

The functionality of the *ErGo* system encompasses definition of roles for system users. Roles define set of function that the particular user may use when interacting with the system. For instance, it is possible to define the role allowing contribution in discussion concerning contract terms but forbidding signing the final version of the contract. The functionality also encompasses the possibility of granting and revoking rights to access the adaptation process for some users or roles.

The adaptation encompasses functionality and data offered by *ErGo* Investment application, *ErGo* Investment Types application and *ErGo* MatchMaker application. Chosen scope of functions is available also through the m-*ErGo* application. For instance, construction supervisor visiting the construction site and experiencing slow motion of ongoing works, may immediately start adaptation process and suggest changes in the planned starting dates of activities.

Modeling Changes. In Case Study 2, the safety inspector submits a change proposition concerning the modification of the process model, suggesting the insertion of a new activity "dangerous chemical removal" before the "demolition" and "clearance" activities. The safety inspector may have provided references to the legal articles regulating dangerous chemical removal from construction site as the reason for the change proposition.

In the *ErGo* system suggestions concerning changes in the construction process are made in form of offers submitted as a part of discussions. Each offer may be described with contextual information in a form of link to referring resources or in a form of attached files. When discussion participants reaches consensus agreed changes are visible in a construction process in the *ErGo* Investments application. Discussions are performed in *ErGo* Investment application. If a discussion concerns selection of new partner or services for construction process, it takes place in *ErGo* MatchMaker application.

7 Conclusions

In this chapter, it has been demonstrated that SOA is a valuable architecture not only at the infrastructural and technical levels but also at the inter-organizational level. SOA-based inter-organizational collaboration is systematically organized around the concept of a service, with organizations mutually providing and consuming services, finally aiming at providing their customers with advanced, but personalized end-user services. As a consequence, application of SOA at the inter-organizational level allows organizations to work in a more user-centric manner, and therefore gain and retain their competitive advantage due to focusing on their core competences.

The solution based on SOA proposed in this chapter supports two main problems of inter-organizational collaboration, namely, flexibility and adaptation. Both the flexible automation of administrative procedures and the adaptation of collaborative processes are tackled by the solution proposed in this chapter. As a consequence, in the context of inter-organizational collaboration, appropriate advanced SOA-based tools and applications, such as the PEOPA platform and the *ErGo* system, improve efficiency and effectiveness of collaborating organizations in highly regulated and competitive environments.

In globalized competitive environments, business sustainability is a major problem. Business environments in which inter-organizational collaboration is fostered by appropriate advanced SOA-based tools and applications are more competitive and supposedly more innovative, which helps to reduce business failures, and thus to increase business sustainability.

The main contributions of this chapter are the following:

1. An in-depth rationale for a service-oriented approach to inter-organizational collaboration.
2. Two SOA based methods: the CMEAP method supporting flexibility, and a method for the adaptation of service protocols supporting process adaptation.
3. Two prototype systems, the PEOPA platform and the *ErGo* system, implementing the proposed methods for the needs of collaborators in the construction sector, followed by a case study illustrating how the prototype systems may support construction processes.

Among future works, performance of the proposed methods, especially their scalability, is still to be evaluated, which is of high importance for open SOA ecosystems in which the number of service providers and consumers is not known in advanced while potentially high.

Further support for flexible automation of administrative procedures should aim at developing appropriate solutions for continuous updating of elementary processes in accordance with the ever changing legislation. It is an open question whether these models should be developed by legislative bodies, e.g., as attachments to the legal acts, or should be developed and distributed by independent third parties on a commercial basis.

Another challenge when it comes to the administrative procedure automation is the issue of discrepancies in operating practices in various public agencies, to implement the same legal provisions. These discrepancies can arise from different organizational structures, various IT infrastructures, and different technical capabilities to access public registries in different agencies.

Support for adaptation of collaborative processes may be further extended by supporting change propagation. Research on change propagation would enable knowledge sharing among groups having similar collaboration patterns. A key research problem concerning change propagation is to understand under which conditions the changes made in one instance of a given service protocol can be propagated to other instances of the same service protocol with a different implementation.

References

1. Abramowicz, W., Haniewicz, K., Kaczmarek, M., Zyskowski, D.: E-marketplace for semantic web services. In: Bouguettaya, A., Krueger, I., Margaria, T. (eds.) Service-Oriented Computing - ICSOC 2008. Lecture Notes in Computer Science, vol. 5364, pp. 271–285. Springer, Berlin (2008). doi:10.1007/978-3-540-89652-4_12
2. Activiti: Activiti BPM Platform. http://www.activiti.org/. Accessed 19 October 2012
3. Adams, M.: Dynamic workflow. In: ter Hofstede, A.M., van der Aalst, W.M.P., Adams, M., Russell, N. (eds.) Modern Business Process Automation: YAWL and its Support Environment, pp. 123–145. Springer, Berlin (2010)

4. Ahuja, M.K., Carley, K.M.: Network structure in virtual organizations. Organ. Sci. **10**(6), 741–757 (1999). doi:10.1287/orsc.10.6.741
5. Aris, BPM Community: Event-driven process chain (EPC). http://www.ariscommunity.com/event-driven-process-chain. Accessed 19 October 2012
6. Barney, J.B.: Firm resources and sustained competitive advantage. J. Manage. **17**(1), 99–120 (1991). doi:10.1177/014920639101700108
7. Barros, A., Dumas, M., Bruza, P.: The move to web service ecosystems. http://www.bptrends.com/publicationfiles/12-05-WP-WebServiceEcosystems-Barros-Dumas.pdf (2005). Accessed 19 October 2012
8. Bichler, M., Lin, K.J.: Service-oriented computing. Computer **39**, 99–101 (2006). doi:10.1109/MC.2006.102
9. Camarinha-Matos, L.M., Afsarmanesh, H., Galeano, N., Molina, A.: Collaborative Networked Organizations: Concepts and Practice in Manufacturing Enterprises. Comput. Ind. Eng. **57**(1), 46–60 (2009). doi:10.1016/j.cie.2008.11.024
10. Camarinha-Matos, L.M., Afsarmanesh, H., Ollus, M.: ECOLEAD and CNO Base Concepts, pp. 3–32. Springer, New York (2008). doi:10.1007/978-0-387-79424-2_1
11. Cellary, W.: Networked virtual organizations: a chance for small and medium sized enterprises on global markets. In: Godart, C., Gronau, N., Sharma, S., Canals, G. (eds.) The 9th IFIP International Conference on e-Business, e-Services and e-Society I3E 2009, Nancy (France), September 23–25, 2009, pp. 73–81. Springer (2009). doi:10.1007/978-3-642-04280-5_7.
12. Cellary, W., Picard, W.: Agile and pro-active public administration as a collaborative networked organization. In: The 4th International Conference on Theory and Practice of Electronic Governance ICEGOV 2010, Beijing (China), October 25–28, 2010, pp. 9–14. ACM (2010).doi:10.1145/1930321.1930324.
13. Cellary, W., Strykowski, S.: E-government based on cloud computing and service-oriented architecture. In: Janowski, T., Davies, J. (eds.) The 3rd International Conference on Theory and Practice of Electronic Governance ICEGOV 2009, Bogota (Colombia), November 10–13, 2009, pp. 5–10. ACM Press (2009). doi:10.1145/1693042.1693045
14. City of Vancouver: Development Permit Process in Vancouver. http://vancouver.ca/commsvcs/planning/landuse2.htm (2010). Accessed 19 October 2012
15. Davulcu, H., Kifer, M., Pokorny, L.R., Ramakrishnan, C.R., Ramakrishnan, I.V., Dawson, S.: Modeling and Analysis of Interactions in Virtual Enterprises. In: Proceedings of the Ninth International Workshop on Research Issues on Data Engineering (RIDE-VE'99), Information Technology for Virtual Enterprises, Sydney, Australia, March 23–24, 1999, pp. 12–18. IEEE Computer Society, Los Alamitos, CA, USA (1999). doi:10.1109/RIDE.1999.758587
16. Demirkan, H., Kauffman, R.J., Vayghan, J.A., Fill, H.G., Karagiannis, D., Maglio, P.P.: Service-oriented technology and management: perspectives on research and practice for the coming decade. Electron. Commer. Res. Appl. **7**(4), 356–376 (2008). doi:10.1016/j.elerap.2008.07.002
17. Economy Watch: USA (United States of America) GDP. http://www.economywatch.com/gdp/world-gdp/usa.html (2010). Accessed 19 October 2012
18. Eniro Polska Sp. z o.o.: Panorama firm. http://panoramafirm.pl/. Accessed 19 October 2012
19. European Commission: Natura 2000 network. http://www.ec.europa.eu/environment/nature/natura2000/index_en.htm (2012). Accessed 19 October 2012
20. Eurostat: Services statistics – short-term developments. http://www.epp.eurostat.ec.europa.eu/statistics-explained/index.php/Services-statistics-short-term-developments (2012). Accessed 19 October 2012
21. Eurostat: E-government online availability. http://www.epp.eurostat.ec.europa.eu/portal/page/portal/product_details/dataset?p_product_code (2012). Accessed 19 October 2012
22. Eurostat: E-government usage by individuals. http://epp.eurostat.ec.europa.eu/portal/page/portal/product_details/dataset?p_product_code=TSDGO330 (2012). Accessed 19 October 2012
23. Georgakopoulos, D., Hornick, M., Sheth, A.: An overview of workflow management: from process modeling to workflow automation infrastructure. In: Distributed and Parallel Databases, pp. 119–153. Kluwer Academic Publishers, New York (1995)

24. Grupa Allegro Sp. z o.o.: Oferia.pl. http://oferia.pl/. Accessed 19 October 2012
25. ITSOA Consortium: ITSOA, Nowe technologie informacyjne dla elektronicznej gospodarki i społeczeństwa informacyjnego oparte na paradygmacie SOA (ang. New information technologies for electronic economy and information society based on service-oriented architecture). http://www.soa.edu.pl/ (2009). Accessed 19 October 2012
26. Jagielski, J.: Administrative law. In: Frankowski, S., Bodnar, A. (eds.) Introduction to Polish Law, Introduction to the Laws of Series, pp. 153–187. Kluwer Law International (2005)
27. Janowski, T., Pardo, T.A., Davies, J.: Government information networks: mapping electronic governance cases through public administration concepts. Gov. Inform. Quart. **29**(Supplement 1), 1–10 (2012). doi:10.1016/j.giq.2011.11.003
28. JBoss: Drools. http://www.jboss.org/drools. Accessed 19 October 2012
29. JBoss community: Hibernate. http://www.hibernate.org/. Accessed 19 October 2012
30. Kanaracus, C.: Gartner: SaaS market to grow 17.9% to $14.5b. http://www.computerworld.com/s/article/9225590/Gartner-SaaS-market-to-grow-17.9-to-14.5B (2012). Accessed 19 October 2012
31. Kobielus, J.: The rhythm of work: a buyer's guide to workflow tools. Netw. World Collab. **12**(42), 12–18 (1995)
32. Krysztofiak, K., Paszkiewicz, Z., Dąbrowski, P., Flotyński, J., Picard, W., Cellary, W.: ErGo: System Project. Tech. rep., Department of Information Technology, Poznań University of Economics, Poznań, Poland (2011)
33. Lave, J., Wenger, E.: Situated Learning: Legitimate Peripheral Participation. Cambridge University Press, Cambridge (1991)
34. Lipnack, J., Stamps, J.: Virtual Teams: Researching across Space, Time, and Organizations with Technology. John Wiley, New York (1997)
35. Lucintel: Global IT services industry analysis 2012–2017: Industry trend, profit and forecast analysis. http://www.researchandmarkets.com/reports/2078480/global_it_services_industry_analysis_2012_2017 (2012). Accessed 19 October 2012
36. Mergers Net sp. z o.o.: Virtual data room, FAQ-Frequently Asked Questions. http://datapoint.pl/en, faq.html (2009). Accessed 19 October 2012
37. Microsoft: Microsoft SQL Server. http://www.microsoft.com/sqlserver/ (2012). Accessed 19 October 2012
38. Ministerstwo Administracji i Cyfryzacji (ang. Ministry of Administration and Digitization): Electronic Platform of Public Administration Services (ePUAP). http://epuap.gov.pl/wps/portal/. Accessed 19 October 2012
39. Oasis SOA Reference Model Technical Committee: Reference Model for Service Oriented Architecture 1.0. OASIS Standard, 12 October 2006, http://docs.oasis-open.org/soa-rm/v1.0/soa-rm.pdf (2006). Accessed 19 October 2012
40. Oasis SOA Reference Model Technical Committee: Reference Architecture Foundation for Service Oriented Architecture Version 1.0. OASIS Committee Draft 02, 14 October 2009, http://docs.oasis-open.org/soa-rm/soa-ra/v1.0/soa-ra-cd-02.pdf (2009). Accessed 19 October 2012
41. OASIS Web Services Business Process Execution Language Technical Committee: Web Services Business Process Execution Language Version 2.0. OASIS Committee Draft, http://docs.oasis-open.org/wsbpel/2.0/wsbpel-v2.0.pdf (2007). Accessed 19 October 2012
42. Object Management Group: Business Process Modeling Notation, version 2.0. http://www.omg.org/spec/BPMN/2.0 (2011). Accessed 19 October 2012
43. OECD: Uptake of e-government services. In: Government at a Glance 2011. OECD Publishing (2011). doi:10.1787/gov-glance-2011-55-en
44. Paszkiewicz, Z., Cellary, W.: Computer supported collaborative processes in virtual organizations. In: Kaynak, E., Harcar, T. (eds.) Advances in Global Management Development. Challenges and opportunities of global business in the new millennium: contemporary issues and future trends, pp. 85–94. IMDA Press, Poznan (2011)
45. Paszkiewicz, Z., Cellary, W.: Computer supported collaboration of SMEs in transnational market. J. Transnat. Manage. **17**(4), 294–313 (2012)

46. Paszkiewicz, Z., Cellary, W.: Computer supported contractor selection for public administration ventures. In: ICEGOV, 6th International Conference on Theory and Practice of Electronic Governance, pp. 332–226. ACM Press, New York (2012)
47. Paszkiewicz, Z., Picard, W.: MAPSS, a Multi-Aspect Partner and Service Selection Method. In: Camarinha-Matos, L.M., Boucher, X., Afsarmanesh, H. (eds.) Collaborative Networks for a Sustainable World: IFIP TC5 WG 5.5 Eleventh IFIP Working Conference on Virtual Enterprises, 11–13 October, 2010, St. Etienne, France, IFIP Advances in Information and Communication Technology, vol. 336, pp. 329–337. Springer, Berlin (2010). doi:10.1007/978-3-642-15961-9_39
48. Paszkiewicz, Z., Picard, W.: Modeling competences in service-oriented virtual organization breeding environments. In: Shen, W., Barthès, J.P.A., Luo, J., Kropf, P.G., Pouly, M., Yong, J., Xue, Y., Ramos, M.P. (eds.) Proceedings of the 2011 15th International Conference on Computer Supported Cooperative Work in Design, CSCWD 2011, June 8–10, 2011, Lausanne, Switzerland, pp. 497–502. IEEE (2011). doi:10.1109/CSCWD.2011.5960118
49. Petrie, C., Bussler, C.: The myth of open web services: the rise of the service parks. IEEE Internet Comput. **12**(3), 95–96 (2008). doi:10.1109/MIC.2008.65
50. Picard, W.: Adaptation of service protocols. Habilitation thesis, Wydawnictwa Uniwersytetu Ekonomicznego w Poznaniu, Poznań, Ploand, (2013)
51. Picard, W.: Semantic modelling of virtual organizations with service network schemata. New Gen. Comput. **30**(2,3), 99–121 (2012). doi:10.1007/s00354-012-0201-0
52. Picard, W., Paszkiewicz, Z., Gabryszak, P., Krysztofiak, K., Cellary, W.: Breeding virtual organizations in a service-oriented architecture environment. In: Ambroszkiewicz, S., Brzeziński, J., Cellary, W., Grzech, A., Zieliński, K. (eds.) SOA Infrastructure Tools – Concepts and Methods, pp. 375–396. Wydawnictwa Uniwersytetu Ekonomicznego w Poznaniu (2010)
53. Porter, M.E.: How competitive forces shape strategy. Harvard Bus. Rev. **57**(2), 137–145 (1979)
54. Porter, M.E.: Competitive advantage: Creating and Sustaining Superior Performance. Free Press, New York (1985)
55. Ricardo, D.: On the Principles of Political Economy and Taxation. John Murray, London (1817). http://www.econlib.org/library/Ricardo/ricP.html. Accessed 19 October 2012
56. Sadiq, S.W., Orlowska, M.E., Sadiq, W.: Specification and validation of process constraints for flexible workflows. Inform. Syst. **30**(5), 349–378 (2005). doi:10.1016/j.is.2004.05.002
57. Spohrer, J., Maglio, P.P.: The emergence of service science: toward systematic service innovations to accelerate co-creation of value. Prod. Oper. Manage. **17**(3), 238–246 (2008)
58. Strykowski, S., Wojciechowski, R.: Ontology-based modeling for automation of administrative procedures. In: Grzech, A., Borzemski, L., Świątek, J., Wilimowska, Z. (eds.) Information Systems Architecture and Technology–Service Oriented Networked Systems, pp. 79–97. Oficyna Wydawnicza Politechniki Wrocławskiej, Wrocław, Poland (2011)
59. Strykowski, S., Wojciechowski, R.: Composable modeling and execution of administrative procedures. In: Kö, A., Leitner, C., Leitold, H., Prosser, A. (eds.) Advancing Democracy, Government and Governance–Joint International Conference on Electronic Government and the Information Systems Perspective, and Electronic Democracy, EGOVIS/EDEM 2012. LNCS, vol. 7452, pp. 52–66. Springer, Berlin (2012)
60. The Apache Software Foundation: Apache axis2/java. http://axis.apache.org/axis2/java/core/ (2012). Accessed 19 October 2012
61. The Eclipse Foundation: Eclipse. http://www.eclipse.org/ (2012). Accessed 19 October 2012
62. U.S. Bureau of Labor Statistics, Business Employment Dynamics: Entrepreneurship and the U.S. economy. http://www.bls.gov/bdm/entrepreneurship/entrepreneurship.htm. Accessed 19 October 2012
63. van der Aalst, W., van Hee, K.: Workflow Management: Models, Methods, and Systems. MIT Press, Cambridge (2002)
64. van Engers, T., Boer, A., Breuker, J., Valente, A., Winkels, R.: Ontologies in the legal domain. In: Digital Government. Integrated Series in Information Systems 17, chap. 13, pp. 233–261. Springer, Berlin (2008)

65. W3C XML Schema Working Group: XML Schema Part 0: Primer, 2nd edn. W3C Recommendation, http://www.w3.org/TR/xmlschema-0/ (2004). Accessed 19 October 2012
66. Wall, Q.: Rethinking SOA governance. http://www.oracle.com/technetwork/articles/entarch/soa-governance-093602.html (2007). Accessed 19 October 2012
67. Wikipedia: Due Diligence. http://en.wikipedia.org/wiki/Due_diligence#Commercial-property (2012). Accessed 19 October 2012
68. Wojciechowski, R., Strykowski, S.: Towards electronic government focused on administrative procedure automation. Bus. Inform. **2**(24), 104–114 (2012)
69. Workflow Management Coalition: Terminology and glossary. Document Number WFMC-TC-1011, Issue 3.0, 1999, http://www.wfmc.org/standards/docs/TC-1011_term_glossary_v3.pdf (1999). Accessed 19 October 2012

Chapter 5
Dependability Infrastructure for SOA Applications

Jerzy Brzeziński, Dariusz Dwornikowski, Anna Kobusińska, Jacek Kobusiński, Michał Sajkowski, Cezary Sobaniec, Michał Szychowiak, Dariusz Wawrzyniak and Paweł T. Wojciechowski

Abstract This chapter describes two tools for improving dependability of SOA-based applications: ReSP (Reliable SOA Platform) and DyMST (Dynamic Management SOA Toolkit). ReSP is a set of modules to improve dependability in respect to availability and reliability, and to some extent safety. It is comprised of the mechanisms of reliable group communication, replication, recovery, and transaction processing. DyMST is a set of components for failure detection, monitoring and autonomic management, and distributed security policy enforcement. In order to show the dependability aspects of real applications and usage of these tools, two case studies from the medical healthcare domain are presented: *Healthcare Integra-*

J. Brzeziński · D. Dwornikowski · A. Kobusińska · J. Kobusiński · M. Sajkowski · C. Sobaniec · M. Szychowiak · D. Wawrzyniak (✉) · P. T. Wojciechowski
Institute of Computing Science, Poznan University of Technology, Poznań, Poland
e-mail: Dariusz.Wawrzyniak@cs.put.poznan.pl

J. Brzeziński
e-mail: jerzy.brzezinski@put.poznan.pl

D. Dwornikowski
e-mail: Dariusz.Dwornikowski@cs.put.poznan.pl

A. Kobusińska
e-mail: Anna.Kobusinska@cs.put.poznan.pl

J. Kobusiński
e-mail: Jacek.Kobusinski@cs.put.poznan.pl

M. Sajkowski
e-mail: Michal.Sajkowski@put.poznan.pl

C. Sobaniec
e-mail: Cezary.Sobaniec@cs.put.poznan.pl

M. Szychowiak
e-mail: Michal.Szychowiak@cs.put.poznan.pl

P. T. Wojciechowski
e-mail: Pawel.T.Wojciechowski@cs.put.poznan.pl

S. Ambroszkiewicz et al. (eds.), *Advanced SOA Tools and Applications*,
Studies in Computational Intelligence 499, DOI: 10.1007/978-3-642-38957-3_5,

tion Platform for the exchange of patients' medical data among various healthcare units, and *Medical Event, and Data Registering Platform* for daily work improvement of medical staff.

1 Introduction

Dependability is a general and complex concept connected with fault tolerance. In the context of computers it mostly concerns distributed systems where the probability of communication failures is relatively high, partial crashes are often hardly detectable, and can lead to misunderstandings. Dependability comprises various attributes relating to system construction and behavior. According to the extensive analysis presented in [5] it covers:

- availability—readiness for correct service,
- reliability—continuity of correct service,
- safety—absence of catastrophic consequences on the user(s) and the environment,
- integrity—absence of improper system alterations,
- maintainability—ability to undergo modifications and repairs.

Dependability is partially addressed in the concept of SOA. One of the main principles of SOA paradigm consists in decoupling of system components and integrating them in the form of loosely coupled services. This implies that the services can operate to some extent independently of one another. On the one hand, possible separation of services limits the consequences of failures, and facilitates system deployment and upgrade, thereby improving dependability in respect to safety and maintainability. On the other hand, the services can independently change their states, in particular fail at arbitrary moments, or lose connection. In contrast to tightly coupled systems that crash as a whole, failures of single services or components may leave other interoperating components in tentative states, consequently leading to uncertainty of system behavior and subsequent results. Vital in this context are mechanisms of improving availability, reliability, and integrity, as well as appropriate tools for system management, capable of monitoring and controlling crucial components.

Fundamental for ensuring dependability is redundancy. Redundancy has multiple forms at different layers of information systems, both in hardware and in software. A straightforward way of incorporating redundancy in SOA-based systems is replication of services (service instances) or system components, with a direct effect of increased availability. However, replication creates a consistency problem—in this case manifesting itself in a divergence between the states of service instances. The problem appears severe if network partitioning is taken into consideration. It has been raised by Eric Brewer [13], then formally analyzed in [45]. Brewer's conjecture—the so-called CAP theorem, states that in a distributed system at most two of the following three properties can be obtained: consistency, availability, and partition tolerance. In a later paper [12] Brewer argues that partitioning is closely related to communication latency (in practice indistinguishable from it) and as such must be tolerated in wide-area systems. Besides, he points out that the properties are continuous rather than binary, which enables tuning the balance between consistency and availability.

There is no approach to replication in the context of SOA that directly addresses this tradeoff. Primary-backup replication scheme [23], applied in [35], *FT-SOAP* [40], *FAWS* [62], and *FV4WS* [75], improves availability in case some replicas crash. It inherently guarantees strong consistency, however, if network partitioning occurs, it may be impossible to designate a new primary replica, which compromises availability. *WS-Replication* [94] uses active replication [96] based on total order multicast, thereby preserves strong consistency. Although the authors emphasize availability, they skip the fact that availability is limited by network partitioning because some replicas may disappear from the consistent view of group members. The concept of *A Middleware for Replicated Web Services* [110] is similar, however, the authors depict the problem of failures only from the client perspective. They do not explain the behavior of the consistency protocol when some replicas crash or become separate. The replication mechanism in *FTWeb* [95] is also based on total order mulitcast and voting, with the same consequences as in the case of *WS-Replication*. Besides, there are some doubts about crucial components (e.g. *WSDispatcher* or *WSInvoker*) whether they are replicated or single point of failure. Strong consistency seems to be also the feature of *Web Services Fault Tolerance Architecture* (WS-FTA) [99]. In order to facilitate consistency maintenance in wide area network, WS-FTA uses hierarchical groups of replicas. The authors mention availability in the presence of network failures, but do not specify their assumptions, i.e. whether the communication stalls only for a short time (the loss of single messages), or the consistency is preserved without access restrictions. *Survivable Web Services* [73] and *Thema* [80] use N-modular redundancy to achieve Byzantine fault tolerance. A response from a service is obtained by voting on the results provided by a number of replica servers. The number of servers depends on the assumed failure threshold. The updates in both systems are to be processed in the same order on each replica, which implies strong consistency. Network partitioning in this case may reduce the number of consistent replica and thereby preclude quorum and consequently availability.

The approach to replication presented in this chapter is two-fold. Firstly, a group communication mechanism is provided as a tool for building internally replicated services, i.e. services organized in the form of replicated components or lower layer items. The primary aim of this replication technique is the improvement of service reliability, which entails preservation of strict consistency at the expense of partition tolerance. Consequently, possible partitioning failures result in the crash stops of some servers, i.e. their exclusion from the group of replicas, thereby compromising availability. The other technique addresses availability taking partitioning into account. Actually, it allows combining optimistic and pessimistic replication by appropriate specification of operations at their submission. Thus, it enables reducing availability at the gain of stronger consistency or vice versa. This technique has been implemented in the form of a reverse proxy server intercepting communication between clients and services. This way, replication can be achieved to some extent transparently to the service to be replicated. Hence, it is suitable for legacy services, implemented without any intention of replication.

Reliability and availability of servers are crucial to dependability, although at the business process layer an important complementary feature is the reliability of interactions between service providers and their clients. The problem of reliable interaction results from possible inconsistency between the server state and its view at the client side, which can appear after server or client crash and subsequent resumption. As a consequence, the state of a business process (understood as a series of interactions between service and its clients) may be incorrect, and consequently further processing may be useless. This requires a mechanism enabling coherent processing despite failures, and may be achieved by using the checkpointing and rollback-recovery techniques. Although these techniques are widely used in the context of general distributed systems, not much work has been done on this approach in the SOA-based systems, where checkpointing and rollback-recovery continues to be an open issue. The idea of logging and recovery has been partially raised in *FT-SOAP* [74], where these mechanisms are used to provide the consistency among replicated services. Another architecture, *WS-Fault Tolerant Architecture (WS-FTA)* [99], also uses replication to ensure reliability, however in case of a given replica failure the logging and recovery mechanism is used to recover its state. The recovery of state of failed replicas is also considered in [95], where a special component, *WSRecovery*, is designed for this purpose. Also in *JTangSynergy* [112] the authors mentioned initializing a newly chosen server with the data and state, but do not explain what data are necessary and how they are gathered for this purpose. The problem of restoring an erroneous state to a consistent state, or restoring to valid operation after an application, or part of it, fails in some way, is partially addressed in [53], where the *Web Service Management System* is proposed.

The above mentioned solutions are focused on providing fault-tolerance of individual services, while the recovery of consistent distributed processing is overlooked or marginalized. In [34], the authors have noticed that while it is relatively easy to make an individual service fault-tolerant, improving fault-tolerance of services collaborating in multiple application scenarios is a challenging task. Therefore, they proposed a framework providing fault-tolerant web services which employs two approaches to recovering from a fault: local recovery mechanism and global recovery mechanism. The first one recovers an individual service, while the latter one applies to the recovery of an application which aggregates a set of service instances to provide the overall capability for the application. Finally, in [6, 7], Barga et al. have developed a framework that provides a comprehensive recovery encompassing data, messages, and the states of web-based applications. The proposed framework masks from users all failures of clients, application servers or database servers. For that reason, in the proposed solution the non-deterministic events are identified, to enable to recover the application component state from an earlier installed state (in an extreme case its initial state) and arrive at the same (abstract) state as before the failure. The key benefit of [6, 7] lies in automating recovery and masking failures to both end users and to application programmers, thus largely relieving them of the need to write explicit code to cope with system failures.

The rollback-recovery mechanism described in this chapter extends the solutions presented in [6, 7, 34]. We develop a solution that recovers a consistent state of distributed processing in case any cooperating component fails (by the component we understand a client-side application, a single service, or an application consisting of several services). The proposed recovery service is specially tailored to the SOA-based systems, and it is transparent to clients and service providers. Though our solution may be used by any kind of service, we target mostly RESTful services.

Another aspect of computer systems, allied with dependability, is security. Despite the differentiation between security and dependability [5], some attributes of these concepts are common. Thus, security encompasses availability (for authorized actions), and integrity in the sense of the absence of unauthorized system alterations, as well as additional attributes like confidentiality, i.e. the absence of unauthorized disclosure of information. Typical security management mechanisms used to provide these attributes include authentication and authorization, access control and information flow control, as well as communication protection. In distributed computing systems, especially large-scale and service-oriented ones, a security policy has become indispensable for efficient management of security requirements. The security policy consists of rules controlling interactions between system components. It allows separating the specification of security requirements from the implementation. However, due to the compound structure of modern service-oriented systems, the policy often suffers from problems, like inconsistencies, which gravely degrade the efficiency of policy execution.

The analysis of requirements for dependability leads to the concept of a toolkit facilitating the construction of dependable services. In fact, it is decomposed into two toolkits:

- ReSP (Reliable SOA Platform)—a set of modules to improve dependability in respect to availability and reliability, and to some extent safety and integrity, comprised of the mechanisms of reliable group communication, replication, recovery, and transaction processing;
- DyMST (Dynamic Management SOA Toolkit)—a set of components for failure detection, monitoring and autonomic management, and distributed security policy enforcement.

In the complex structure of SOA-based solutions these toolkits operate on lower levels constituted by hardware, system software and network infrastructure. They are not aimed at application software, i.e. faults (and subsequent failures) in the implementation of services or business processes. Consequently, the crash-recovery failure model is assumed. In other words, these toolkits are not suitable for Byzantine (arbitrary) failures, especially N-modular redundancy and closely related N-version programming, often considered in the context of SOA and Web Services [26, 47].

Although the concept of ReSP and DyMST is not limited to any technology or model of distributed services, it is inspired by RESTful web services. Wherever possible, the specific features of the RESTful model are exploited, especially the explicit semantics of operations—HTTP methods in this case. The implementations of these toolkits are also mainly provided for RESTful web services, which fills the gap in dependability solutions for SOA because the great majority of them is provided for the Web Services [3, 26, 40, 47, 73, 75, 80, 94, 95, 99, 110, 114]. RESTful web services are addressed in [38]. However, the authors concentrate only on the client side, and assume that a number of equivalent endpoints for a given RESTful service are available. Consequently, they do not describe an infrastructure for fault-tolerant services, but limit their work to the specification of client behavior in case an error occurs when accessing a service.

To show the dependability aspects of real applications, the medical healthcare domain is chosen, where achieving dependability is of crucial importance. To this end, two case studies are analyzed:

- *Healthcare Integration Platform* for the exchange of patients' medical data among various healthcare units,
- *Medical Event, and Data Registering Platform* for daily work improvement of medical staff.

2 RESTful Web Services

The *REpresentational State Transfer (REST)* [42, 43] is an architecture style for building distributed systems that are efficient and scalable. The style evolved from the experiences with the World Wide Web service but is more general and can be applied to other contexts. The fundamental assumptions of REST are following:

Client-server architecture—the server has a clear interface, and can be developed independently of the clients.

Statelessness—the service is stateful itself but does not maintain any state information associated with specific clients. In other words: statelessness regards the state of interaction with clients. This way visibility of interactions is improved, as well as reliability and scalability of the system. Transparency is improved because all requests are self-descriptive—they do not rely on previous interactions, and therefore can be handled by any service instance. The system is more reliable because recovery process after a crash is greatly simplified. Scalability is improved because the server does not need to maintain additional data structures for individual clients.

Caching—responses from the servers can be cached by clients if the server allows this, which can substantially reduce the number of requests sent by clients.

Uniform interface—all services are accessed using the same, standardized interface, which improves transparency, and allows more independent development of both client and server side of services.

The most important assumption concerning uniform interface because both communicating sides must use it. Uniform interface covers the following aspects:

Resources—clients interact with the service referring to resources. A resource is any addressable piece of information of any size. However, the clients do not access the resources directly: they are manipulated through representations. This way the logical structure of information is decoupled from its internal data structure and implementation, which allows for more flexible and independent development of clients and servers.

Resource identifiers—every resource has its unique identifier, which is then used for accessing it using the uniform interface (using a standardized set of operations). The resources can change their state but the semantics of a resource associated with a given ID should remain the same.

Self-descriptive messages—the requests sent to the server contain all information necessary to properly interpret and handle them, which is a direct consequence of the assumption of statelessness of services.

Resource relationships—representations of resources may contain references to other resources (like links in Web pages), which helps the client to discover the service structure.

The presented assumptions of REST are very general and do not refer to any specific technology that may be used to implement it. However, the predominant implementation environment for REST architecture style is the HTTP protocol [41]. The application of REST rules to the context of the HTTP protocol is not straightforward but good proposals are given in [91]. The services built according to REST and using the HTTP protocol are called "RESTful web services", and are characterized by the following properties:

- Resources are identified using Uniform Resource Identifiers (URI).
- Operations on resources are specified using methods of the HTTP protocol, namely: GET for reading the resource, PUT for updating, POST for creating, and DELETE for removal.
- Representations can use any data format indicated by means of the MIME standard.

RESTful web services have many advantages, just to name a few:

Simplicity—the services use just the HTTP protocol and do not rely on other specific standards. The simplest client of such services is a web browser. The simplicity also applies to the implementation level where the simplest tools are sufficient to build client and server applications.

Performance—contrary to SOAP protocol, the requests used in RESTful web services do not use an XML envelope for transport, but instead use directly the most appropriate format in a given context.

The use of existing Web infrastructure—the HTTP protocol is well-known, broadly implemented, thoroughly verified, and used for over 20 years now. The same applies to the whole infrastructure of the Web with its proxy servers and accompanying standards. RESTful web services build upon these foundations, simply extending them.

Transparency—the proper use of the HTTP protocol with its uniform interface results in semantic clarity of operations on resources. This helps building monitoring and intermediary services which extend the functions of existing services.

SOAP-based web services also use the HTTP protocol but they treat it as a pure transport protocol: the messages can be also sent using SMTP or a queuing system with a JMS interface. In case of the HTTP protocol (the most widely used one) the messages are just tunneled using a single POST method, which results in reduced efficiency. The independence of SOAP and protocols used for communication means that many additional standards, protocols and conventions cannot be used directly with SOAP because they are related to a specific communication protocol (e.g. HTTP). As a consequence there are numerous accompanying standards and protocols for SOAP-based web services, collectively called WS-*, defining additional functionality.

3 Group Communication

The Web can provide a common, language-independent execution platform for interoperable and loosely-coupled services. The programmers can use this platform to develop Web-based applications by composing (or connecting) web services into seamless and robust distributed systems. The key programming abstractions required to realize this scenario are offered by *group communication* (described in this Sect. 3) and *atomic transactions* (described in the next Sect. 4). Unfortunately, REST currently lacks sufficient support of these programming abstractions, which provided motivation for our work. Below we characterize the group communication semantics. Then, we describe our implementation of group communication abstractions in REST.

Group communication systems provide various primitives, e.g. for unicast and broadcast within a group of processes, for maintaining group membership, for maintaining virtual (or view) synchrony, and for detecting process failures. Below we characterize broadcast primitives which are essential for group communication.

The simplest broadcast primitive is *Unreliable Broadcast*, which allows a message to be sent to all processes in a group, guaranteeing that if the process which sent a message is correct, then all processes in the group will eventually deliver the message. However, if the sender crashes during the broadcast, then some processes in the group may not deliver the message. Obviously, this primitive is not much useful in systems in which failures may occur and should be tolerated. *Regular Reliable Broadcast* solves this problem; it guarantees the following properties:

- *Validity*: if a correct process broadcasts a message, then it eventually delivers the message;
- *Agreement*: if a correct process delivers a message, then all correct processes eventually deliver the message;
- *Uniform Integrity*: for any message, every process delivers the message at most once, and only if the message was previously broadcast.

Note that Regular Reliable Broadcast allows executions in which a faulty process delivers a message but no correct process delivers the message. *Uniform Reliable Broadcast* is a stronger version of Reliable Broadcast, which satisfies the Validity and Uniform Integrity properties defined above but replaces the Agreement property with the following:

- *Uniform agreement*: if a process (correct or not) delivers a message, then all correct processes will eventually deliver the message.

The Regular and Uniform Reliable Broadcast primitives provide a basis for stronger broadcast primitives, which have additional properties, e.g.:

- *FIFO order*: this property guarantees that messages sent by a process are delivered in the order of sending;
- *Causal order*: this property means that if some process has sent a message m_1 that caused sending of another message m_2, then each process in a group will deliver m_1 before m_2;
- *Total order*: this property means that messages sent by any different processes in a group will be delivered by all processes in the group in the same order (note that this property does not guarantee FIFO).

It is important to emphasize that group communication systems are fully distributed—they do not depend on any central server. Therefore, there are no single points of failure, nor performance bottleneck. These features are obviously desirable but they require complex protocols for *distributed agreement*. In the past 20+ years, many group communication systems have been implemented (e.g. JGroups [90] and Spread [100]; see also [79] for other references). They often use different protocols. Unfortunately, they also have quite different APIs, which are language dependent and not easy to use. Moreover, many of these systems are *monolithic*, i.e. it is not possible to easily replace their protocols or add new features. Using the API of these systems to develop a distributed Web-based application makes the code of this application neither easily reusable nor interoperable with other web services, which is a counterexample to the openness of the Web. Moreover, none of the system that we know offers a REST/HTTP-based interface.

In the world of SOAP-based web services, group communication has been standardized, albeit it does not offer the semantics which we described earlier. For example, *WS-BaseNotification* [84] and *WS-BrokeredNotification* [85] are two OASIS standards specifying the protocols for one-to-many communication of SOAP messages in the *publish-subscribe* model. In this model, the system users can create *topics* of messages, to which the message recipients (or *consumers*) can subscribe. The standards allow the users to have a separate *subscriber* that subscribes a number of consumers to a given topic. When a message sender (or a *publisher*) publishes a message on a given topic, the message is propagated to all consumers who subscribed (or have been subscribed) to that topic and whose subscription remains active. The *WS-BrokeredNotification* standard also introduces a *broker*, who is responsible for recording published messages of a given topic, and resending them to all *consumers* that have subscribed to that topic. However, the WS-BaseNotification and WS-BrokeredNotification standards focus on the information exchange protocol only, leaving the issues of reliable communication to "a delivery mechanism for transmission", where transmission properties are unspecified: "depending on the actual delivery mechanism, this transmission might be reliable or might be done on a best-effort basis" [84].

When it comes to RESTful approaches to the implementation of Web-based distributed applications, group communication solutions that offer all associated properties and guarantees are unknown to us. At the moment there is a lack of standards in the domain of group communication intended for the REST style. This provided motivation for development of RESTGroups, described in this chapter. *RESTGroups* [68, 69] is a module of the ReSP programming toolkit which allows the programmers to easily implement Web-based applications that require broadcast in a group of distributed RESTful web services (with the broadcast semantics described above). RESTGroups is itself a service. Thus, distributed applications developed using our tool conform to the Service-Oriented Architecture paradigm. The closest work to RESTGroups is *WS-Multicast* [94] that has been designed as a broadcasting service for SOAP-based web services. It is built on top of the JGroups group communication system [90]. WS-Multicast uses its own transport layer module for message communication based on SOAP. A WSDL interface has been defined, making WS-Multicast a web service itself. Since WS-Multicast only replaces the transport layer of the JGroups system, leaving the rest of the protocol stack unchanged, all the assurances offered by JGroups remain in place. Nonetheless, as was noted, the use of SOAP involves sizable cost stemming from the character of this protocol. Thus, in the final version of WS-Multicast, parsing XML data was avoided.

3.1 RESTGroups

RESTGroups [68, 69] is a group communication *front-end* for RESTful web services. The current implementation is based on Spread [4, 100]—a popular group communication toolkit implementing protocols for reliable, ordered multicasts and group membership. Spread is a monolithic system, consisting of a daemon program, client libraries, and a system monitor. Spread's API consists of many functions with bindings available for several programming languages: C/C++, Java, Perl, Python, and Ruby. The programmers can use this API to implement their distributed applications. On the other hand, RESTGroups represents group communication services provided by Spread as Web resources, addressed by URIs. Thus, web services and their clients can use our group communication system to broadcast among themselves (with the broadcast guarantees described earlier) in the same style as they communicate point-to-point. Moreover, since firewalls usually do not block the HTTP protocol, RESTGroups supports communication across firewalls. The system has been implemented; see the project web page [61].

RESTGroups has a small but powerful API that consists of just four methods of the HTTP protocol: GET, POST, PUT, and DELETE. They can be used for:

- detection of malfunctioning/crashed distributed processes,
- reliable point-to-point transfer of messages,

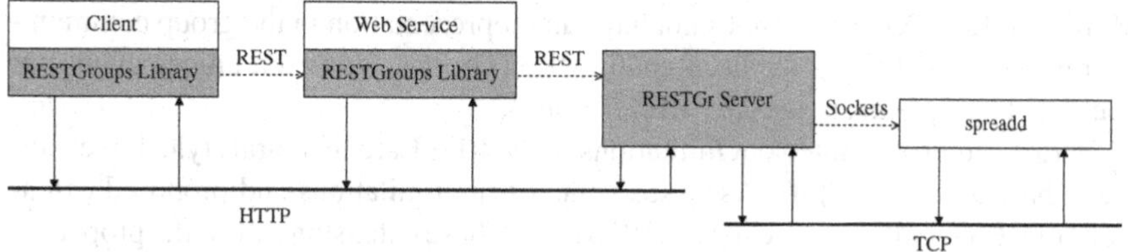

Fig. 1 The RESTGroups system

- formation of distributed processes into groups, the structure of which can change at runtime,
- reliable message multicasting with a wide range of guarantees concerning delivery of messages to group members (e.g. causally-, FIFO- and totally-ordered message delivery).

3.2 Main Components

Web-based distributed applications developed using RESTGroups consist of four types of communicating components (see Fig. 1): Web Service, Client, RESTGroups Server (RESTGr Server in short), and spreadd (which is a daemon of Spread). The Client and Web Service components are respectively, a user-defined RESTful client and a RESTful web service. The RESTGr Server acts as a *proxy* between Web Service and group communication protocols; the protocols are implemented by spreadd. The communication between Client and Web Service, as well as between Web Service and RESTGr Server uses the REST/HTTP style. On the other hand, RESTGr Server and spreadd communicate using TCP and may or may not run on the same machine.

Group communication services (provided by Spread) are represented as Web resources identified by URIs. Instead of calling Spread methods, a user-defined Web Service can invoke only four methods of the HTTP protocol (i.e. GET, PUT, POST, or DELETE). Then, a suitable HTTP request, possibly containing an XML document, is sent to RESTGr Server that translates it into a group communication call to spreadd. A crash of RESTGr Server results in the disconnection of all Web Services that are using this server. They can establish connection with another RESTGr Server which is available within the same group. In Sect. 3.5, we present the architecture of a system in which the RESTGroups components are replicated for resilience.

3.3 Statelessness

RESTGr Servers are almost *stateless*—they only store data that are necessary to maintain group communication sessions for the connected Web Service replicas.

Moreover, RESTGr Server does not have any representation in the group communication system that forms the back-end of RESTGroups. However, unique client IDs generated by Spread are used by RESTGroups.

Various authors pointed out limitations of the REST architectural style. For example, Khare and Taylor [66] discussed some of the limitations and proposed extensions of REST, collectively called *ARRESTED*. These extensions allow the properties required by distributed and decentralized systems to be modelled. Similarly to Khare and Taylor, we are not bound by the rules of the original model since REST cannot model group communication well (as the RESTGr Server is not 100 % stateless). Our original goal was rather to design a REST-inspired interface to the group communication service, albeit sacrificing strict conformance to the original REST model.

3.4 Application Programming Interface

Below we explain the semantics of the RESTGroups API. A complete description of the API with examples of its use is in the User Guide, available from the project web page [61]. The following methods of the HTTP protocol are used, where resources represent some group communication services or data structures (such as a mailbox):

- GET is used to perform a query on a resource, e.g. to retrieve messages from the mailbox (in a blocking or non-blocking manner);
- PUT is used to create a new resource, e.g. to extend a process group with a new process; the server responds with a status indicating success or failure;
- POST is used to update existing resources, e.g. to connect to the server on system start-up (this operation is executed only once) or to send/broadcast a new message;
- DELETE is used to remove a resource, e.g. to remove a process from a process group; in some cases, the update and delete actions may be performed with POST operations as well.

The following guarantees of message delivery are supported (based on the protocols implemented by Spread):

- `unreliable`—no guarantee of message delivery,
- `reliable`—reliable broadcast,
- `fifo`—fifo broadcast (first-in-first-out),
- `causal`—causal broadcast, consistent with Lamport's definition of causality,
- `safe`—total order broadcast,
- `agreed`—total order broadcast that is consistent with causal broadcast, ie. messages are delivered to all recipients in the same order, and the order agrees with the causal relation between messages.

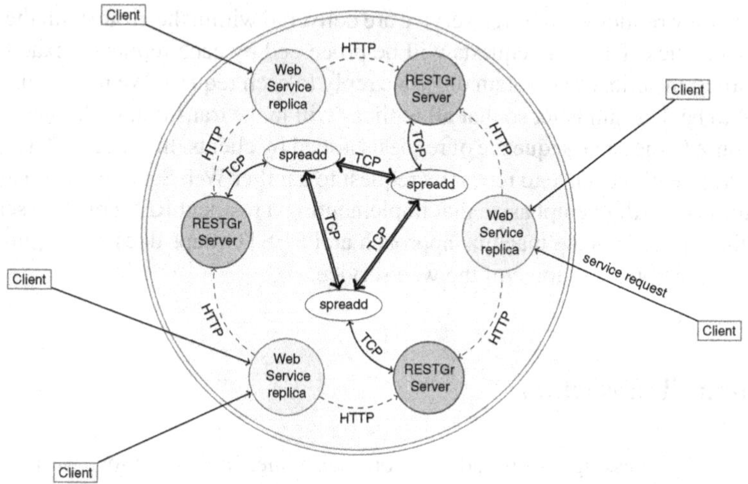

Fig. 2 Replication of a RESTful web service using RESTGroups

3.5 Example Application: Service Replication

Figure 2 presents a Web-based distributed application that uses RESTGroups for pessimistic replication of a RESTful web service to make it tolerant to server crashes. Service replication will be described in detail in Sect. 5. Below we only describe the use of RESTGroups for replication. There are three service replicas (each one called *Web Service replica*) perceived by the clients as a single web service, represented as a large circle. Clients can issue REST/HTTP requests to any of them. Each Web Service replica connects to two RESTGr Servers using HTTP, so it can tolerate a crash of one server. Each RESTGr Server has its own spreadd, so that partial failure of the Spread group communication system used as the back-end is also tolerated.

In general, replicating a web service to tolerate at most $\lfloor (n-1)/2 \rfloor$ machine crashes requires the following steps:

- spawning n Spread daemons (spreadd) on n independent machines;
- spawning n RESTGroups servers on different machines; each server communicates only with one Spread daemon (usually located on the same machine);
- spawning n instances of the RESTful web service on different machines (in this example application, they would run on the same machines as Spread daemons); each service replica can communicate with one or many RESTGroups servers.

The service developers can use the replicated state machine approach [96, 109] to implement a resilient RESTful web service in the following way. After system start up, a group is created which all Web Service replicas must join. A client can issue a request to any known replica which then forwards the request to the RESTGr Server that is alive; the latter broadcasts the request in the group. All client requests issued to (any replica of) the web service are delivered within the group with the total order semantics. Thus the requests will be processed by each replica in exactly the same order; the client will obtain only one reply to each request. We require the web service to be deterministic, so that all replicas will make transition to the same state in response to the same sequence of requests issued by clients. In the case of a replica crash, the client may have to repeat its request to another Web Service replica after a timeout.It is worth to emphasize that implementing a resilient RESTful web service using the replicated state machine approach and RESTGroups does not require any changes to the business logic of the web service.

4 Atomic Transactions

Transaction processing is a broad and complex issue. In this section, we focus on atomic transactions. An *atomic transaction* (or *transaction*, in short) is a series of operations that have to be executed atomically, i.e. they should either be executed completely and successfully or not at all. Alternative approaches to define a unit of work (e.g. Sagas [93]) do not guarantee atomicity. Atomic transactions is a very useful programming abstraction.That's why transaction processing is a valuable extension to REST.

Although there are many interesting proposals of systems and design patterns that introduce transactions to REST (we describe some examples below), none of them has gained wide acceptance among the community. In most cases such systems or patterns arguably break some of the REST style principles. For example, the client-server communication is constrained by no client context being stored on the server between requests. This *statelessness* constraint—a key requirement in relation to RESTful web services—is often a subject of discussions about interpretation. Developers of systems (including some transaction systems described later) often work around the statelessness constraint by giving the session state a resource identifier on the server side. An interpretation proposed in [24] disallows this approach and claims that such a design cannot be called RESTful. But it might be REST with some exceptions, and this "with exceptions" approach is probably right for most enterprise architectures.

Pessimistic Transactions

One of the first proposals of atomic transactions in REST is described by JBoss [82]. It is an extension of JAX RS—a popular Java API for RESTful web services, with atomic transactions based on exclusive locks represented as resources on the server side. Contrary to this approach, Atomic REST, described in this chapter, adopted a novel architecture, introducing separate services (mediators) that are responsible for the execution of transactions, and using server proxies that allow the services to remain unaware of transaction processing.

A similar approach to [82] is represented by RETRO [77]—a transaction model that defines many fine-grained resources for transaction processing, with a choice of exclusive and shared locks. We are not aware of any RETRO implementation announced yet. Some authors [92] pointed out drawbacks of this model: cluttering the business representations with transactional entities and the complexity that makes programming cumbersome.

Optimistic Transactions

Optimistic concurrency control [10] fits REST better than pessimistic transactions because it increases availability of a web service by decreasing resource blocking. Below we discuss a few example realizations of optimistic transactions in REST.

The most common solution for providing atomic transactions to REST is using the POST method to execute a batch (set) of operations. The concurrency control is optimistic since the data is cached by a client, and the consistency of the cache is checked during commit-time. The main advantage of the overloaded POST-based solution is its simplicity. On the other hand it is often criticized because it does not respect the semantics of uniform interface methods [42]: POST should create a resource, not execute any operations. Moreover, the mechanism is quite limited, e.g. contrary to Atomic REST, it does not allow transactions that span many services.

Overloading the POST method is used, among other systems, in the cloud computing environments, such as Microsoft Windows Azure [51]. It offers structured storage in the form of tables with a REST-compliant API, enabling to perform a transaction across entities stored within the same table and partition. An application can atomically perform multiple Create/Update/Delete operations across multiple entities in a single batch request to the storage system, as long as the entities in the batch have the same partition key value and are in the same table [81]. Thus, the high scalability and accessibility of the service is achieved by introducing the limitation on the set of resources that may be included in one transaction.

A simple design pattern that provides transactions in REST is described in [91]. A new transaction is created by sending a POST to a factory resource. Once the transaction is created successfully, we can access it as a "gateway" to the main service, sending all possible HTTP requests to a variety of resources. The pattern is simple and seems to be effective, but is it RESTful? In the same book, the authors emphasize the difference between application state and resource state. A user transaction is, in fact, an application state, therefore it should not be maintained by the server. Exposing it as resources does not change anything. In fact, the authors admit that their proposal is not "the official RESTful or resource-oriented way to handle transactions"—it is just "the best one they could think up". On the other hand, even if the pattern breaks the statelessness constraint of REST, it is a clean concept that can work successfully for a variety of services.

This section shows that the existing transaction processing patterns are either too limited or do not produce generic, reusable clients. On the other hand, service-centric products often break the REST statelessness constraint. When designing Atomic

REST, we used several ideas from the work described above but at the same time our approach is novel.

4.1 Atomic REST

Atomic REST [70] is a lightweight transaction processing system developed for RESTful web services. Contrary to other similar approaches, web services do not require any changes to be used with Atomic REST. Moreover, the system is completely transparent to non-transactional clients. This has been achieved by introducing an overlay network of mediators and proxy servers, and restricting transactions to a batched set of REST/HTTP operations (or requests) on Web resources addressed by URIs. The system guarantees isolation of concurrent transactions, and a limited form of atomicity (which will be defined later in this chapter). If a transaction commit fails due to system failures, compensations are used, if possible, to rollback the respective transaction and restore service state. The system (with one mediator only) has been implemented; see the project web page [59].

While other proposals of transactions in REST are mostly software design patterns or libraries to be used by the client/server developers, the Atomic REST approach is different: most of the transaction processing work is done by separate services—proxies and mediators—communicating using an overlay network. In particular, web services do not require any changes to be used with our system. Moreover, the system is completely transparent to the clients that do not require transactions. These two features enable straightforward use of Atomic REST to develop distributed Web-based applications using existing RESTful web services.

A *distributed transaction* in Atomic REST (or a *transaction*, in short) is a batch of REST operations (or requests) to be executed using Web resources maintained by servers and addressed by URIs. Thus, from a client view-point a transactional request does not differ much from an ordinary HTTP request. This means that clients are able to cache transaction responses, which fulfills one of the REST architectural constraints, ie. *cacheable responses*. Execution of concurrent transactions satisfies the isolation property and a weak form of atomicity, described in Sect. 4.3.

Batching of transaction operations restricts transactions to be rather short and non-interactive (similarly to, e.g. Sinfonia [1]). Hence the time when resources are blocked by a client is reduced to a minimum. This means that our system could be deployed on the Web and platforms, such as those provided for cloud computing, in which dependencies between network nodes should be avoided and the request processing time has to be short.

Compensation

To be able to use existing Web hosts that normally do not support versioning of Web resources, transaction resources are currently modified in-place, with a simple compensation mechanism. The compensation mechanism is based on the symmetry of HTTP operations. *Compensating* an operation on some resource requires executing a complementary operation on this resource. Transaction compensation could be provided by a user or done automatically whenever possible. For the latter, the following table presents REST/HTTP and complementary operations that are used by Atomic REST for compensation.

Request	Compensating request
GET	no compensation needed
PUT (modification)	PUT
PUT (creation)	DELETE
POST (creation)	DELETE
DELETE	PUT

4.2 Main Components

We can define four components of an example RESTful web service that uses Atomic REST for distributed atomic operations:

- *Server* provides a user-defined RESTful web service, executing client requests and returning results, without being aware of Atomic REST;
- *Client* is a user-defined client, with or without knowledge of Atomic REST;
- *Mediator* (or *Transaction Manager*) is a web service managing transaction execution on behalf of the client;
- *Proxy* is a server's *façade*, intercepting messages addressed to the server and handling Atomic REST-specific requests; the proxy enables a RESTful web service to support transactions without any changes in its code.

We can explain the *modus operandi* of the Atomic REST system using two examples of interaction patterns presented in Figs. 3 and 4.

A client can submit several requests, to be executed by many servers, as a single transaction. For example, in Fig. 3 there is a single mediator and two clients which

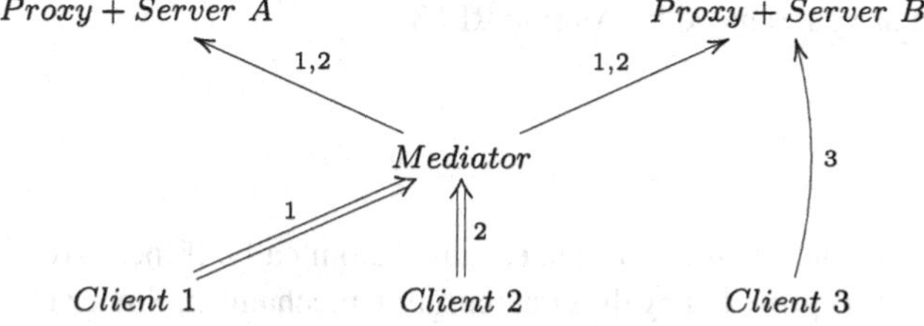

Fig. 3 An example interaction pattern of Atomic REST (single mediator)

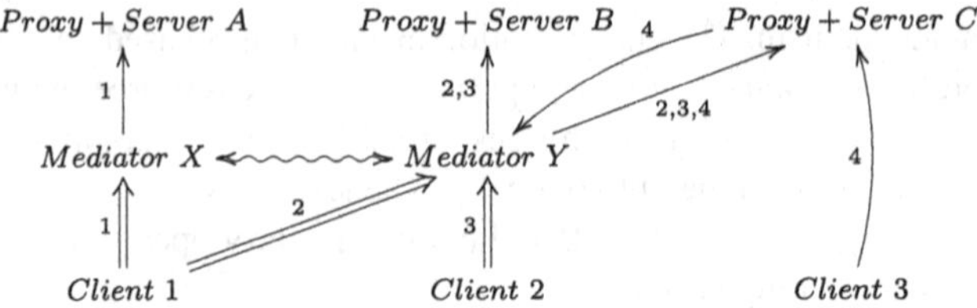

Fig. 4 An example interaction pattern of Atomic REST (several mediators)

submit their respective transactions, 1 and 2, to the mediator for execution. The transactions request some resources on servers A and B. The mediator executes transactions sequentially, first 1, then 2. Thus, isolation is satisfied. At the same time, client 3 submits a non-transactional request to server B that does not conflict with the transactions and is handled by server B normally.

In Fig. 4, there is an example with many mediators. Introducing many mediators supports privacy and load balancing. Each server gets transactions only from its trusted (single) mediator, e.g. server A trusts only mediator X. Each mediator can handle many servers. Mediators could be replicated for fault-tolerance if required, e.g. using the RESTGroups system described earlier. Client 1 executes transactions 1 and 2 using, respectively, mediator X and Y, while client 2 uses only Y. At the same time client 3 submits a non-transactional request to server C. Since the request conflicts with the transactions, it is forwarded to B's mediator Y as a transaction containing only a single request. At the end, all results will be returned to the clients. In order to agree upon the order of transaction execution, mediators communicate using a coordination protocol described below.

4.3 Properties

In database systems [10], atomicity and isolation are usually defined as follows. *Atomicity* guarantees that either all transaction operations are performed, or none of them (e.g. if some failures occur; this is called all-or-nothing guarantee). *Isolation* defines how operations of concurrent transactions interact with each other in terms of reading results written by another transaction. There can be defined provides serializable level (excluding fatal failures). Below we characterize the atomicity and isolation that are guaranteed by Atomic REST.

Atomicity

If there are no errors, transactions are executed atomically. If there are errors, transaction atomicity is provided by the compensation mechanism. Since automatic compensation of transaction operations is not always possible (either due to HTTP-bound issues or application semantics), the Atomic REST system cannot guarantee atomicity in all cases. This is acceptable since Atomic REST is no more tolerant to failures than an average web service that can fail at any time. Therefore, in case of some failures, some transaction operations may or may not have been executed or compensated; we call such failures *fatal*. Thus, the client should always check the results returned by the system, and in case of some error messages, execute a suitable action. For example, the client could repeat transaction operations. Specifically, the PUT, DELETE and GET methods are *idempotent* methods, and so they can be repeated many times.

Implementing stronger semantics of atomicity would require significant changes in the code of web services, such as resource multiversioning and the two-phase commitment (2PC) or three-phase commitment (3PC) protocols for the mediator-server communication. Multiversioning would allow transaction operations to be executed on shadow copies of resources, made public on transaction commit and rejected on transaction abort. However, we think that supporting existing web services by Atomic REST compensates the drawback of a weaker atomicity semantics. Moreover, we intend Atomic REST transactions to be a mechanism for increasing expressiveness in Web programming rather than for implementing fault-tolerant web services.

Isolation

Atomic REST guarantees serializability, which is the strongest form of isolation: the non-faulty concurrent execution of transactions is similar to the case when transactions are executed serially. In case of fatal failures the property can be guaranteed *up to* the fatal failure.

5 Replication

Replication can be achieved in a variety of ways depending on the approach to different aspects of its implementation or application. The presented approach is two-fold. Replication can be deeply embedded in the service itself, e.g. tightly integrated with the service implementation (semantic approach [32]). For this purpose, a group communication mechanism has been developed (see Sect. 3) that provides a reliable broadcast protocol. On top of this protocol multiple servers can cooperate with one another constituting so called replicated state machine. The primary aim of this replication technique is the improvement of service reliability, which entails preservation of strict consistency at the expense of partition tolerance.

Alternatively, replication can be set up as an external mechanism intercepting interactions between clients and a service, especially a service not designed to work in a replicated form. This approach, called syntactic [32], is applied in ReREST—a component working as an HTTP proxy. ReREST is aimed at combining optimistic and pessimistic replication by appropriate specification of operations at their submission. Thus, it enables reducing availability at the gain of stronger consistency or vice versa.

5.1 Replication Architecture

The architecture of replication mechanism is illustrated in Fig. 5. A replicated service is accessed by a client through a proxy. For this purpose, a modular proxy server for HTTP protocol, called MProxy [60], has been developed. Appropriate replication modules of MProxy are responsible for the communication with one another and with the real server (service provider). This way, MProxy can represent one replica. However, it can also provide access to multiple replicas organized in the form of replicated state machine (denoted by clouds in Fig. 5). A replicated state machine is intended for replication over LAN or co-located servers. In contrast, due to efficiency and possible communication failures, ReREST is more suitable for geographically separated replicas exposed to network partitioning. Actually ReREST is a module for MProxy.

MProxy

MProxy is a modular, highly efficient proxy server for HTTP protocol. It is written in pure Java, and is built around Apache HTTPComponents—a framework for constructing services and applications communicating by means of HTTP protocol. The implementation uses an asynchronous processing model based on Java NIO interfaces, which makes possible handling of thousands of simultaneous connections in a resource efficient manner. The proxy server is modular: it is very easy to implement

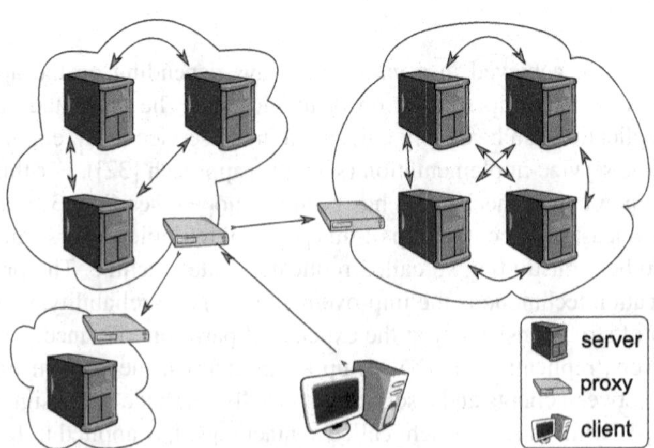

Fig. 5 Replication architecture

a new module for specific processing of client requests and server responses. The modules can form a pipeline providing compound processing. From within the module it is possible to communicate with external services, and by that means form a complex distributed application. MProxy can be used as an integration platform for different solutions enhancing RESTful web services or general web applications.

Replication Modules

Replication mechanisms of RESTful web services have been implemented as modules of the MProxy server. There are different modules providing different approaches to replication. The early implementations followed the "eager" pessimistic approach, which instantly forwarded requests to all other replicas and responded to the client after collecting confirmations from replicas. This module was build for performance comparisons only, and the main effort was concentrated on implementation of a module combining both pessimistic and optimistic approach (described in the next subsection).

There is an additional module related to a specific form of replication—the caching module. The caching mechanism is designed specifically for the purpose of handling RESTful web services, therefore it differs from a typical caching proxy (like Squid or Varnish). The caching can be more aggressive because all updates are assumed to be explicit (they pass through the HTTP channel visible to the proxy), thus making the validation cache contents redundant. Additional service description providing information on resource dependencies enables further optimization of necessary invalidation of cached data items. There is also an additional consistency manager module for coordinated invalidation of a group of distributed caches.

5.2 *Pessimistic and Optimistic Approach*

ReREST supports both optimistic and pessimistic replication. As a matter of fact it allows combining these strategies, thus joining the benefits of both high availability and predictability of system behavior [8]. High availability is a straightforward effect of optimistic replication, because optimistic strategies allow for access to replicas without instant synchronization. However, they expose to the risk of losing consistency of replicas, thereby involving tentativeness of the replica state and subsequent results of interaction. On the other hand, consistent state of replicas is crucial to some operations when firm results are expected. Just to meet this requirement, pessimistic replication is incorporated.

Understanding the difference between pessimistic and optimistic replication requires detailed analysis of the operation life cycle—especially modifications—from its issue to completion (see Fig. 6), and its influence on the replica state. An operation is *issued* at the client side and sent to the server holding a replica. After the operation is *accepted* by the server it should be *scheduled*, i.e. arranged within

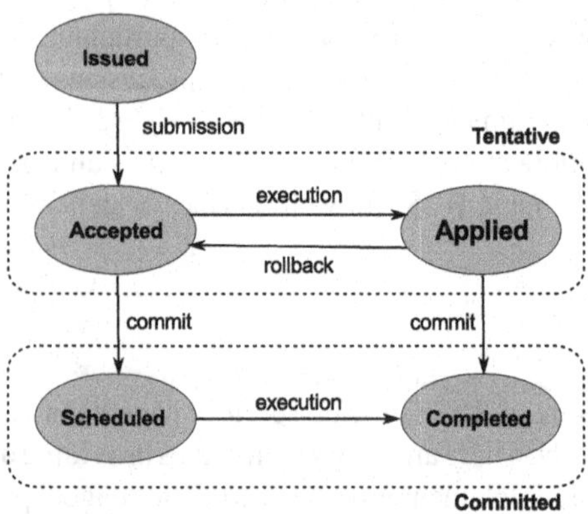

Fig. 6 States of modification processing

the set of operations concurrently accepted by other replica servers. The operations are executed according to the schedule, generating a sequence of *committed* (stable) states. This way the operation that has been executed by the accepting server becomes *completed*. This sequence of operation states is characteristic of pessimistic replication. Optimistic replication, in contrast, allows an operation to be executed before scheduling, at risk of conflicts, i.e. concurrent execution of non-commutative modifications. Operations executed in this manner are termed *applied*, and the states resulting from their execution— *tentative*. If the optimistic execution does not cause conflicts, the operation becomes *completed*, otherwise the conflict must be resolved.

The combination of optimistic and pessimistic replication consists in coexistence of committed and tentative states of replicas. The requirements about the replica state at which a given operation can be executed are specified at the issue of the operation. To this end, two general submission modes are distinguished: optimistic and pessimistic. Moreover, modifications are specified as either synchronous or asynchronous, and read-only operations either as synchronized or immediate. This way two orthogonal classifications of submission modes are achieved as presented in Table 1 for modifications and Table 2 for read operations.

The synchronous mode for modifications (Table 1) means that after issuing an operation the client waits for its execution. Depending on whether the operation is pessimistic or optimistic, it must be completed in a committed state or can only be executed (become applied) in a tentative state. As it is easy to notice, there is

Table 1 Submission modes for modifications

	Pessimistic	Optimistic
Synchronous	The modification must be committed	previously accepted modifications must be executed before
Asynchronous	Modification is submitted and acknowledged	

Table 2 Submission modes for read operations

	Pessimistic	Optimistic
Synchronized	Modifications accepted before the read must be committed	Modifications accepted before the read must be executed
Immediate	Uncommitted operations must be retracted	The current state

no distinction between the pessimistic and the optimistic mode for asynchronous modifications. Since the interaction ends before the modification is executed, the system should postpone subsequent state changes till the operation can be completed, i.e. executed in an appropriate order as if it were issued in the pessimistic mode.

The difference between the pessimistic and the optimistic mode for read operations (Table 2) lies in the state of replica to be read. For pessimistic read operations the state must be committed, while for optimistic ones it may be tentative. The orthogonal distinction between synchronized and immediate reads, in turn, concerns the treatment of uncommitted modifications. The synchronized mode requires all previously accepted modifications to be included in the state of replica, while the immediate mode expects the current state.

6 Recovery

Failures of the SOA components lead to the limitations in the availability of services, and thus affect the reliability of the whole system. Restoring normal work of the system requires failed components to resume their work in a finite time after a failure occurrs. However, in general, the state of the component resuming the work may differ from its state before the failure. As a consequence, the state of a business process (understood as a processing consisting of interactions between service and its clients) after a failure occurrence may be incorrect, and in the result the further processing of such a business process may be stopped. Such a situation is undesirable in the SOA systems, where the reliability aspect is particularly important. Therefore, to increase the reliability of a service-oriented system, the design of a mechanism enabling the continuation of processing despite failures is an important and current research objective.

One approach to solving this problem, and to fully masking failure occurrence is backward-recovery. In general, this approach relies on the idea of periodically saving (during the failure-free execution) the current state of the system in the non-volatile memory in the form of so-called checkpoints, to be able to restore the error-free system state from the saved data in the case of failures.

A wide range of checkpointing and rollback-recovery techniques for general distributed systems and databases have been explored in the literature [39, 49, 64, 67, 71, 103]. Depending on when checkpoints are taken, existing approaches can be divided into *synchronous* (*coordinated*) *checkpointing* [39, 71], *asynchronous* (*uncoordinated, independent*) *checkpointing* [103, 105] and *communication-induced checkpointing* [39, 88]. In synchronous (coordinated) checkpointing strategy, all servers coordinate the moment of taking their checkpoints, in order to form a consistent global state [71]. As a result, in the case of failure and recovery, every server restarts from its most recent checkpoint. As a consequence, no more than one checkpoint per server has to be kept in stable storage, this way reducing the storage overhead and eliminating the need for garbage collection. The main disadvantage of synchronous checkpointing, however, is unavoidable synchronisation overhead of failure-free executions, which may influence the scalability of the system, and can be unacceptable for some applications. To eliminate the overhead introduced by synchronous checkpointing and to make checkpointing scalable, an asynchronous checkpointing approach has been proposed. In this approach, checkpoints can be saved independently, at the point of time the most convenient for each server, and without the requirement of cooperation with other servers [65, 103, 105]. However, there is a potential a risk that no consistent global checkpoint can be formed from local checkpoints in this case. This is a problem known as the *domino effect* [89]. The domino effect can be avoided with a communication-induced (quasi-asynchronous) checkpointing, which allows servers considerable autonomy in deciding when to take checkpoints. A server can thus, take a checkpoint at times when saving a state would incur a small overhead [88]. Checkpointing may be combined with log-based rollback-recovery protocols, which save messages sent and received by each server. If the server fails, the log can be used to replay the progress of the server after the most recent checkpoint in order to reconstruct the state it had before the failure. This has the advantage in that the server recovery results in a more recent checkpoint of the server's state than checkpointing alone can provide. Log-based recovery has been widely studied in the context of process based systems with asynchronous message passing [39], where three types of these protocols are considered: *pessimistic* [11, 63], *optimistic* [64, 101, 102] and *causal* [2].

Regardless of the applied approach to checkpointing and message-logging, the direct use of these solutions in general distributed systems faces in the context of the service-oriented architectures a number of problems arising from such systems specific characteristics, among which are the autonomy of the service providers, dynamic nature and longevity of interactions, as well as the inherent constant interaction with the outside world. For example, in the classical solutions using the mechanisms of synchronous and asynchronous checkpointing [39] one either has to control when checkpoints are taken, or to appropriately choose the checkpoint which will be used during the recovery process. In the case of the SOA systems such solutions cannot be applied, because of the autonomy of the services, which is expressed among the others in the implementation of services' own fault-tolerant policies. As a result a service sometimes cannot be forced to take a checkpoint, or to rollback, it may also refuse to inform other services on checkpoints it has taken. Another limitation, which arises from the fact that every service invocation may result in irrevocable changes, is the necessity of applying so-called output-commit protocols [39], in which the pessimistic approach is used, and checkpoints are taken every time when the external interaction is performed. The assumed SOA model also imposes certain restrictions on the rollback-recovery of services. The failure of one service can not affect the availability of other services taking part in the processing. This means that the rollback-recovery of one service should neither cause the cascading rollback of other services, nor influence their state.

Consequently, the existing rollback-recovery solutions have to be revised, and specially profiled for the SOA environments to efficiently meet their requirements, and to take advantage of their specifics. This problem has been recently investigated, and some solutions have been proposed [27, 95, 113]. However, most of existing solutions are acceptable only for simple application scenarios, and they usually do not consider nested invocation dependencies. If nested interactions are not allowed, then checkpointing the local state of some chosen replica of a single service is sufficient for the correct recovery. Such a simplified recovery approach has been commonly adopted in practice by business process execution engines, such as BPEL [87]. However, in the complex application scenarios, where nested interactions are used for service composition, distributed checkpointing is necessary to maintain a correct global state useful for recovery. A noticeable fault tolerance framework with distributed checkpointing for services has been proposed in [34]. Unfortunately, while offering very strict consistency of the recovered processing state, this proposal requires complex fault detection and costly global recovery coordination. It is clear based on the past experience [44] that many applications could benefit from less restrictive consistency guarantees, allowing the recovery of the processing state in a more efficient way.

It is thus reasonable to still search for new rollback-recovery solutions, appropriate for service-oriented systems. Therefore, in the course of the IT-SOA project, the business process rollback-recovery framework ReServE (Reliable Service Environment), being a part of a ReSP (Reliable SOA Platform), was proposed.

ReServE aims to provide fault tolerance of the SOA systems based on the RESTfulWeb services. The primary method used by the proposed framework is message logging. The proposed framework ensures that in the case of failure of one or more system components, a coherent state of a business process (i.e. one that could be achieved in the failure-free processing) is transparently recovered and that it is consistently perceived by the business process participants: clients and service providers.

Rollback-recovery carried out through the ReServE service is transparent to processing participants, and does not require either their intervention, or semantic knowledge related with the processing. While the proposed framework can be used in any SOA environment, it is particularly well-suited for processing, which does not have the transactional character, and where clients applications do not use the business process engines.

6.1 General Idea

The idea behind the proposed ReServE framework is based on two main assumptions. First, the purpose of the framework is to recover the consistent state of the processing in the service-oriented systems in the case of failure of any of processing participants (clients or services). The second assumption is to impose the minimal requirements on the functionality of the business process participants that use the proposed framework. The usage of the ReServE should be as transparent to clients and services taking part in the processing as possible.

Essentially a system state is said to be consistent, if it could have occurred during the failure-free preceding execution of the system from its initial state regardless of the relative speeds of individual processes. This ensures that the total execution of the system is equivalent to some possible failure-free execution. A theoretical foundations for finding consistent global checkpoints in general distributed systems are presented in [64, 76, 106–108]. In the context of the SOA systems, we will guarantee the recovery of the consistent SOA state if all the effects of clients' invocations performed by the service before the failure will be reflected in the state of the recovered service, and the results of the execution of these invocations will be reflected in the state of the client. To meet these assumptions, the proposed rollback-recovery framework takes advantage of the fact that the processing in the service-oriented systems is based on the exchanging of messages. Therefore, ReServE intercepts the communication between participants of the processing (clients and services) and logs messages exchanged among them. As a result, the intercepted messages reflect the complete history of communication, which is then used to recover the system state in the case of failure of any client or service.

The correctness of the above idea requires remembering also the order of messages issued by the participants of the business process. However, since in the SOA environment a service autonomy is one of the most important characteristics, it is assumed that the participants of the business process may have their own reliability policies, and may use different mechanisms that provide fault-tolerance. We will denote by a recovery point an abstraction allowing the partial reconstruction of a service state after the failure occurrence, but we do not make any assumptions on how and when such a recovery points are made (recovery points may take the form of logs, checkpoints, replicas or other mechanisms). We only assume that for each service there is always at least one recovery point available (e.g. the initial state of a service). Each service takes recovery points independently, and in general it may take no recovery points at all. Since in the recovery point the state of the service (or the client) resulting from the processing of certain requests may be saved, therefore, during the recovery procedure carried out by the ReServE framework, the reprocessing of such requests should be omitted. In other words, only those messages, the processing of which is not reflected in services (clients) recovered state, should be processed again during recovery. Such messages have to be found and performed, which is also the task of the proposed framework.

Other additional assumptions we make are the following: it is assumed that both clients and services are expected to be piece-wise deterministic, i.e. they should generate the same results (in particular, the same URIs for a new resource) in the result of multiple repetition of the same requests, assuming the same initial state. Also, the crash-recovery model of failures is assumed [48]. System components can fail at arbitrary moments, but every failure is eventually detected, for example by the failure detection service FaDE [21]. The failed service becomes temporally unavailable until it is restored.

6.2 ReServE Architecture

The distributed ReServE framework is designed in the modular way. It includes the distributed *Recovery Management Units* (RMU). Due to a distribution od the RMUs we avoid the existance of a single point of failure, which occured in the centralized version of the ReServE [29–31]. Other modules of the proposed ReServE service are: client and service proxy servers, called the *Client Intermediary Module* (CIM), and the *Service Intermediary Module* (SIM).

Additionally, we assume that in the case of a service failure, an external failure detection service, like FaDE, is responsible for monitoring the services involved in the processing. In the case of failure of any of them, FaDE informs the failed service proxy server on this fact. In this way, after the resumption of service work, the appropriate service rollback-recovery procedure is started.

The overall concept of the ReServE framework is presented in Fig. 7.

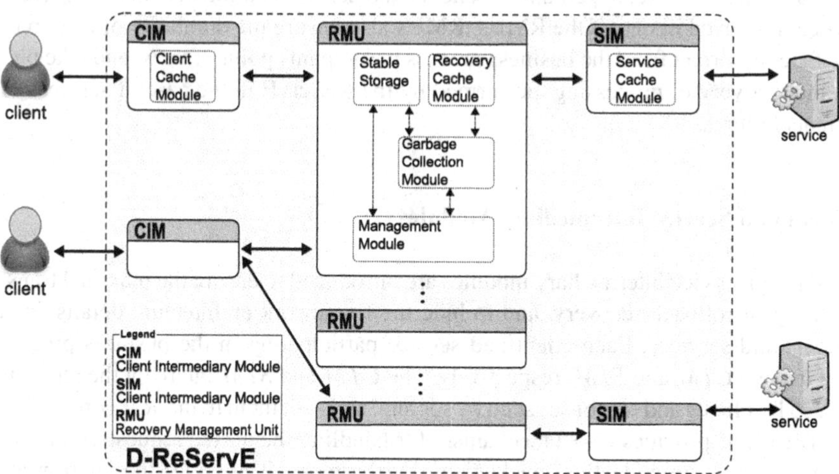

Fig. 7 ReServE architecture

Recovery Management Module

The *RMU*s are the backbone of the proposed framework. Each *RMU* includes the following components:

- *Management Module*—manages the execution of specific operations associated with saving the state of communication and rollback-recovery actions
- *Stable Storage*—stores the data necessary to recover the state of a distributed processing; it is assumed that the data in the Stable Storage are stored in the database. Currently at least 8.4.3 version of PostgreSQL database is supported.
- *Garbage Collection Module*—monitors the status of Stable Storage and removes unnecessary data, in order to ensure the high performance of the ReServE service
- *Recovery Cache Module* (*RCM*)—the volatile memory that stores pairs: <service_URI, RMU_URI>. The first element denotes the service address, and the second one is the address of the RMU in which this services is registered. Such an information is obtained when the RMU gets from the client a requests sent to a service not registered in the given RMU module. In such a case the RMU asks the service's SIM to respond with it's default RMU URI. SIM sends back a message containing the address of the RMU it is registered in.

Each service is registered in the one, chosen RMU module. It is assumed, however, that one RMU module can be used by many services. The RMU, in which the service is registered is called the default (master) RMU for this service. In turn, the client can be registered simultaneously in many RMUs. The client's default (master) RMU is the one, in which the client is registered at the beginning of the processing of the business process. The client's default RMU stores the information on the other RMUs, the client contacted with while invoking services.

Each message exchanged among clients and services taking part in the business process is saved in one of the RMUs. RMUs also ensure the durability of resources that are important from the business process participants point of view, and take part in the recovery of processing, performed by the ReServE in the case of services or clients failures.

Client and Service Intermediary Modules

Client and service intermediary modules are introduced to ensure the maximal transparency of rollback-recovery, and to hide the framework architecture details from clients and services. Each client and service participating in the business process has its own CIM and SIM, respectively. The CIM and SIM intercept the requests issued by clients and responses send by servers, and sent them to the RMU modules.

The CIM provides also a mechanism for handling the nested requests, and supports the garbage collection mechanism. Furthermore, it provides the appropriate interface, designed directly for the client, to access the RMU functions. Among such functions are: the ability to store the state of client's processing, the acquisition of the response to the recently issued request, the acquisition of a list of active conversations, and the list of RMU modules, with which the client had contacted.

A single CIM should be used by a single client only. The CIM consists of the client cache module, which is developed in order to increase processing efficiency. When the client calls the services registered in the RMU modules other than its default one, then the addresses of RMUs in which the requested services are registered are saved in the client cache module.

The Client Intermediary Module is implemented using the MProxy caching module. In order to install the CIM at the client side, the configuration script must be run. Such a script includes among other things, the address of the client proxy default RMU.

The last module of the ReServE framework is the Service Intermediary Module (also referred to as service proxy, SIM). The primary function of Service Intermediary Module, is to monitor the service status and to react in the case of its eventual failure. It is the SIM that is responsible for initiating and managing the service rollback-recovery procedure.

SIM works in two modes: normal, and recovery one. The normal mode is carried out when there are no failures. During the normal processing, the SPS manages the clients requests sent by the m to the service, captures generated responses and returns them to the RMU module. Each response is expanded by the unique, within the RMU domain, service identifier. In this mode, the service proxy server is equipped with a cache memory, used to store responses in the case of expiration of timeout limit.

Service proxy server also makes the detailed analysis of the requests directed to the service. It filtrates the outdated requests, i.e. the requests sent before the failure, but received by the service after its rollback-recovery. Acceptance of such requests could result in the inconsistency. In order to filtrate requests, the additional information attached to requests by the RMU is used.

The SIM is implemented using the MProxy caching module. The installation of the SIM is similar to the installation of the CIM—the configuration script, including the RMU address, is run.

6.3 ReServE Functionality

Failure-free execution

The request issued by a client to a chosen service is intercepted by the client's CIM and forwarded to the client's master RMU. If the required service is registered in the RMU, the request is saved in the stable storage and then forwarded to the service through its SIM. Otherwise, client's master RMU obtains the URI of requested service RMU from its SIM, and sends back this information to the CIM, which reissues the request to a proper URI. The service performs request and sends the response back to RMU. The response is saved in the stable storage and forwarded to the client through its CIM.

The distributed nature of the RMU denotes that such a history of communication is dispersed among many RMUs. The mechanism of history of communication is mainly used during the recovery of a distributed processing. However, it can also be used to ensure the idempotence of all requests. If the client?s request is received by the RMU, and response to such a request has already been saved in the RMU, then there is no need to send this request to the service once again, as the already saved response can be sent to the client immediately. With this solution, the same message may be sent multiple times, (i.e. the message with the same identification number), without the danger of multiple service invocations.

In the RMU modules the garbage collection mechanism has been applied in order to periodically clear unnecessary messages from the Stable Storage. The request is claimed to be unnecessary when there is certainty that it will be never used during the recovery of a service. The response, on the other hand, is unnecessary when the client will never re-send request which generated this response. Therefore the process of marking messages as ready to be removed engages both parties of a business process. Client saves its state in RMU's stable storage along with the information about the progress of processing. The progress is expressed by the identifier of the last message included in the state. It is assumed by the garbage collecting protocol that all messages stored in the client's state are ready to be deleted from RMU's stable storage. Taking into account the service side—the decision on messages that can be removed from the RMU's stable storage are taken on the basis of a metadata associated with each recovery point taken by the service. According to such metadata, the messages that will not further take part in the recovery process are chosen. Choosing such messages is realised periodically by the service proxy server transparently to the service.

The service proxy server contains a service cache module. In this module, the last responses send by the service are saved. These responses may be used by the RMU module, if the waiting timeout error occurs. In the case of such an error, the RMU module sends a request to the service proxy server to re-send the service responses, and the service proxy replies by sending the responses saved in its cache module (thus avoiding the need to send the request again to the service). If the RMU request can not be realized (e.g. the required responses are already removed from the service cache module), the RMU initiates a rollback-recovery of a service.

The occurrence of client's or service's failure

In the case of client's application failure, for some clients the last response from the service may be enough for recovery (according to the HATEOAS principle of Resource Oriented Architecture). Since in general the client communicates with many RMUs, the information on the last response obtained by the client from services before its failure has to be agreed, based on the information from each of them. The client's CIM gathers the information on the obtained responses from all RMUs it contacted before the client's failure (the list of such RMUs is stored by the client's master RMU), and sends a response with the highest identifier to the client. The client then proceeds with the execution.

In turn, when the service fails, an external failure detection service, like FaDE, notifies the SIM on this fact, so it can start the rollback-recovery process. In order to correctly recover the service and processing state, we must first identify the last externally visible state of the failed service. Therefore, SIM starts with getting the information on available recovery points taken by such a failed service. The service returns SIM the list of available recovery points along with the information on the identifier of the last message contained in each of them. Afterwards, SIM asks RMU for an identifier of the oldest request which has not received a response yet. The obtained identifier is used in the process of determination of the recovery point to which the service should be rolled back. The service rolls back its state to the designated recovery point and informs SIM about this fact. Afterwards, SIM and RMU cooperate to resend to the service a sequence of requests, for which the responses were marked with Response-Id identifiers greater than the Response–Id value saved in the recovery point to which the service rolled back. Such requests are reexecuted by the service, in the same order as before the failure. Lastly, requests which were directed to the service before the failure, but have not yet received a response, are sent.

7 Management

Systems created according to the SOA paradigm can be characterized as loosely-coupled, complex and dynamic. Independent services in such systems cooperate with each other by forming sophisticated business processes. Business orientation of SOA systems forces them to emphasize quality of service and reliability in order to fulfill users' requirements. In these systems the task of management is challenging because the systems are complex. Various scenarios of business process executions, dynamic services orchestration and composition result in overall difficulty of managing such systems. One of the problems of management in such systems is the aspect of configuration of the management system itself. Managed resources can appear and disappear continuously, dynamic resources can be created and destroyed on demand. For such systems a new approach is needed, an approach that will, to some extent, support automatic configuration of the management system, making it able to detect new resources and automatically configure itself to handle them. By doing so it will take over at least some responsibilities of human administrators and simplify their work. This section presents a *Dynamic Management SOA Toolkit* management platform that focuses on self-configuration and auto-discovery of managed resources.

Heterogeneous distributed service oriented systems, such as SOA, tend to grow in the number of components taking part in the processing, as well as in complexity. Services are built with the use of various technologies and according to different paradigms, and models. On the lower level, software and hardware components provide building blocks and running environment for these services. Infrastructure plays a role of a foundation for such systems to provide storage, communication, virtualization, and processing power. In such multi-layered systems, management becomes a challenging task. One of the most difficult aspects that arises is the aspect of configuration. Let us imagine a system composed of possibly hundreds of elements (services, software and infrastructure components), each providing a set of sensors and effectors. Manual configuration of such a system is time consuming and error prone. If the system is dynamic in nature, i.e. its topology can change during the execution, the task of configuration of the management system becomes almost impossible to carry out. One of the solutions to this problem is equipping the system with self-configuration and self-management capabilities. However, in order to do so, the management system has to know how to automatically configure itself. This means it has to either know *a priori* which managed resources are present in the system, or be able to construct this kind of knowledge dynamically. The first approach is used in most of management systems where configuration is static. Such systems are as good as the functionality they provide out of the box. The latter approach, where resources in the system are discovered dynamically, and represented by means of knowledge, is more flexible and in fact solves the problem stated before. First of all, such systems are much more extensible;knowledge once discovered can be reused later by other components. Second, such systems can consume knowledge, which opens the door to more sophisticated self-management abilities, such as self-optimization and self-healing.

7.1 M^3

M^3 (*Metrics, Monitoring, Management*) is a distributed management platform intended for management of loosely-coupled environments. Being a platform means that M^3 itself is not a management application, it neither enforces nor makes policies and decisions that affect the whole system. Instead it provides means for external applications that implement such mechanisms and treat M^3 as an underlying source of monitoring data, knowledge about the components building up the system, and a set of effectors to act on it. From an autonomic computing perspective, M^3 platform can be seen as an implementation of Monitoring, Execution and Knowledge parts of the *MAPEK* loop of the autonomic element.

M^3 follows the SOA paradigm; all its components are in fact loosely-coupled RESTful web [42] services with well-defined interfaces. Such an approach is needed in order for M^3 to be used by external management applications. Since SOA systems are of great complexity and dynamism, management platforms should take this issue under consideration. Therefore M^3 is equipped with some mechanisms that ease human administrators in their work, and try to replace them in part of their responsibilities. We believe that configuration of the management system is the most time consuming task. That is why M^3 focuses strongly on being self-configurable. It does that by providing mechanisms of discovering its surrounding and automatically configuring itself in order to fullfill this task. This autodiscovery feature provides M^3 the means to uncover, store and consume knowledge about the managed system. The knowledge can later be searched and used by higher level management applica-

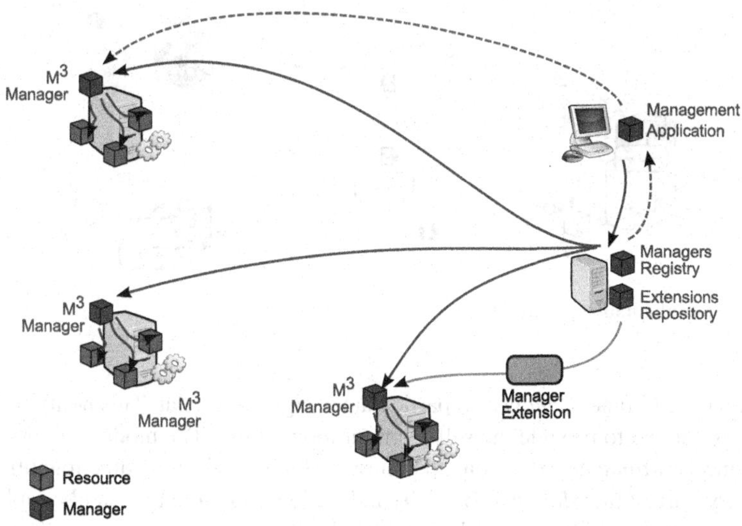

Fig. 8 Architecture of M^3

tions that enforce global planning policies or implement intelligent mechanisms of proactive management [14, 22].

M^3 architecture

The architecture is shown in Fig. 8. *Managers* are tied to one or more managed resources. They register with the *Registry* that can be described as a central knowledge repository for the whole M^3platform. *Managers* communicate with the *Registry* by uploading knowledge about their surroundings . In general this information contains locations of managed resources, monitoring and management capabilities, i.e. sensors and effectors. Combining knowledge from *Managers*, *Registry* stores a complete *picture* of the whole managed system. In order to keep the knowledge database up to date, the *Registry* sustains a simple pulse mechanism with all active *Managers* in the system. When a *Manager* registers or disconnects, *Registry* updates its knowledge repository accordingly. Similarly *Managers* are obliged to monitor its surrounding for occurrences of new resources and inform the *Registry* about this fact. The *Registry* also provides a search feature, which can be used by external management application to search the knowledge database for managed resources, sensors and effectors.

The component, often combined with the *Registry*, is the extensions *Repository*. It is a simple service storing all of the extensions that are used for self-configuration purposes of *Managers*. The *Repository* is also responsible for sending extensions to the requesting *Managers*. *Extensions* can be thought of as plugins that extend *Manager's* functionality with new effectors, sensors and event internal components.

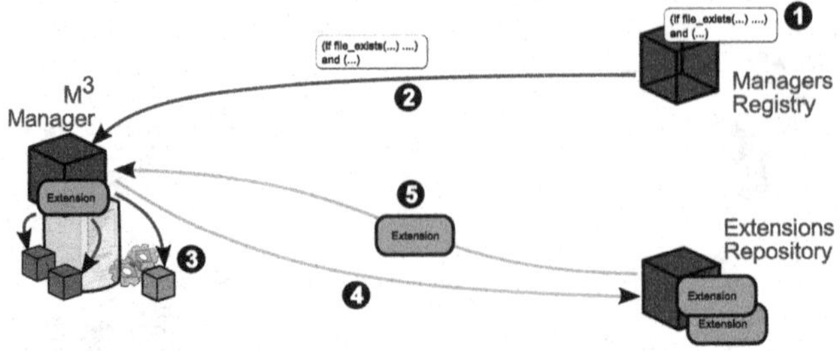

Fig. 9 Self-configuration pattern in M^3

Note that M^3 (and *Manager*, in particular) is highly modular. This nonfunctional feature is needed to provide the self-configuration ability. The modularity does not only concern *Manager* extensions that enrich *Managers* in new functionalities. In fact every part of the *Manager* is a fully independent module that can be replaced, upgraded or uninstalled without the need for downtime.

7.2 Self-Configuration

Figure 9 presents a self-configuration pattern used by M^3. In step (1) a definition written in a RDL is submitted to the *Registry* to update the knowledge used to detect new resources. The submission can be done either by a human administrator or a higher level management application. The definition is then forwarded to all *Managers* in the system (2), in order to update their capability of detecting new resources. When the *Manager* installs a new definition, it can detect a new resource (3). After discovering the new resource the *Manager* contacts the *Repository* to check for existence of a matching extension to the newly discovered resource (4). The *Repository* sends the extension to the *Manager* that can now install it and be ready to handle a request for new sensors and effectors associated with the new resource (5). Of course the *Manager* also uploads information about a new resource to the *Registry*. This information is combined with the knowledge stored in the *Registry*.

Resource Detection Language is used by *Managers* to detect resources in the system. RDL definitions are uploaded with the corresponding extension that is combined with the resource that is detected by the definition. Upon uploading the extension to the *Registry*, the definition is extracted from the extension and uploaded to all *Managers* in the system. *Managers* update their detection database with the new RDL document, process and execute it. Those *Managers* that get positive detection results communicate with the *Repository* to download appropriate extension to handle the newly detected resource. RDL is a declarative language that can be thought of as a simple logic rule based language. It provides simple operators that perform certain checks. Currently, supported operations can perform checks for an existence of file in the file system (*has-file*), check if a process is running (*proc-running*) or check for open socket (*socket-open*). Since outcomes of the operations are boolean, operations can be chained by logical AND, OR, XOR operators, as well as negations—NOT. This way RDL is parsimonious, yet powerful enough to detect any resource. A simple example is presented in below. If the whole block evaluates to *true*, it means that the resource (in this case "*apache*") is detected. The way the extension is associated with the corresponding definition is described in *META-INF* directory inside the *jar* file, where the definition body has to be provided in a *m3-precondition.txt* file. Upon uploading an extension, *Repository* reads *META-INF/m3-precondition.txt* file and propagates its content to all active *Managers* in the system. After receiving the content, *Managers* execute the definition and if the resource is detected, they download the associated extension.

```
(AND (isos-linux)
    (OR // Block1
        (AND    (has-file "/etc/init.d/apache")
                (proc-running "httpd")
                (socket-open "tcp:80") ) // Block2
        (AND    (has-file "/etc/init.d/apache2")
                (proc-running "httpd2")
                (socket-open "tcp:80") ) ) )
```

In the above example a check for the existence of Apache HTTPd server in the system is carried out. First clause *isos-linux* checks whether the operating system is of Linux family, and if either *Block1* or *Block2* are true. *Block1* checks for existence of a file */etc/init.d/apache*, if the process *httpd* is running and whether port 80 is open. Another block is similar, but checks for alternative Apache installation. RDL is continuously developed, the next versions will provide new operators and mechanisms to extend the language with custom modules.

In order to detect a resource, a definition in RDL should be provided. From this point of view, M^3 relies partially on the knowledge of a human expert. Notice, however, that the definition has to be provided only once; a higher level management application that uses the platform can store such definitions in a case-study database and reuse them if appropriate. The problem of configuration is still present, yet it does not involve human assistance as much as in a legacy management system. RDL definitions can be stored in a central server and extensions combined with them can be reused for different kinds of managed resources.

7.3 M^3 Extensions

It is possible to extend the functionality of M^3 by developing and installing *Manager* extensions. Extensions are automatically uploaded and deployed in *Managers* if particular resources managed by the extension are detected. The range of functionalities provided by extensions is very wide, including:

- possibility to define managed resource,
- discovery of managed resources,

- defining effectors for resources,
- defining metrics for sensors that return performance values,
- defining sensors for resources.

Currently M^3 is equipped with extensions that can be divided into four categories:

Sensors are extensions that provide the functionality for collecting various monitoring data, as well as provide definitions of metrics. An example can be an extension that collects CPU usage from the operating system and provides average CPU usage metric for the last hour.

*Effector*s provide endpoints for controlling the environment in which a *Manager* is placed. In collaboration with sensors they can create control loops in order to continuously make impact on the environment. An example can be an effector that is used to control the number of threads assigned to a Web application, depending on the current and the predicted number of requests.

Consumers are all the extensions that take metrics values as their input, in order to process these values in some way. In M^3 there are several built-in consumers, responsible for storing values in a database, file, memory, alert handling, and graphing.

Internals are the extensions that provide mechanisms for the *Manager* itself; they provide self-configuration management, RESTful API and various utilities. The *Manager* is built in such a way that its every part can be dynamically replaced, even its internals. Such approach facilitates all tasks connected to upgrading the *Manager*.

It is worth noting that every sensor is in fact divided into two parts. One part is an adaptor that is responsible for collecting data, and the second part is the module that calculates various metrics on the collected data. Such an approach allows for defining new metrics by the administrator without replacing the whole sensor. To achieve this, a special metrics definition format is used—MDL (*Metrics Definition Language)* [20], which is capable of expressing metrics in a human readable language.

7.4 FADE: Failure Detector Service

This section describes a more sophisticated extension, FADE (*Failure Detection Service*) [21, 36], which can be used with M^3. This kind of extension can also work as a standalone fully fledged SOA service. M^3 is flexible enough to incorporate FADEeither as a dedicated extension or detect the presence of its standalone version and register it as a managed resource.

The starting point of many approaches to counteract the negative events occurring in the system is to detect irregularities. Irregularities are detected with a mechanism known as a distributed failure detector. This mechanism consists of software modules located on each node of the system. In order to obtain information about the status of other nodes, modules exchange messages between each other. In the simplest possible scenario, each of the modules in certain moments of time sends a control message called heartbeat to other modules, thus indicating that the node is working properly. Absence of such messages for a specified time justifies the suspicion that the sender has crashed.

Design and implementation of failure detection service in SOA systems involves the solution of new problems arising from different client applications interaction model and the potentially large variety of scenarios for the use of this service. A layered model of the SOA system also opens new possibilities when it comes to detecting failures at different levels of action (node, a group of services, service). The possibility of detection of new classes of errors, specific to SOA remains an open question.

FADE is a distributed service that monitors the availability of nodes, and services, and the various resources provided by these services. It provides a wide range of built-in adjustable failure detection mechanisms, which allow to tune up detection activity for the environment in which it was launched. Moreover, the failure detection service allows the user to specify preferences considering the speed and accuracy of the detection. Distributed and decentralized nature of the service increases its reliability and resilience to the network failures. In addition, it is possible to register client requests for callback notification of a specific event occurrence in the context of failure detection.

Figure 10 presents a sample FADE service usage scenario. In this case, the FADE service consists of three nodes. The nodes communicate with each other in order to exchange information about connection topology and monitored service states. The FADE nodes monitor three services: A, B and C. It is worth noting that

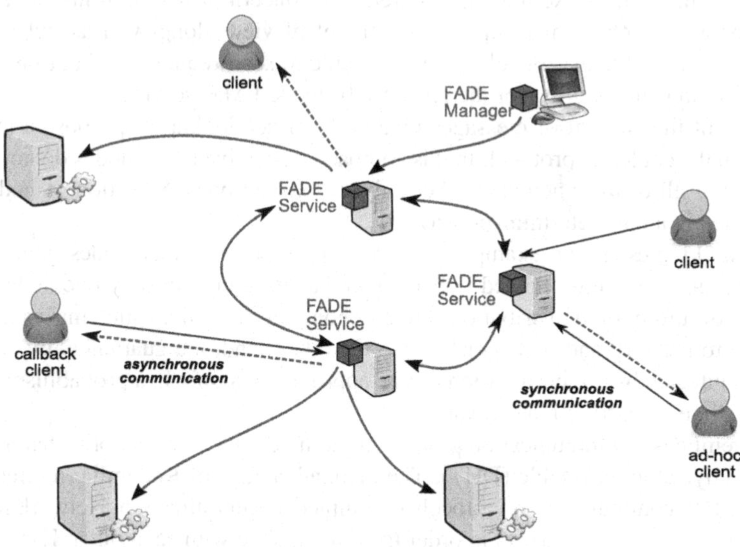

Fig. 10 FADE service sample scenario

FADE node can monitor services that provide SOAP or REST type interfaces. It is transparent to the client, which node monitors which service. FADEdistinguishes three types of clients: standard, ad-hoc and callback, all of which use FADE service in different manner. In general, clients can inquire FADE service about the status of the monitored services. Depending on the type of request, appropriate action is performed.

Communication

Assuming a large number of monitored services, responsibility for the monitoring process is distributed among the different FADE service nodes. Thus, even a large number of services can be monitored at the same time. Moreover, the presumption of flexibility according to realized approach enforced introducing a possibility to request service state from every single detector node, even the one that do not directly monitor the service. Thus a communication mechanism between the FADE nodes is needed, so that they can exchange information about monitored services. Finally two models of communication are available. The first of them is based on epidemic (gossiping) algorithms. The other one is an adaptation of implementation of distributed hash tables—Kademlia [78]. Choice of the communication mechanism should be done at the stage of the FADE service configuration and should be consistent for all the running nodes. The exchange of information between nodes is transparent from the perspective of the service user.

First of proposed models of communications is based on gossiping. The client should get the answer about a service state from any node, so suitable mechanisms should be introduced. Additional requirements concern providing transparency of the proposed mechanism from a client's point of view along with asynchronous communication. Finally, the client must be able to send requests without blocking the processing and waiting for the response from the FADE service.

To limit the amount of messages related to service lookup wepropose a model based on the epidemic protocol. In this approach respective FADE nodes do not send requests to all of their neighbors, but only to selected ones. Selection of nodes is performed in a non-deterministic way.

Figure 11 presents an example scenario, in which particular nodes gain some information (are being infected by it), which has its source in only one node. The process of information distribution is totally non-deterministic, which makes it very durable to network and node crashes. On the other hand, the guarantee, that every node is infected by the information can be approximated only in a probabilistic way and the time of distribution can vary.

Kademlia is a communication protocol used in peer-to-peer networks that implements a dynamic hash table (DHT). The main advantage of Kademlia in reference to other P2P communication protocols is complete decentralization. Network nodes do not need any proxy servers in order to communicate with each other. Each node receives information about other nodes from its direct neighbors.

The FADE uses monitored service URIs as keys for a distributed hash table. Values stored in the hash table are current service states. The FADE service assures that each monitored service has a unique identifier. The identifiers are 160-bit binary keys, they are calculated from service URIs.

The situation is more complex with reference to failure detectors. There may exist a single detector accessible via multiple addresses, thus generating a unique identifier from an URI would be ambiguous. A solution is proposed, that each detector upon start-up generates a random identifier. The identifier is attached to every message the detector sends. This provides unique and unambiguous identifiers for FADE service nodes.

Assuming that services and FADE nodes are identified with keys of similar structure and each service is supposed to be monitored by a few detectors (the number is algorithm parameter), the detectors that are responsible for monitoring the service are those that have identifiers similar to the monitored services identifier (according to some metric). This makes FADE service resistant to FADE node crashes, and the probability of losing information about the monitored service is extremely low.

Monitoring

Monitoring of services is done by cyclic exchange of messages between the FADE node and monitored service. In this model FADE works in active mode; its task is to send a monitoring message. According to the assumptions relating to compliance FADE service with the REST paradigm, the HTTP protocol is used.

The monitoring mechanism is implemented as a failure detector . At the moment it is possible to use two models of the detector, depending on the configuration:

binary failure detector A binary failure detector [25], gives only two different answers, which means that the service is running or not. This means that the detector monitors the set of services. The absence, or too long waiting time for a response make the failure detector start suspecting the service. In the case of a failure, the values returned by it should be interpreted as follows: 1—there is a suspicion that the service is not available, 0—otherwise.

accrual failure detector An accrual failure detector [52] returns a numeric value that can be interpreted as the degree of probability that the service is unavailable.

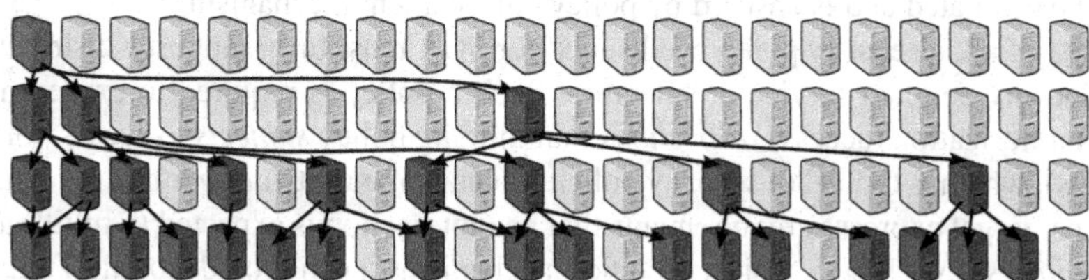

Fig. 11 FADE epidemic communication

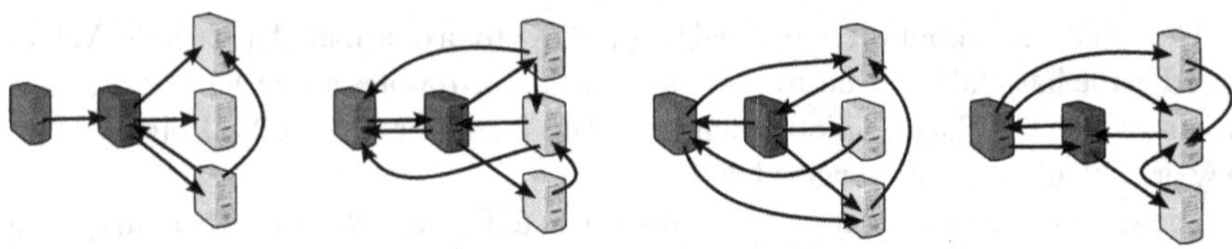

Fig. 12 FADE Kademlia communication

The higher the value returned by the detector, the greater the likelihood that the monitored service crashed. The decision whether the monitored service is unavailable is left to the user, which can interpret the results returned by the failure detector depending on its own preferences in the context of speed and accuracy of detection.

Integration with M^3

FADE can work as a standalone service or be an embedded extension in a *Manager*. In both situations the integration with the M^3 is carried out using the RESTful interface provided by the failure detection service. *Managers* have built-in rules for detection of any FADE that is either installed on the same host or can be fed with their URL address from the *Repository*. Upon the discovery, a suitable extension is automatically downloaded from the *Repository* and new sensors and effectors connected with FADE are accessible.

8 Security

As SOA systems are often built upon multiple development technologies and span over different control domains, a security policy adapted for such a diverse and heterogeneous environment is required to manage security related problems of SOA interactions. In general a policy represents some constraints or conditions of the access to and use of a service or any other entity managed by any participant. Each of those constraints or conditions is expressed in the form of a policy assertion which may be evaluated and is ensured by policy enforcement mechanisms.

First of all, a security policy well suited for SOA must fully support a distributed and loosely coupled service architecture. Such a policy is required to express not only authorization (access) *restrictions* (authorizing policy *subjects* to access policy *targets*), which is typical of security policies, but also security *obligations* (by which a policy target expresses requirements that any interaction is expected to fulfill) and *capabilities* (by which a subject specifies mechanisms it can provide to fulfill target-side requirements) that must be satisfied by all participants of the service interaction (e.g. the communication protection will be set up by negotiating obligations and capabilities). The ability to express obligations and capabilities is imperative for true SOA environments, where interactions (possibly nested) can be established automatically and dynamically without any human intervention or coordination.

Since large scale SOA systems are often federations of smaller autonomic systems, we also need the ability to express in any autonomic system the security policies concerning other federated subsystems, acquire and incorporate security policies from another federated subsystem, define distributed trust relationships and delegation of security rights between the subsystems.

Many policy frameworks and languages have been proposed in literature to date. Some of them have reached their maturity and received real-world implementations, e.g. XACML [37], Ponder [28], SecPAL [9] and ORCA [19]. Among them the ORCA framework [19] is currently the only one offering full support for SOA-specific requirements.

8.1 Security Policy

A service provider may define conditions under which the service is provided. These conditions may include such constraints as the requirement of successful authentication of the service consumer before processing his/her request or confidentiality protection of the content of a response message, etc. On the other hand, service consumers can impose some constrains on the service interactions. They may specify their own capabilities and requirements concerning the protection of the interaction. The set of all these conditions constitutes a security policy of a service. A policy is expressed in the form of policy assertions representing particular security conditions. Actually, the notion of policy can be successfully extended beyond the scope of security, into more general dependability policy (including such additional aspects of processing dependability as reliability—see WS-ReliableMessagingPolicy [86]).

In general, the most fundamental entities related to security are:

- a target that is to be protected (e.g. resource or service);
- a policy which defines what protection must be applied to targets (governed at a Policy Administration Point);
- policy subject performing actions on targets (e.g. service consumer issuing access requests);
- mechanisms by means of which targets are protected.

Since in SOA we consider independent applications interacting across different administration domains, it is important for a policy to contain all information necessary to ensure the required degree of security without scarifying application interoperability. In our opinion, the following types of policy rules must be supported:

- Restrictions—rules granting subjects specific rights with respect to targets. They define access control policy.

- Obligations—by which a target expresses what requirements a subject is expected to fulfill in order to accept his/her requests, and *vice versa*. This may include for example, a requirement to ensure communication confidentiality with one of accepted cryptographic algorithms or to authenticate the request with an acknowledged signature standard.
- Capabilities—by which a subject provides a list of alternative mechanisms it can provide to fulfill requirements. For example a subject informs that it is able to provide confidentiality protection with only a given subset of cryptographic algorithms or is using only selected signature standards.

In spite of access control, usually provided by current policy languages, obligations and capabilities are also necessary in SOA-based communication to allow both service participants to agree on one of possible alternative security requirements. Expressing obligations and capabilities of both sides of the service interaction is imperative for automatic context negotiation in veritable SOA environments, where interactions (very likely nested) are established dynamically, without any human intervention or coordination, and where security requirements may probably be independent and separate from the service interface, and can change autonomously.

Selected problems in security policy languages

In large size policies some problems (or *conflicts*) can arise. Their occurrence usually follows from imperfect management of the policy during its growth and adjustment, especially for large and complex policies for large-scale and loosely-coupled distributed systems. Hence their extreme significance for SOA-based environments.

One of the problems a security policy must face is *confused deputy*. This problem arises with nested invocations. A subject delegates to another subject (a service, typically) his/her own access rights to a given resource. Then, this service can be forced, by the interaction invoker, to operate on a different resource than it was originally intended, using the delegated access rights [50].

Another problem is *ambient authority*, which is somehow similar to the first problem. Here, every nested invocation invoked by a subject contains delegated access rights. When a subject delegates too many access rights to a target, then it is possible that some service will use those delegated access rights to some malicious behavior [72].

The next problem is *excess authority*. A nested invocation issued by a subject contains delegated access rights. When a subject delegates to a service the access rights to multiple targets, then it is possible that the service will use these delegated access rights without further control, up to some evil behavior.

The last problem involves *modality conflicts*. These are inconsistencies in the policy specification which may arise when two or more rules—referring to the same subjects, actions and targets—lead to ambiguous policy decisions. Typically, modality conflicts concern access control (restriction rules). Two or more restriction rules may apply to a given interaction, some of them allowing the interaction, while some others—denying it. In SOA, modality conflicts can also occur between obligation and capability rules used to negotiate protection mechanisms for particular interactions. That kind of modality conflicts occurs when one side of a given interaction requires (by its obligations) the other to use some mechanisms which cannot be offered by the other side (i.e. are not allowed by its own capabilities).

The occurrence of the above conflicts prevents the policy enforcement mechanisms from making consistent policy decisions. Therefore, suitable work has been carried out in ORCA to detect and resolve any of those conflicts [15–17].

8.2 ORCA Framework

The architecture of the ORCA policy execution framework is shown in Fig. 13.

The Policy Decision Point (PDP) makes decisions about granting access to resources, issued by policy subjects. Moreover, it is charged with establishing security contracts between interaction participants. The *security contract* is a set of negotiated communication protection mechanisms, used to securely perform the interaction. This set consists of a choice of authentication, confidentiality and integrity protection mechanisms (e.g. cryptographic algorithms) and their parameters (e.g. encryption/decryption keys). The security contract is established by the PDP according to the capability and obligation policy rules of both interaction participants.

Policy enforcement would be implemented in a SOA environment most likely as intermediaries which intervene in the communication (interactions). The pipelined processing of interaction messages fully conforms to the SOA architecture. The Policy Enforcement Point (PEP) can then be treated as a filter that performs protection of the input/output interaction messages. In input and output processing pipelines a number of policy assertions can be added or removed to/from interaction messages.

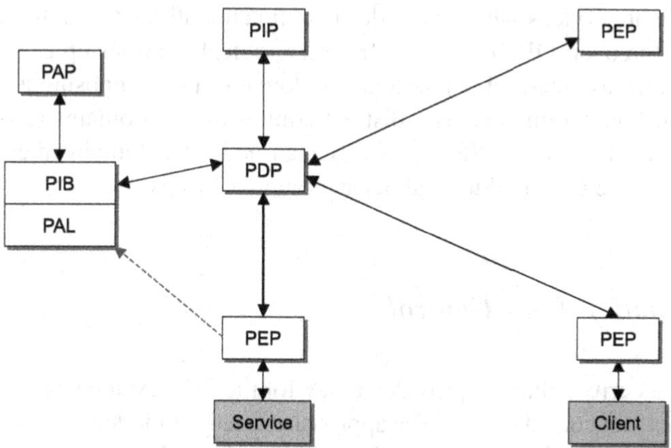

Fig. 13 Architecture of a distributed security policy framework

The PEPs are also responsible for generating service access requests to the PDP and executing access decisions received from the PDP. Typically, the authorization decision can be either "allow access", "deny access", or "no policies apply to this access request". If the PDP is unable to evaluate the policies or to retrieve required information for some reason, it may return an error instead of an authorization decision.

Two PEP intermediaries are involved in each interaction. The client-side PEP is responsible for extracting—from a received interaction request—all the information necessary for the PDP to find appropriate policy rules related to that request. Such information includes, for instance, the subject identity and attributes like authentication credentials or certificates. Then the PEP is responsible for enforcing the security contract imposed by the PDP. The service-side PEP is responsible for requesting access control decisions from the PDP (following from restriction policy rules processed by the PDP). Obviously, it also enforces the service-side part of the security contract obligations returned by the PDP.

The Policy Administration Point (PAP) is used to define policy rules concerning access to resources. Rules defined by PAP are stored in policy repository—Policy Information Base (PIB). The Policy Information Point (PIP) is used for acquiring additional information, unavailable locally for the PDP (in a local PIB), required to resolve proper decisions. Typically, PIP obtains information from other administration domains. This can be useful in federated systems (or Virtual Organizations) where some part of knowledge needed to take proper local decisions can be distributed across multiple autonomous systems. PIP may be e.g. an external Identity Provider for Federated Identity Management (FIdM) or an external Trusted Authority acting as a source of Federated Access control Management data (FAccM). Finally, the Policy Audit Log (PAL) repository plays an important role in policy management, keeping log trails about policy enforcement events.

This separation of basic policy elements isolates policy information from implementation, significantly simplifying policy system management.

In ORCA, the following entities: PIP, PDP, PIB and PAL are implemented as SOA-compliant services—they provide their functionalities through well-defined interfaces. Moreover, ORCA allows for a hierarchical relationship of several PDP services, governing fine-grained system sub-domains and composing a larger SOA environment. Interactions between distinct administration domains (top-level) are supported with the use of PIP entities. Nested and inter-domain interactions are supported with the use of additional delegation policy rules.

8.3 Information Flow Control

Most of the security solutions provided today for the SOA systems only take access control issues into consideration. The approaches used for modeling security restrictions suffer from lack of information dissemination control after access to this information has been authorized. Allowing a service uncontrolled use of any received confidential information (e.g. when further invoking other services) causes a serious risk of unauthorized information dissemination. Thus, information flow control (IFC [33]) mechanisms are an imperative security requirement which must be addressed where any sensitive information is processed in an SOA environment. Informally, IFC requires the system to ensure noninterference [46]. The noninterference paradigm defines the authorized information flow in the following way: low security level outputs of a service may not be directly affected by its high security level (confidential) inputs. Low level data can never be influenced by high level data in such a way that someone is able to determine the high level data from the low level output. The real world case of using IFC for securing information often arises in the context of HIS (Health Information Systems). Let us consider a hospital using SOA-based software to manage its day-to-day activities and patients' medical data. This information is often highly confidential and must be released only to authorized medical personnel. Several hospital departments, such as cardiology, accounting or main pharmacy, share different information regarding patients and their Electronic Health Records. However, specific information categories must be released only to the authorized departments and under strict control, e.g. personal treatment data must not be available to the accounting department. It is the responsibility of IFC to ensure this.

The existing approaches to enforcing IFC in distributed systems impose strong requirements on internal control of service implementation [83, 97, 98, 104, 111]. Since one of fundamental properties of the SOA model is autonomy of services, provided by third party vendors and being unlikely under any internal control, these approaches are of limited use in service-oriented environments and, in practice, have never been successfully applied there. The ORCA framework incorporates a novel concept of lightweight IFC aimed at controlling service-oriented interactions without sacrificing the autonomy of services. Compared to existing solutions, this proposal offers full compliance with the SOA processing model and does not require any source code analysis. The information flow control can be easily managed in an SOA system by using a security policy language. The concept of lightweight IFC is closer to obligation and capability concepts for controlling communication between services rather than to pure access control.

The lightweight IFC implementation in ORCA is based on labeling messages and services with a security level and information category [18]. The security level represents the degree of desired data confidentiality, while the information category determines the possible scope of data usage. Only services with security labels corresponding to message labels are allowed to process messages. Labeling is governed by a new type of security policy rules, the IFC policy. The PDP grants similar labels (clearance) to particular services, according to the IFC policy.

9 Case Studies

Nowadays, we observe an enormous growth in the use of computer systems in the domain of medical healthcare. Such systems are present in all fields of health provision, and substantially improve medical practice. They range from those that directly support human lives, assist diagnosis and deliver certain treatments, to administrative systems that are not directly related to the health outcomes.

Regardless of the purpose, to provide safe and solid healthcare service, the system must conform to a suitable level of reliability and availability. The improper processing (e.g. suspension) is inadmissible as it could be life-threatening. The healthcare system should also protect and guarantee the privacy and integrity of both system users, and medical data. Especially patients' medical records, called Electronic Health Record (EHR) [54], containing the history of care, orders, prescriptions and patients' tests results, are particularly important. Their inappropriate use may violate the privacy and confidentiality of the personal health information. Therefore, individual medical data should be protected and access to this data granted only to the authorized users.

Consequently, the dependability is indispensable or even critical property for all healthcare systems, both in the context of their functionality, and provided medical data. As a result, the healthcare domain is chosen as an application area for the developed infrastructure.

In this chapter we present two healthcare systems: the Healthcare Integration Platform and the Medical Event and Data Registering Platform. The first platform exchanges and integrates medical information (especially patients' medical data) originating from different healthcare information systems, and offers access to it for authorized medical staff, regardless of their current location. The platform utilizes some concepts contained in the Integrating Healthcare Enterprise (IHE) profiles [56–58], combined with existing Electronic Health Record standards in order to maintain the high level of healthcare units interoperability. In turn, the latter solution—Medical Event and Data Registering Platform—provides the availability, reliability and consistency of patients' medical data. It supports primary care physicians in everyday tasks like management of patient visits and maintenance of their medical data. Mobile technology support offers additional opportunities increasing mobility of physicians.

In the following sections, we explain how dependability of the proposed platforms can be increased by applying mechanisms described in the previous sections.

9.1 Healthcare Integration Platform

There are many complex Hospital Information Systems (HIS) that use Electronic Health Records (EHR). However, the ways of storing EHR may differ significantly depending on the system used. Thus, in order to provide unified and effective integration of existing HIS systems, the way of exchanging and sharing data securely has to be defined, and the reference information model needs to be standardized. To meet these expectations, the Healthcare Integration Platform (HIP) was proposed.

HIP allows the exchange of patients' medical data among various healthcare units while respecting the existing law regulations. Users of the proposed platform have in-depth knowledge on the patient, based on the information from different healthcare units.

Due to law requirements, medical data should be stored in the place of their origin. To fulfill this requirement the Healthcare Integration Platform creates an index which stores references to the original data locations instead of a creating centralized data repository. Moreover, the indexed data needs to preserve their context, thus the platform assumes that a document is an indivisible set of data. Indexed documents can be further described using metadata and categorized by creating specific subsets. These subsets can be shared by the patient or healthcare unit in an independent manner, and according to the determined rules. A person requesting access to these data will have to meet the authorization policy in order to access a specific data set.

As a result, HIP can be seen as a distributed repository of medical data. Such a repository includes information on medical data location, as well as some additional metadata that describes medical data. The platform supports querying, accessing and collecting information from various original sources.

The Healthcare Integration Platform is based on the SOA paradigm in order to increase the HIP scalability, and to improve its efficiency. Within the HIP, the appropriate set of indexing, searching, and acquiring data services is distinguished. The set consists of Source, Index, Registry, Authorization and Mediator services.

The Source web service performs several tasks within the platform: it maintains patient's personal data, makes medical documentation available and submits information about it. During processessing medical data, the Source service creates HL7 CDA compliant documents [55], which are meant to be stored locally. Therefore, the Source service is supposed to be located at an appropriate healthcare unit and closely integrated with its HIS system.

The HIP platform may contain several Source services, therefore there is a need to index all shared data. Thus, each Source service submits data concerning patients and documents to appropriate indexing services within the platform. Two such services are distinguished: Index and Registry.

The Index service plays a critical role in the proposed HIP Platform. It stores and provides information about patient's personal data. It is also responsible for merging data of the same patient that originate from various sources. With this service unavailable, it would be impossible to determine the identity of a new patient in the system, and thus, no new document related to that patient could be submitted. Beyond doubt, this is a service that requires additional reliability mechanisms.

The next service, the Registry, is a metadata manager and index of data locations, and it acts as a catalog for the Source services. Similarly to the Index, it is a centralized service used by many healthcare units. To ensure the extensibility of the proposed platform, many Registries may be built. The only assumption in that matter is that a single Source should be assigned to only one Registry. Due to personal data separation and protection requirements the Index and Registry could not be combined and need to operate as independent services.

Authorization is a simple web service with tools to manage the permission data required to secure the access to the platform and protect information from an unauthorized use. The service enables a patient to grant access rights to selected information

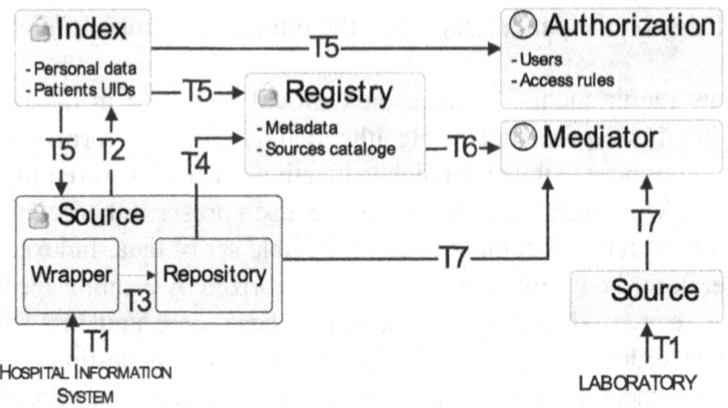

Fig. 14 Healthcare integration platform architecture

of users, clinicians or the whole medical units. Access rights may be granted permanently or for specified time and restrict access only to a selected Source and/or type of data.

Finally, the Mediator service is an entry point to the data shared through the platform. It defines an API for the applications connecting with the platform, and performs requests by appropriate interaction with other services to search and view medical data.

Fig. 14 shows the interactions among services belonging to the HIP platform. A description of these interactions follows.

The Source service obtains data from medical units with the use of a dedicated wrapper. Its task is to extract data from a specific type of HIS (T1) and to convert it into documents in an adopted format. As mentioned earlier, a document is an indivisible part of data exchanged within the platform and its format is consistent with HL7 CDA with partially structured data as a minimum level of detail.

Furthermore, the Source service submits information about new patients (T2) to the Index service in order to identify them among different healthcare providers. The information is sent with the use of HL7 compliant messages, which ensures the interoperability with many information systems.

The Source service may act either as an independent document repository or a mediator in retrieving documents directly from HIS inside a healthcare facility (T3) in order to ensure accessibility of patient's data. For each new document, the service is responsible for generating metadata and submitting it to the Registry service (T4). Such a process of managing documentation is based on the IHE XDS profile [56].

The Index service is required to cross-reference patient's identity and bind his/her documents from different facilities.The service provides consistency of such identification by sharing UIDs among all the services within the platform (T5). Managing patients identity cross-referencing is based on IHE PIX profile [57].

The Registry service is used to gather all metadata about desirable documents (T6), in particular the addresses of the Source services. The Sources services are in turn used to retrieve document sets from them (T7). Last but not least, the retrieved documents are verified against rules provided by the Authorization service.

9.2 Medical Event and Data Registering Platform

Medical Event and Data Registering Platform (MEDReP) improves the daily work of physicians and nurses working in clinics, hospitals, and engaged in private medical practices, by providing the functionality that allows medical staff to schedule patients' visits, and supports the overall process and documentation. Moreover, MEDReP allows staff to record patients' personal data, the history of their diseases, and any other text data defined by users. The platform also supports the management of the healthcare unit structure and resources.

Medical Event and Data Registering Platform is composed of Management, EHR Directory, and Scheduler services, which implement MEDReP's business logic. It supports various platforms from desktop PC to mobile tablets by providing thin client-side applications. Moreover, MEDReP can be integrated with external systems in order to export/import data.

Management service defines the healthcare unit structure, human resources, and stores administrative data. The service is responsible for managing both these data, and access to them. Therefore, the service records patient's personal data, contact numbers, and insurance information, as well as the information on the medical staff (physicians, nurses, midwives), their medical specialties, and the structure of their employment (work at various clinics and healthcare units). Since each patient is under the medical care of a given physician, they register at appropriate healthcare units, and fulfill a declaration, which determines physicians involved in the patients' treatment. Such a relationship is also recorded in the Management service. As a result, only part of the medical staff have access to personal and medical patients' data. When a chosen physician is absent, his/her replacement also needs access rights to the patient's data. The fact of substitution and assigning temporary access rights to another physician in the case of a doctor's absence is also recorded in the Management service.

Another MEDReP's service is the EHR Directory service (EHRDir), which stores the medical data, among which are patients' EHR and detailed information on patients' visits. The patient's medical data is stored in the form of entries defined by physicians. The platform provides default templates to store EHR records, but the templates can be also redefined and extended in order to allow collecting all necessary data. Additionally, with each single visit, extra data can be stored, such as issued prescriptions and referrals to specialists or a list of conducted medical treatments. Based on these data, non-modifiable documents can be created to record the conducted medical treatment. The EHRDir service also provides medical dictionaries, such as registry of medical procedures (ICD-9-CM), diagnoses (ICD-10) or dictionary of drug names and their dosages.

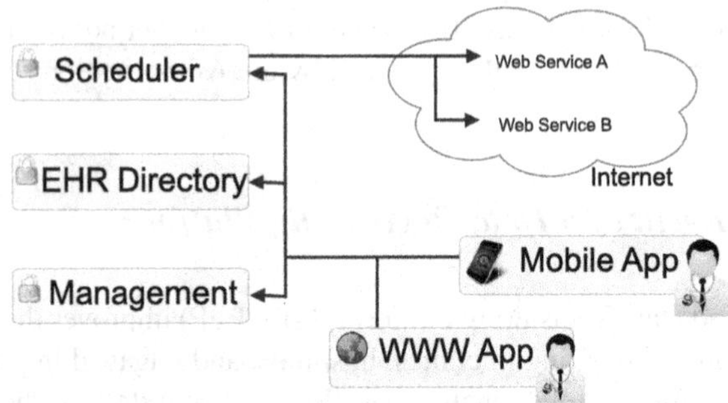

Fig. 15 Medical event and data registering platform

Finally, the Scheduler service provides an electronic calendar. The calendar stores information about the clinic opening hours and admission hours of individual physicians, and is used to plan visits of individual patients. Fig. 15 presents the overall architecture of Medical Event and Data Registering Platform and shows the connections between the above described services.

The services of MEDReP platform are designed according to the REST approach. They easily integrate with other services and systems, and facilitate the adaptation of client applications so that these applications can use and benefit from the proposed platform. Two reference client applications were developed along with the MEDReP platform: a web application, and a mobile application, designed for devices running the Android system.

The web application provides the full functionality of the MEDReP platform. It is designed for administrators, registration employees, and medical staff. Depending on role, the client possesses, the application provides various options of MEDReP.

The mobile application was created mainly for medical staff performing their duties outside the office. With the use of the proposed application physicians, nurses or midwives obtain the access to the system resources during home visits. The application allows them to browse the patient's medical history (in an offline mode—without the need to connect to the Internet), and to fill the patient's medical records at the visit. The data from the device automatically synchronize with the MEDReP platform, so when the physician or another medical staff member returns to the healthcare unit office, he/she has an immediate preview of the information (also the option to correct and extend it).

9.3 Dependability Issues

The previous chapters presented various mechanisms and solutions that increase dependability of SOA-based IT systems. Healthcare is a good example of a domain where dependability plays a crucial role. Both platforms characterized above use

ReSP and DyMST, described in the previous sections, in order to increase dependability by means of the mechanisms offered by these tools.

With reference to medical data, the availability of data and access time are very important factors. Certain services should operate in the 24/7/365 manner, which requires the data to be available all the time without any disturbances. In order to increase the efficiency and availability of Healthcare Integration Platform, the replication mechanism described in Sect. 5 is applied. Since the Mediator service issues the short-term read operations to Registry, and long operations associated with network communication, service is replicated at the server level. As a result, the user queries are processed faster and the availability of Registry service is increased.

The security requirements in case of sharing medical data, often described as fragile data, are very high. Moreover, the HIP internal services must exchange messages in secure way. As described in Sect. 8, ORCA Framework offers lightweight information flow control mechanism easily managed in the SOA-based system by using a security policy language. What is worth emphasizing, by using ORCA the services retain their autonomy. By using this framework it was possible to separate the security mechanism from a business logic. Defined security policies precisely describe the restrictions, obligations and capabilities of each service and each role.

In this case, to attain high reliability of healthcare systems, the mechanisms that allow management and failure detection should be provided to the end user. Therefore, HIP internal services are monitored and managed using a DyMST platform. DyMST allows administrators to monitor the status of every single system component, and to react quickly in the case of its failure. Due to using sensors and effectors described in Sect. 7, the configuration and maintenance of the platform is much easier and more accurate in terms of QoS. Failure detection service FADE is used to monitor crucial HIP internal services, thus in case of failure the proper corrective actions are implemented.

The services of MEDReP platform should ensure high data integrity, availability, reliability and consistency. To this end, the MEDReP platform can be integrated with the modules of ReSP and DyMST tools as stated above.

The MEDReP platform is used to collect and record the medical data. In general, such information is relevant and sensitive (e.g. the results of specific medical results), and often difficult or even impossible to reproduce. Therefore, the loss of such data may not only impede the process of diagnosis and future medical treatment of the patient in the case of his/her illness, but also expose the patient to a danger of multiple non-indifferent performance of medical examinations (e.g. taking X-ray pictures several times). Hence, it is necessary to ensure that the data recorded in the MEDReP platform are neither destroyed, nor lost in the case of failure of MEDReP's individual components.

In order to support reliable recording of medical data by the MEDReP platform, the ReServE service can be used. ReServE stores the operations performed in the system, and enables MEDReP to recover the state of services in case they fail. In order to take advantage of the ReServE service, each service of MEDReP's platform should be equipped with the CIM and SIM modules, and the client-side applications (both web and mobile) should be provided with CIMs, as described in Sect. 6.

Medical data in the MEDReP platform can be found in three services: Management service (organizational data), EHRDir service (EHR data) and Scheduler service (work plans, schedule of patient visits). Each MEDReP service is an independent component, but in order to obtain complete information, it is necessary to combine data from two (or even three) services. As an example, let us consider the patient's medical history. The medical data stored in EHRDir does not contain any personal data (the patient is described only by an abstract identifier). Therefore to find out to which patient the given medical history belongs to, the data from Management and EHRDir services should be merged using the given identifier. Thus, to provide the reliability of MEDReP, all platform services should be correct.

On the other hand, operations performed on one service may result in calling other services (which we denote as a nested call), e.g. en entry can be added to the EHRDir service, and simultaneously the visit can be registered in the Schedule service. Also in such a situation all services of the platform should operate properly. The occurrence of failures causing loss of data stored in one of services could lead to an inconsistent state of the whole system. If the information on one of the visits is lost in the EHRDir, and at the same time, the Schedule service shows that such a visit took place, a system user will be confused. The ReServE service described in Sect. 6 can be used to address the problem of inconsistency of data stored in different services of MEDReP platform.

Medical work requires constant access to current medical information related with providing medical care to patients. In this context, it is vital to ensure constant, high availability and performance of services provided by the MEDReP platform. In order to ensure these features, a ReREST module of ReSP is used. The replication mechanism enables MEDReP to duplicate services, while maintaining their state consistency. Therefore, failures of individual replicas will not affect the reliability and efficiency of the system. The usage of ReREST in context of the MEDReP platform requires each replica to have a proxy server. However, there is no need for a central management service. The Management service has a dominant number of read operations. Since the majority of data saved in the Management service do not change, in order to increase the availability of services of such characteristics, the optimistic replication of Management service is used.

10 Summary

In this chapter, we have discussed the problem of increasing dependability of SOA-based systems. Since dependability issues are particularly important and imperative in SOA, we have proposed two toolkits: the Reliable SOA Platform (ReSP), and the Dynamic Management SOA Toolkit (DyMST), which provide solutions to improve selected aspects of dependabilityin the considered SOA-based environments.

ReSP is aimed at improving availability, reliability, and integrity of service-oriented systems and applications, by using the mechanisms of group communication, replication, checkpointing, rollback-recovery, and transactional processing techniques. Although these mechanisms are well-known and widely used in general distributed systems, in the proposed solution we tailored them in order to take into account the specific features of SOA systems. In turn, Dynamic Management SOA Toolkit (DyMST) monitors and controls crucial components of SOA-based systems. DyMST consists of components which provide failure detection, monitoring and autonomic management, and distributed security policy enforcement.

In the chapter we have illustrated the practical use of the proposed ReSP and DyMST solutions in the context of medical healthcare systems: the *Healthcare Integration Platform* and the *Medical Event, and Data Registering Platform*.

References

1. Aguilera, M.K., Merchant, A., Shah, M., Veitch, A., Karamanolis, C.: Sinfonia: a new paradigm for building scalable distributed systems. In: Proceedings of SOSP' 07: The 21st ACM Symposium on Operating Systems Principles (SOSP), pp. 159–174 (2007)
2. Alvisi, L., Marzullo, K.: Message logging: pessimistic, optimistic, causal and optimal. IEEE Trans. Softw. Eng **24**(2), 149–159 (1998)
3. Alwagait, E., Ghandeharizadeh, S.: Dew: a dependable web services framework. In: Proceedings of the 14th International Workshop on Research Issues on Data Engineering: Web Services for e-Commerce and e-Government Applications (2004)
4. Amir, Y., Stanton, J.: The Spread wide area group communication system. Technical Report, CNDS-98-4, Department of Computer Science, Johns Hopkins University (1998)
5. Avizienis, A., Laprie, J.C., Randell, B., Landwehr, C.: Basic concepts and taxonomy of dependable and secure computing. IEEE Trans. Dependable and Secure Comput. **1**(1), 11–33 (2004). doi:10.1109/TDSC.2004.2
6. Barga, R., Chen, S., Lomet, D.: Improving logging and recovery performance in phoenix/APP. In: International Conference on Data Engineering (ICDE) (2004)
7. Barga, R.S., Lomet, D.B., Shegalov, G., Weikum, G.: Recovery guarantees for internet applications. ACM Trans. Internet Technol. **4**(3), 289–328 (2004)
8. Bazydło, M., Francuzik, S., Sobaniec, C., Wawrzyniak, D.: Combining optimistic and pessimistic replication. In: Proceedings of the 9th International Conference on Parallel Processing and Applied Mathematics (PPAM 2011). Lecture Notes in Computer Science, vol. 7203, pp. 20–29. Springer, Toru (2012)
9. Becker, M.Y., Fournet, C., Gordon, A.D.: SecPAL: design and semantics of a decentralized authorization language. Technical Report, MSR-TR-2006-120, Microsoft Research (2006)
10. Bernstein, P.A., Newcomer, E.: Principles of Transaction Processing. Morgan Kaufmann, Amsterdam (2009)
11. Borg, A., Baumbach, J., Glazer., S.: A message system supporting faulttolerance. In: Proceedings of the Symposium on Operating Systems Principles, pp. 90–99 (1983)
12. Brewer, E.: CAP twelve years later: How the "rules" have changed. Computer **45**(2), 23–29 (2012). doi:10.1109/MC.2012.37
13. Brewer, E.A.: Towards robust distributed systems (abstract). In: PODC '00: Proceedings of the Nineteenth Annual ACM Symposium on Principles of Distributed Computing, p. 7. ACM, New York (2000). http://doi.acm.org/10.1145/343477.343502
14. Brodecki, B., Brzeziński, J., Dwornikowski, D., Kobusiński, J., Sajkowski, M., Sasak, P., Szychowiak, M.: Selected aspects of management in SOA. In: Ambroszkiewicz, S., Brzeziński, J., Cellary, W., Grzech, A., Zieliński K. (eds.) SOA Infrastructure Tools: Concepts and Methods. UEP (2010)

15. Brodecki, B., Brzeziński, J., Sasak, P., Szychowiak, M.: Modality conflict discovery for SOA security policies. In: Temam, O., Yew, P.C., Zang B. (eds.) Advanced Parallel Processing Technologies 2011 (APPT 2011), Lecture Notes in Computer Science, vol. 6965, pp. 112–126. Springer, Shanghai (2011). doi:10.1007/978-3-642-24151-2-9
16. Brodecki, B., Brzeziński, J., Sasak, P., Szychowiak, M.: ModCon algorithm for discovering security policy conflicts. In: 6th Joint Workshop on Information Security—JWIS 2011. Kaohsiung, Taiwan (2011)
17. Brodecki, B., Brzeziński, J., Sasak, P., Szychowiak, M.: Consistency maintenance of modern security policies. In: Thilagam, P.S., Pais, A.R., Chandrasekaran, K., Balakrishnan N. (eds.) Advanced Computing, Networking and Security (ADCONS 2011), Lecture Notes in Computer Science, vol. 7135, pp. 472–477. Springer, Mangalore (2012). doi:10.1007/978-3-642-29280-4-55
18. Brodecki, B., Kalewski, M., Sasak, P., Szychowiak, M.: Lightweight information flow control for web services. In: Wyrzykowski, R., Dongarra, J., Karczewski, K., Wasniewski J. (eds.) Proceedings of the 9th International Conference on Parallel Processing and Applied Mathematics (PPAM 2011), vol. 7204, pp. 608–617. Springer, Toruń (2012). doi:10.1007/978-3-642-31500-8-63
19. Brodecki, B., Sasak, P., Szychowiak, M.: Security policy definition framework for SOA-based systems. In: Vossen, G., Long, D.D.E., Yu J.X. (eds.) 10th International Conference on Web Information Systems Engineering (WISE 2009), Lecture Notes in Computer Science, vol. 5802, pp. 589–596. Springer, Poznań (2009). doi:10.1007/978-3-642-04409-0-57
20. Brzeziński, J., Dwornikowski, D., Kalewski, M., Sajkowski, M., Pawlak, T.: MDL: metrics definition language. In: Nguyen, N.T., Kim, C.G., Janiak A. (eds.) Proceedings of 3rd International Conference, ACIIDS 2011, Lecture Notes in Artificial Intelligence, vol. 6591/2011, pp. 248–257. Springer, Poznań (2011)
21. Brzeziński, J., Dwornikowski, D., Kobusiński, J.: FADE: RESTful service for failure detection in SOA environment. In: Malyshkin, V. (ed.) Parallel Computing Technologies, Lecture Notes in Computer Science, vol. 6873, pp. 238–243. Springer, Berlin (2011)
22. Brzeziński, J., Dwornikowski, D., Sajkowski, M.: Koncepcje zarządzania w SOA. In: Niemir, D., Stroiński, M., Węglarz J. (eds.) Nauka w obliczu społeczeństwa cyfrowego. I Konferencja i3: internet—infrastruktury—innowacje, pp. 222–232. Ośrodek Wydawnictw Naukowych (2010)
23. Budhiraja, N., Marzullo, K., Schneider, F.B., Toueg, S.: The primary-backup approach. In: Mullender, S. (ed.) Distributed Systems, 2nd edn., pp. 199–216. Addison-Wesley, Reading (1993)
24. Carlyle, B.: The REST statelessness constraint. http://soundadvice.id.au/blog/2009/06/13/#stateless (2009)
25. Chandra, T.D., Toueg, S.: Unreliable failure detectors for reliable distributed systems. J. ACM **43**(2), 225–267 (1996)
26. Chen, Y.: WS-mediator for improving dependability of service composition. Ph.D. thesis, Newcastle University, Newcastle upon Tyne, UK (2008)
27. Chen, J.Y., Wang, Y.J., Xiao, Y.: SOA-based service recovery framework. In: Proceedings of the 9th International Conference on Web-Age, Information Management, pp. 629–635 (2008)
28. Damianou, N.C., Dulay, N., Lupu, E., Sloman, M.: Ponder: A language for specifying security and management policies for distributed system. Technical Report, Imperial College of Science, Technology and Medicine; Department of Computing, London (2000)
29. Danilecki, A., Hołenko, M., Kobusińska, A., Szychowiak, M., Zierhoffer, P.: The reliability service for service oriented architectures. In: Proceedings of the 3rd Workshop on Design for Reliability (DFR'11), pp. 33–38. Heraklion, Crete (2011)
30. Danilecki, A., Hołenko, M., Kobusińska, A., Szychowiak, M., Zierhoffer, P.: ReServE service: An approach to increase reliability in service oriented systems. In: Malyshkin, V. (ed.) Proceedings of the 11th International Conference on Parallel Computing Technologies, Lecture Notes in Computer Science, vol. 6873, pp. 244–256. Springer, Berlin (2011)

31. Danilecki, A., Kobusińska, A.: Message logging for external support of web services recovery. In: Proceedings of the IADIS International Conference on Collaborative Technologies, pp. 199–203. Freiburg, Germany (2010)
32. Davidson, S.B., Garcia-Molina, H., Skeen, D.: Consistency in partitioned networks. ACM Comput. Surv. **17**(3), 341–370 (1985)
33. Denning, D.E.: A lattice model of secure information flow. Commun. ACM **19**, 236–243 (1976). http://doi.acm.org/10.1145/360051.360056
34. Dialani, V., Miles, S., Moreau, L., Roure, D.D., Luck, M.: Transparent fault tolerance for web services based architectures. In: Proceedings of the 8th International Euro-Par Conference (Euro-Par 2002), Lecture Notes in Computer Science, vol. 2400, pp. 889–898. Springer, Paderborn (2002)
35. Dustdar, S., Juszczyk, L.: Dynamic replication and synchronization of web services for high availability in mobile ad-hoc networks. SOCA **1**(1), 19–33 (2007)
36. Dwornikowski, D., Kobusińska, A., Kobusiński, J.: Failure detection in a RESTful way. In: Proceedings of the 11th International Conference on Parallel Processing and Applied Mathematics, Lecture Notes in Computer Science. Springer, Toruń (2012)
37. Moses T. (ed): eXtensible Access Control Markup Language (XACML) version 2.0, OASIS Open (2005)
38. Edstrom, J., Tilevich, E.: Reusable and extensible fault tolerance for RESTful applications. In: 2012 IEEE 11th International Conference on Trust, Security and Privacy in Computing and Communications (TrustCom), pp. 737–744 (2012). doi:10.1109/TrustCom.2012.244
39. Elmootazbellah, N., Elnozahy, L.A., Wang, Y.M., Johnson, D.: A survey of rollback-recovery protocols in message-passing systems. ACM Comput. Surv. **34**(3), 375–408 (2002)
40. Fang, C.L., Liang, D., Lin, F., Lin, C.C.: Fault tolerant web services. J. Syst. Architect. **53**(1), 21–38 (2007). doi:10.1016/j.sysarc.2006.06.001. http://www.sciencedirect.com/science/article/pii/S1383762106000609
41. Fielding, R., Gettys, J., Mogul, J., Frystyk, H., Masinter, L., Leach, P., Berners-Lee, T.: Hypertext Transfer Protocol—HTTP/1.1. Internet Engineering Task Force (1999). RFC 2616 (Draft Standard). Updated by RFC 2817
42. Fielding, R.T.: Architectural styles and the design of network-based software architectures. Ph.D. thesis, University of California, Irvine (2000)
43. Fielding, R.T., Taylor, R.N.: Principled design of the modern Web architecture. ACM Trans. Internet Technol. (TOIT) **2**(2), 115–150 (2002)
44. Friedman, R., Vitenberg, R., Chockler, G.: On the composability of consistency conditions. Inf. Process. Lett. **86**(4), 169–176 (2003)
45. Gilbert, S., Lynch, N.: Brewer's conjecture and the feasibility of consistent, available, partition-tolerant web services. SIGACT News **33**(2), 51–59 (2002). http://doi.acm.org/10.1145/564585.564601
46. Goguen, J.A., Meseguer, J.: Security policies and security models. In: Proceedings of the IEEE Symposium on Security and Privacy (1982). http://doi.ieeecomputersociety.org/10.1109/SP.1982.10014
47. Gonalves, E., Rubira, C.: Archmeds: An infrastructure for dependable service-oriented architectures. In: 2010 17th IEEE International Conference and Workshops on Engineering of Computer Based Systems (ECBS), pp. 371–378 (2010). doi:10.1109/ECBS.2010.51
48. Guerraoui, R., Rodrigues, L.: Introduction to Distributed Algorithms. Springer, Heidelberg (2004)
49. Gurevich, Y., Soparkar, N., Wallace, C.: Formalizing database recovery. J. Univs. Comput. Sci. **3**(4), 320–340 (1997)
50. Hardy, N.: The confused deputy. ACM SIGOPS Oper. Syst. Rev. **22**(4), 36–38 (1988). doi:10.1145/54289.871709
51. Haridas, J., Nilakantan, N., Calder, B.: Windows azure table. Microsoft (2009)
52. Hayashibara, N., Défago, X., Yared, R., Katayama, T.: The φ accrual failure detector. In: Proceedings of the 23rd IEEE International Symposium on Reliable Distributed Systems (SRDS '04), pp. 66–78. IEEE Computer Society (2004). http://csdl.computer.org/comp/proceedings/srds/2004/2239/00/22390066abs.htm

53. He, W.: Recovery in web service applications. IEEE International Conference on E-Technology, E-Commerce, and E-Services, pp. 25–28 (2004). http://doi.ieeecomputersociety.org/10.1109/EEE.2004.1287284
54. ISO 20514:2005: Health informatics—Electronic health record—Definition, scope, and context. International Organization for Standardization, Geneva, Switzerland
55. Health Level Seven International. http://www.hl7.org/
56. IHE IT infrastructure technical framework, vol. 1, Integration Profiles (2009a)
57. IHE IT infrastructure technical framework, vol. 2a, Transactions Part A (2009b)
58. Integrating the healthcare enterprise. http://www.ihe.net/
59. IT-SOA research network: Atomic REST. http://www.it-soa.eu/atomicrest (2011)
60. IT-SOA research network: MProxy—modular proxy server (2011). http://www.it-soa.eu/mproxy/
61. IT-SOA research network: RESTGroups. http://www.it-soa.eu/restgroups (2011)
62. Jayasinghe, D.: FAWS for SOAP-based web services (2005). http://www.ibm.com/developerworks/webservices/library/ws-faws/
63. Johnson, D., Zwaenepoel, W.: Sender-based message logging. 17th Annual International Symposium on Fault-Tolerant, Computing, pp. 14–19 (1987)
64. Johnson, D., Zwaenepoel, W.: Recovery in distributed systems using optimistic message logging and checkpointing. J. Algorithms **11**, 462–491 (1990)
65. Juang, T., Venkatesan, S.: Crash recovery with little overhead. In: Proceedings of the 11th Conference on Distributed, Computing Systems, pp. 454–461 (1991)
66. Khare, R., Taylor, R.N.: Extending the representational state transfer (REST) architectural style for decentralized systems. In: ICSE '04: Proceedings of the 26th International Conference on Software Engineering, pp. 428–437. IEEE Computer Society, Washington (2004)
67. Kim, J.L., Park, T.: An efficient protocol for checkpointing recovery in distributed systems. IEEE Trans. Parallel Distrib. Syst. **4**(8), 955–960 (1993)
68. Kobus, T., Wojciechowski, P.T.: A 90 % RESTful group communication service. In: Proceedings of DCDP '10: The 1st International Workshop on Decentralized Coordination of Distributed Processes, Amsterdam, The Netherlands (2010). An extended abstract appeared in EPTCS, vol. 27. A full version published as Technical Report RA-2/10, Institute of Computing Science, Poznań University of Technology
69. Kobus, T., Wojciechowski, P.T.: RESTGroups for resilient Web services. In: Proceedings of SOFSEM '12: The 38th International Conference on Current Trends in Theory and Practice of Computer Science: Software & Web Engineering Track, vol. 7147, LNCS, pp. 505–517 (2012)
70. Kochman, S., Wojciechowski, P.T., Kmieciak, M.: Batched transactions for RESTful web services. In: Harth, A., Koch N. (eds.) Proceedings of Composable Web '11: The 3rd International Workshop on Lightweight Integration on the Web (co-located with ICWE '11: The 11th International Conference on Web Engineering, Paphos, Cyprus), Lecture Notes in Computer Science, vol. 7059, pp. 86–98. Springer, Berlin (2011)
71. Koo, R., Toueg, S.: Checkpointing and rollback-recovery for distributed systems. IEEE Trans. Softw. Eng. **13**(1), 23–31 (1987)
72. Li, J., Karp, A.H.: Access control for the services oriented architecture. In: Proceedings of the 2007 ACM Workshop on Secure Web Services, pp. 9–17. ACM (2007). doi:10.1145/1314418.1314421
73. Li, W., He, J., Ma, Q., Yen, I.L., Bastani, F., Paul, R.: A framework to support survivable web services. In: IPDPS '05: Proceedings of the 19th IEEE International Parallel and Distributed Processing Symposium (IPDPS'05)—Papers, p. 93.2. IEEE Computer Society, Washington (2005). http://dx.doi.org/10.1109/IPDPS.2005.27
74. Liang, D., Fang, C.L., Chen, C., Lin, F.: Fault tolerant web service. In: Proceedings of 10th Asia-Pacific Software Engineering Conference, pp. 310–319 (2003). doi:10.1109/APSEC.2003.1254385
75. Liu, L., Wu, Z., Ma, Z., Wei, W.: A fault-tolerant framework for web services. In: WRI World Congress on Software Engineering, WCSE '09, vol. 3, pp. 138–142 (2009). doi:10.1109/WCSE.2009.211

76. Mannivanan, D., Netzer, R., Singhal, M.: Finding consistent global checkpoints in a distributed computation. IEEE Trans. Parallel Distrib. Syst. **8**(6), 623–627 (1997)
77. Marinos, A., Razavi, A., Moschoyiannis, S., Krause, P.: RETRO: A consistent and recoverable RESTful transaction model. In: Proceedings of ICWS '09: The 7th IEEE International Conference on Web Services (2009)
78. Maymounkov, P., Mazières, D.: Kademlia: A peer-to-peer information system based on the xor metric. In: Revised Papers from the First International Workshop on Peer-to-Peer Systems, IPTPS '01, pp. 53–65. Springer, London (2002)
79. Mena, S., Schiper, A., Wojciechowski, P.T.: A step towards a new generation of group communication systems. In: Endler, M., Schmidt D. (eds.) Proceedings of Middleware '03: The 4th ACM/IFIP/USENIX International Middleware Conference (Rio de Janeiro, Brazil), Lecture Notes in Computer Science, vol. 2672, pp. 414–432. Springer, Heidelberg (2003)
80. Merideth, M.G., Iyengar, A., Mikalsen, T.A., Tai, S., Rouvellou, I., Narasimhan, P.: Thema: byzantine-fault-tolerant middleware for web-service applications. In: SRDS, pp. 131–142. IEEE Computer Society (2005). http://dblp.uni-trier.de/db/conf/srds/srds2005.html#MeridethIMTRN05
81. Microsoft: Windows Azure—team blog. http://blogs.msdn.com/windowsazure (2008–11)
82. Musgrove, M.: Transactional support for JAX RS based applications. http://community.jboss.org/wiki/TransactionalsupportforJAXRSbasedapplications (2009)
83. Myers, A.C., Liskov, B.: Protecting privacy using the decentralized label model. ACM Trans. Softw. Eng. Methodol. **9**, 410–442 (2000). http://doi.acm.org/10.1145/363516.363526
84. OASIS: Web Services Base Notification 1.3 (WS-BaseNotification) (2006)
85. OASIS: Web Services Brokered Notification 1.3 (WS-BrokeredNotification) (2006)
86. OASIS: Web services reliable messaging (WS-ReliableMessaging) version 1.1 (2008).http://docs.oasis-open.org/ws-rx/wsrm/v1.1/wsrm.html
87. Oracle Corporation: Oracle BPEL Process Manager (2009). http://www.oracle.com/technology/products/ias/bpel/
88. Plank, J., Beck, M., Kingsley, G.: Compiler assisted memory exclusion for fast checkpointing, pp. 62–67. IEEE Technical Committee on Operating Systems Newsletter, Fault Tolerance (1995, special issue)
89. Randell, B.: System structure for software fault tolerance. IEEE Trans. Software Eng. **1**(2), 221–232 (1975)
90. Red Hat: JGroups—a toolkit for reliable multicast communication. http://www.jgroups.org/ (2009)
91. Richardson, L., Ruby, S.: RESTful Web Services. O'Reilly, Media (2007)
92. Rotem-Gal-Oz, A.: Transactions are bad for REST. http://www.rgoarchitects.com/nblog/2009/06/15/TransactionsAreBadForREST.aspx (2009)
93. Rotem-Gal-Oz, A., Bruno, E., Dahan, U.: SOA Patterns, chapter 5.4 Saga. Manning Publications Co., Greenwich (2007)
94. Salas, J., Perez-Sorrosal, F., Martínez, M.P., Jiménez-Peris, R.: WS-Replication: a framework for highly available web services (Edinburgh, Scotland). In: Proceedings of the 15th International Conference on World Wide Web, pp. 357–366. ACM Press, New York (2006)
95. Santos, G.T., Lung, L.C., Montez, C.: FTWeb: a fault tolerant infrastructure for web services. In: Proceedings of 9th IEEE International EDOC Enterprise Computing Conference, pp. 95–105 (2005)
96. Schneider, F.: Implementing fault tolerant services using the state machine approach: a tutorial. ACM Comput. Surv. **22**(4), 299–319 (1990)
97. She, W., Yen, I.L., Thuraisingham, B., Bertino, E.: The SCIFC model for information flow control in web service composition. In: Proceedings of the 2009 IEEE International Conference on Web Services, ICWS '09, pp. 1–8. IEEE Computer Society, Washington (2009). http://dx.doi.org/10.1109/ICWS.2009.13
98. Smith, G., Volpano, D.: Secure information flow in a multi-threaded imperative language. In: Proceedings of the 25th ACM SIGPLAN-SIGACT Symposium on Principles of Programming Languages, POPL '98, pp. 355–364. ACM, New York (1998). http://doi.acm.org/10.1145/268946.268975

99. Souza, J.L.R., Siqueira, F.: Providing dependability for web services. In: SAC '08: Proceedings of the 2008 ACM Symposium on Applied Computing, pp. 2207–2211. ACM, Fortaleza (2008)
100. Spread Concepts LLC: The Spread toolkit. http://www.spread.org/ (2006)
101. Storm, R., Yemini, S.: Optimistic recovery in distributed systems. ACM Trans. Comput. Syst. **3**(3), 204–226 (1985)
102. Venkatesan, S.: Efficient algorithms for optimistic crash recovery. Distrib. Comput. **8**(2), 105–114 (1994)
103. Venkatesan, S., Juang, T., Alagar, S.: Optimistic crash recovery without changing application messages. IEEE Trans. Parallel Distrib. Syst. **8**(3), 263–271 (1997)
104. Volpano, D.M., Smith, G.: A type-based approach to program security. In: Proceedings of the 7th International Joint Conference CAAP/FASE on Theory and Practice of Software Development, TAPSOFT '97, pp. 607–621. Springer, London (1997)
105. Wang, M., Fuchs, W.: Optimistic message logging for independent checkpointing in message-passing systems. In: Proceedings of the Symposium on Reliable Distributed Systems, pp. 147–154 (1992)
106. Wang, Y.M.: Maximum and minimum consistent global checkpoints and their applications. In: Proceedings of IEEE the 14th Symposium on Reliable Distributed Systems, pp. 86–95 (1995)
107. Wang, Y.M.: Consistent global checkpoints that contain a given set of local checkpoints. IEEE Trans. Comput. **46**(4), 456–468 (1997). http://dx.doi.org/10.1109/12.588059
108. Wang, Y.M.: Consistent global checkpoints that contain a given set of local checkpoints. IEEE Trans. Comput. **46**(4), 456–468 (1997)
109. Wiesmann, M., Pedone, F., Schiper, A., Kemme, B., Alonso, G.: Understanding replication in databases and distributed systems. In: Proceedings of ICDCS '00: the 20th IEEE International Conference on Distributed, Computing Systems, pp. 464–474 (2000)
110. Ye, X., Shen, Y.: A middleware for replicated web services. In: Proceedings of the IEEE International Conference on Web Services (ICWS'05), pp. 631–638. IEEE Computer Society, Washington (2005)
111. Yildiz, U., Godard, C.: Information flow control with decentralized service composition. In: 2009 IEEE International Conference on Web Services (ICWS 2009), pp. 9–17 (2007)
112. Yin, J., Chen, H., Deng, S., Wu, Z., Pu, C.: A dependable esb framework for service integration. Internet Computing, IEEE **13**(2), 26–34 (2009). doi:10.1109/MIC.2009.26
113. Zhao, W.: A lightweight fault tolerance framework for web services. In: Proceedings of IEEE/WIC/ACM International Conference on Web, Intelligence, pp. 542–548 (2007)
114. Zhao, W., Zhang, H., Chai, H.: A lightweight fault tolerance framework for web services. Web Intell. Agent Syst. **7**(3), 255–268 (2009). doi:10.3233/WIA-2009-0167, http://dx.doi.org/10.3233/WIA-2009-0167

Chapter 6
Implementation, Deployment and Governance of SOA Adaptive Systems

R. Brzoza-Woch, Ł. Czekierda, J. Długopolski, P. Nawrocki, M. Psiuk, T. Szydło, W. Zaborowski, K. Zieliński and D. Żmuda

Abstract This chapter introduces a pragmatic methodology for adding and managing adaptability in multiple layers of the SOA application execution infrastructure. Adaptability mechanisms and techniques are investigated by referring to the MAPE-K pattern, which is viewed as the most representative solution for adaptive and autonomous systems. The SOA solution stack developed by IBM is selected as the basis for the application execution infrastructure model. This makes the proposed concepts easier to understand, while not detracting from their general nature. The adaptability aspect is considered in a broad context, with attempts to address, in a uniform way, all SOA applications composed of software services (Virtual Services) and hardware components (Real World Services). The proposed methology is supported by the AS3 Studio package which is a complete suite of tools providing extensions of SOA systems with adaptability features. This methodology is presented as a crucial part of the governance process of SOA applications. Finally, a case study which illustrates the proposed approach is described.

1 Introduction

Service-oriented applications operate in dynamic business environments. These applications should therefore become highly flexible and adaptive, as they need to adequately identify and react to various changes observed in the execution environment [78]. Adaptation is the relation between a system and its environ-

R. Brzoza-Woch · Ł. Czekierda · J. Długopolski · P. Nawrocki · M. Psiuk · T. Szydło · W. Zaborowski · K. Zieliński(✉) · D. Żmuda
Faculty of Computer Science, Electronics and Telecommunications,
Department of Computer Science, AGH University of Science and Technology,
al. A. Mickiewicza 30, 30-059 Krakow, Poland
e-mail: kz@agh.edu.pl

S. Ambroszkiewicz et al. (eds.), *Advanced SOA Tools and Applications*,
Studies in Computational Intelligence 499, DOI: 10.1007/978-3-642-38957-3_6,

ment where change is provoked to facilitate proper operation of the system in the environment.

The adaptation process of SOA applications should be investigated in the context of the widely accepted characteristics of service orientation proposed by Thomas Erl [20] which refer to the following eight principles:

- Standardized Service Contract,
- Service Loose Coupling,
- Service Abstraction,
- Service Reusability,
- Service Autonomy,
- Service Statelessness,
- Service Discoverability,
- Service Composability.

It is evident that, taken together, these principles allow for easier adaptation of SOA applications by system integrators during the (re-)development process and at runtime. Standardized Service Contract principles mean that services within the same service inventory remain in compliance with the same contract design standards which simplifies applications integration. Moreover, the service contract is usually developed separately from the service logic and provides the sole means of accessing service functionality and resources. This allows for independent creation of a contract and service implementation, reducing unintentional dependencies between services. This feature is referred to as Service Loose Coupling and supports adaptability, since each service can be controlled separately. Service contracts contain only essential information about services, thus enabling their categorization. This principle, known as Service Abstraction, makes searching for suitable services more efficient. The Service Autonomy principle means that services should exercise a higher level of control over their underlying runtime execution environment, which implies more predictive behavior. The ability of services to function autonomously is achieved by reducing shared access to service resources and increasing their physical isolation. It simplifies adaptation performed on the resource allocation level. The autonomy of individual services is also especially important for adaptation of SOA applications performed at the service composition or integration level. This feature is complemented by the Statelessness principle, which implies that services should minimize resource consumption by deferring the management of state information when necessary. Additionally, services are supplemented with communication-oriented metadata with which they can be effectively discovered and interpreted. This is an important prerequisite of dynamic structural adaptation.

Service orientation principles allow for effective development of adaptive SOA applications. Unfortunately, SOA is mainly focused on design techniques which support developers in constructing services. It does not encompass runtime aspects of service operation, i.e. how to manage and maintain services. Without proper management business objectives cannot be met as it is impossible to specify goals and determine whether they are, in fact, reached.

SOA applications and services should satisfy Service Level Agreements despite changing execution conditions such as system load, number of users, etc. This is why loose coupling and composition features inherent in SOA should be extended to cover adaptive behavior [10], ensuring self-adaptation of the system at the runtime. This aspect is often referred to as compositional adaptation [77], enabling software to modify its structure and behavior dynamically in response to changes in its execution environment. Adaptive systems are a remedy for the complexity of computing environments, expressed not only by the number of connected hardware and software components but also by the growing space of configuration parameters and management strategies offered by middleware technologies and virtualized computational infrastructures. Better exploitation of modern IT infrastructures is a prerequisite of achieving the required QoS (Quality of Service) and end-user satisfaction, characterized by QoE (Quality of Experience).

Implementing an adaptive SOA system remains a challenging issue: the adaptation process requires suitable mechanisms to be built into applications or the execution environment itself. Satisfying service orientation principles means that self-adaption of SOA systems can be considered a self-contained aspect, introduced during development or at runtime. Adaptive applications can be developed in the process of transforming a preexisting application and hardware components (which does not involve changes in the application's business services). This is consistent with the fact that the investigated adaptive systems remain business-agnostic and instead focus on ensuring the required nonfunctional parameters.

SOA application development and deployment should be considered in the context of the SOA Solution Stack (S3) proposed by IBM [3], which provides a detailed architectural definition of SOA. This allows for clear separation of different adaptability mechanisms referencing particular layers of the S3 model and assigning the requested adaptability extensions.

The goal of this chapter is to propose a pragmatic methodology for adding and managing adaptability aspects in multiple layers of the S3 stack, as well as to present the Adaptive S3 (AS3) Studio package which is a complete suite of tools supporting extensions of SOA systems with adaptability features. This methodology is considered part of the governance process of SOA applications. The adaptability mechanisms and techniques are investigated by referring to the MAPE-K pattern [1] as the most representative for adaptive and autonomous systems.

The adaptability aspect is considered in a broad context, with attempts to address, in a uniform way, all SOA applications composed of software services (also called Virtual Services) and hardware components (Real World Services) [35]. Such an approach is justified by the increasing importance of pervasive systems [67], bringing interaction from enterprise systems back to the real world. In this context the adaptive behavior of Real Word Services is considered a critical element, combining adaptive interaction, adaptive composition and task automation by involving knowledge of user profiles, intentions and previous usage patterns.

2 Development of Adaptive Systems: Motivation

This section describes the fundamentals of adaptive system development, referring to the state of the art in this domain. The presented analysis is performed in the context of the SOA Governance process definition and requirements to precisely identify the place and role of adaptive systems in SOA applications. Adaptability mechanisms are evaluated by taking into account service orientation principles to better illustrate the challenges involved in SOA adaptive system implementation.

According to [51] the goal of SOA Governance is to ensure reliable long-term operation of a SOA. More specifically, it provides the ability to guarantee SOA adaptability and integrity as well as check services for capability, security and strategic business alignment. Its overall goal is SOA Compliance, i.e. compliance with legal, normative and intra-company regulations respectively. SOA Governance includes the identification of a decision-making authority for the definition and modification of business processes that are supported by SOA and the requirements for service levels and performance including access rights to the services. It also defines the way how reusable services are defined, designed, accessed, executed, and maintained as well as the determination of service ownership and cost-allocation in a shared-service organization. This definition includes all crucial tasks and activities of SOA Governance and structures them into organizations, processes, policies and metrics. In addition, this definition covers all important aspects of IT Governance and specializes them in the context of SOA. SOA Governance defines the organizational structure of SOA and provides a way to implement it within an existing corporate structure.

Numerous models for SOA Governance have been proposed so far. Ten of them are investigated and compared in [51]. Each emphasizes different aspects, including service lifecycle management and organizational change [8]. The result of this study is the TEXO Governance Framework [32], compiled on the basis of the existing frameworks. The processes provided by this framework (depicted in Fig. 1) have been grouped into five phases: design, deployment, delivery, monitoring, and change.

The design phase covers all strategic aspects related to the operation of a service marketplace, enabling services to be purchased and traded. The development and deployment of services as well as selection of third-party services belong to the deployment phase. The delivery phase addresses all aspects of service and infrastructure operations. It is closely coupled with the monitoring phase as both phases occur concurrently. The monitoring phase covers all aspects of service and infrastructure monitoring. The change phase contains all processes and tasks required to adjust and change the infrastructure and services traded in the marketplace.

The presented division of governance phases leads to the conclusion that adaptation processes primarily concern the monitoring and change phases of the Governance Framework. The activities most relevant to the adaption process are highlighted in Fig. 1. Adaptation can be performed manually (by a system administrator) or automatically. System administrators may specify high-level policies which determine how the system should adjust its behavior at runtime in order to meet the specified requirements. Administrators are consequently relieved from dealing with low-level

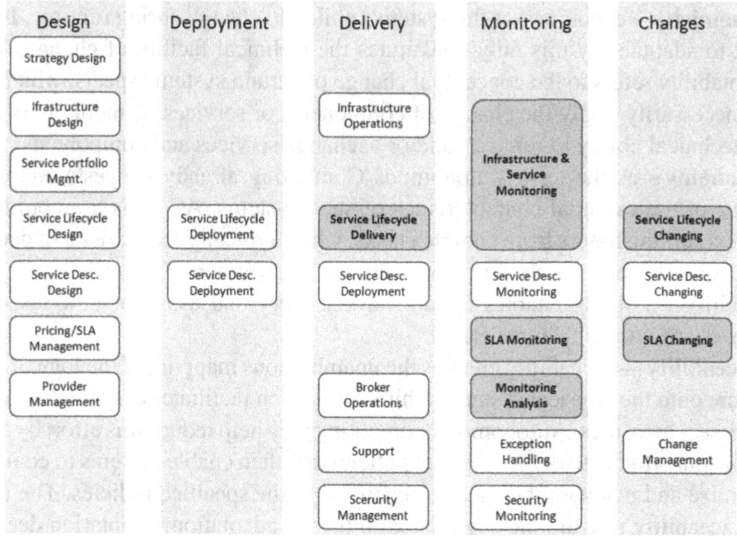

Fig. 1 SOA governance processes [32]

aspects of system operation. This goal is absolutely desirable, leading to adaptive systems which operate without (or with limited) human intervention. One of the emerging paradigms aimed at reducing effort involved in deploying and maintaining complex computer systems is called Autonomous Computing (AC) [37]. This paradigm might be applied in a very natural way to the development of adaptive SOA systems.

Autonomous Computing applications share some common properties [39] enabling them to properly apply this paradigm. Such properties aim to clarify the relation between adaptability and AC, and can be summarized as follows:

- **Adaptability**—the core concept behind adaptability is the general ability to change a system's observable behavior, structure or realization. This requirement is amplified by automatic adaptation that enables a system to decide about adaptation by itself, in contrast to ordinary adaptation, which is decided upon and triggered by the system's environment (e.g. users or administrators).
- **Awareness**—closely related to the adaptation and the execution context. It is a prerequisite of automatic adaptation. The term "context" is defined as sufficiently exact characterization of the situations in which a system might find itself by means of perceivable information relevant for the adaptation of the system. Awareness has two aspects: self-awareness (enabling a system to observe its own system model, state, etc.) and awareness of the environment.
- **Monitoring**—since monitoring is often regarded as a prerequisite of discovery and response to emerging events, it constitutes a system awareness. Monitoring indicates the system's state and thus characterizes a situation in which adaptation is necessary.

- **Dynamicity**—encompasses the system's ability to change during runtime. In contrast to adaptability this only constitutes the technical facility of change. While adaptability refers to the conceptual change of certain system aspects, which does not necessarily imply the change of components or services, dynamicity is about the technical ability to remove, add or exchange services and components.
- **Autonomy**—as the term Autonomous Computing already suggests, autonomy is one of the essential characteristics of such systems. AC aims at unburdening human administrators from complex tasks, which typically require a lot of decision making and problem solving without human intervention.
- **Mobility**—mobility enables dynamical discovery and usage of new resources, recovery of crucial features etc.
- **Traceability**—traceability enables the unambiguous mapping of the logical architecture onto the physical system architecture which facilitates easy deployment of necessary measures. Autonomous Computing may help reduce this effort by allowing administrators to define abstract policies and then enable systems to configure, optimize and maintain themselves according to the specified policies. The notion of traceability is again closely related to that of adaptation: adaptation decisions are also based on an abstract system model.

The relation between AC properties and the service orientation principles is presented in Table 1. The AC adaptability is supported by all principles, but most significantly by P2 and P4. AC monitoring is strongly related to P7—Service Discoverability, as it is a prerequisite of monitoring activity. AC dynamicity is very much related to P8 and P2. Service composability represents more flexible relation than integration and allows for services to be interconnected at runtime. The next AC property, mobility, is supported by P4 and partially by several other principles. This property is very much in line with service orientation. The same concerns AC Traceability which is supported to some extend by P3. Services have abstract descriptions which allows for abstract model construction and mapping of the logical system architecture onto the physical one. Unfortunately, support for specification of abstract policies which

Table 1 Autonomic Computing Properties vs. Service Orientation Principles

Autonomic Computing	Service Orientation Principles							
Properties	P1	P2	P3	P4	P5	P6	P7	P8
Adaptability	+	++	+	++	+	+	+	+
Awareness								
Monitoring	+						++	
Dynamicity		+			+	+	+	++
Autonomy								
Mobility	+	+		++		+	+	
Traceability		++						

P1-Standardized Service Contract, *P2*-Service Loose Coupling, *P3*-Service Abstraction, *P4*-Service Reusability, *P5*-Service Autonomy, *P6*-Service Statelessness, *P7*-Service Discoverability, *P8*-Service Composability

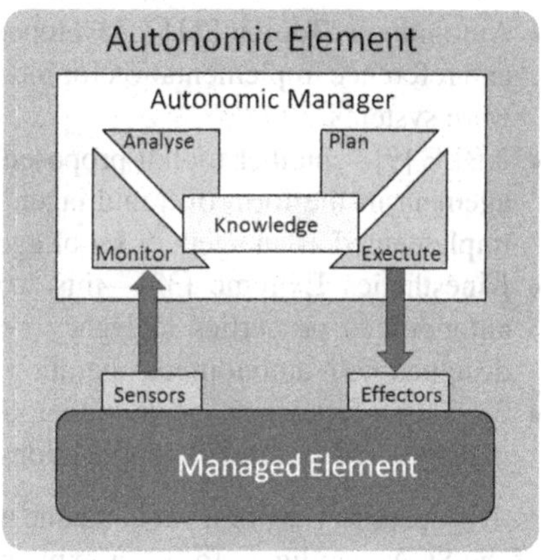

Fig. 2 IBM's MAPE-K reference model for autonomous control loops [1]

would enable systems to configure, optimize and maintain themselves according to the specified policies, is not offered directly by service orientation principles.

Two AC properties—Awareness and Autonomy—have no direct service orientation counterparts. Awareness is related to the ability of a system to observe its own system model, state, etc. This feature is not required by any service orientation principles. The AC Autonomy property has a different meaning than P5—Service Autonomy. While the former concept concerns the ability to make decisions without human intervention, P5 instead refers to isolation of the execution environment of a service. This consideration leads to an important observation, namely that implementation of SOA self-adaptive systems requires both AC Awareness and AC Autonomy. Support for AC Traceability should also be provided where possible.

To achieve autonomous computing IBM has suggested a reference model for autonomous control loops [1], which is sometimes called the MAPE-K (Monitor, Analyse, Plan, Execute, Knowledge) loop and is depicted in Fig. 2. This model is used to express the architectural aspects of autonomous systems.

In the MAPE-K autonomous loop the managed element represents any software or hardware resource that is given autonomous behaviour by coupling it with an autonomous manager. This element is equipped with: (i) Sensors, often called probes or gauges, which collect information about the managed element, and (ii) Effectors, which carry out changes in the managed element.

The data collected by the sensors allows the autonomous manager to monitor the managed element and execute changes through effectors. The autonomous manager is a software component that can be configured by human administrators using high-level goals and uses the monitored data from sensors and internal knowledge of the system to plan and execute the low-level actions deemed necessary to achieve high-level goals. The internal knowledge of the system is often an architectural model of the managed element. The goals are usually expressed using event-condition-action (ECA) policies, goal policies or utility function policies [78].

There are many diverse implementations of the MAPE-K loop:

- Autonomous Toolkit [31]—developed by IBM. It provides a practical framework and reference implementation for incorporating autonomous capabilities into software systems.
- ABLE [9]—another toolkit proposed by IBM which provides autonomous management in the form of a multiagent architecture: each autonomous manager is implemented as an agent or set of agents,
- Kinesthetics Extreme [33]—this work was driven by the problem of adding autonomous properties to legacy systems, i.e. existing systems that were not designed with autonomous operation in mind.
- 2K [40]—represents an autonomous middleware framework and offers self-management features for applications built on top of this framework.

It is necessary to point out that none of these systems addresses or takes advantage of the SOA paradigm. To better explain the technical aspects of the MAPE-K loop activities, we will consider them in more detail.

Monitoring

Monitoring involves capturing the managed element state or properties of the environment. Two types of monitoring can be distinguished: (i) Active Monitoring which requires instrumentation of software or hardware at some level, for example by modifying and adding code to the implementation of the application or the operating system in order to capture function or system calls; (ii) Passive Monitoring which relies on already built-in system capabilities for presenting information about their operation, e.g. system logs, load monitors, etc.

Analysis

Analysis is rather straightforward and depends on the planning activity specified below. It may concern data filtering or recognition of the managed element state.

Planning

Planning takes into account the monitoring data from sensors to produce a series of changes to be effected on the managed element. ECA rules, already mentioned in this section, that directly produce adaptation plans from specific event combinations, could be used in the simplest case. Rule-based planning determines the actions to take when an event occurs and certain conditions are met. This type of planning (referred to as Policy-Based Adaptation Planning) usually does not take into account system history and is therefore stateless. More advanced planning known as Architectural Models acknowledges a model of the system in the form of a connected component network. This allows users to ascribe constraints and properties to individual components and connectors. Violation of these constrains triggers adaptation actions. The third type of planning involves the Process-Coordination Approach where adaptation tasks result from defining the coordination of processes executed in the managed elements.

Execution

This activity does not merit special attention as it concerns technical issues related to execution of adaptation tasks. The managed element effectors are used for this purpose.

The presented analysis of self-adaptive systems design with special attention to AC Systems led to the definition of an adaptive systems space. This space covers three directions: (i) how the adaptation process is executed, (ii) where the adaptation mechanisms are located, (iii) when the adaptability mechanisms are added to the system. Definition of the adaptive systems space (depicted in Fig. 3) enables us to show where the adaptive SOA system investigated in this chapter can be located. The properties of such a space could be summarized as follows: the adaptation process is executed automatically; adaptation mechanisms are located in the middleware or infrastructure layer; adaptation mechanisms are added during deployment or at runtime.

The reference model which explains the role of Infrastructure and Middleware Layers in Adaptive SOA is shown in Fig. 4. The Infrastructure Layer is interpreted in a rather broad sense as it combines not only physical resources such as typical servers connected to computer networks, but also small physical devices (Real-World Devices) such as mobile phones, intelligent sensor networks, etc. with pre-installed

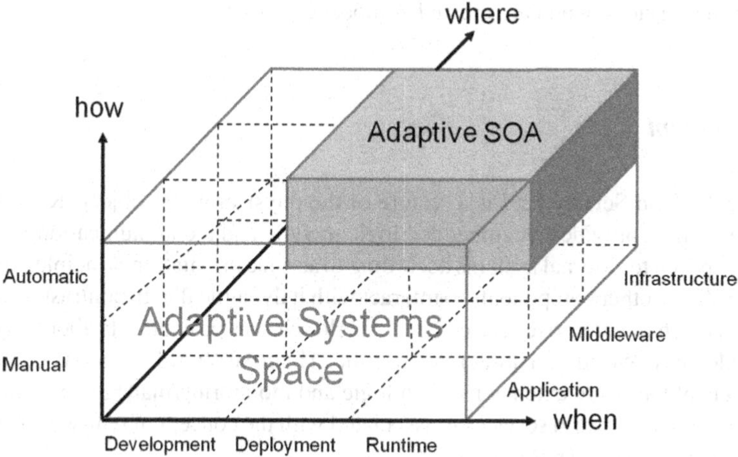

Fig. 3 Adaptive SOA vs Adaptive Systems Space

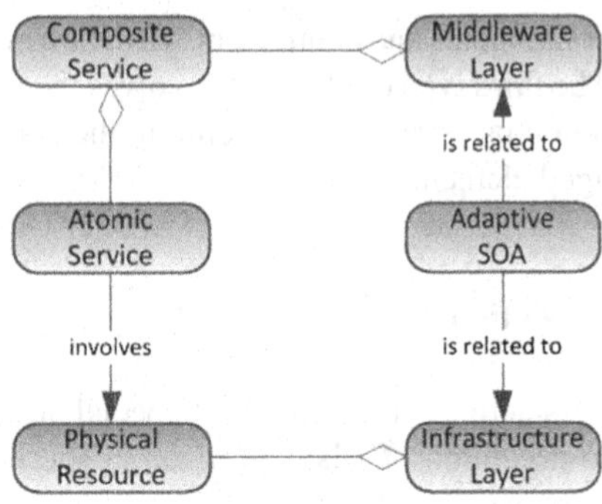

Fig. 4 Concept map showing Adaptive SOA

systems or embedded software/firmware. The Infrastructure Layer could be exposed as a set of virtual resources for higher layers, e.g. Middleware.

The Middleware Layer provides all of the system services that are relevant to the service orientation principles and their implementation. From an abstract point of view the Middleware Layer serves as a container for Composite and Atomic Services. These services require physical resources to perform their function.

3 Real-World Service

As shown in Fig. 5, a physical resource can be used by a computer or a Real-World Device. In the first case, through a computer, Virtual Services can be provided. In the second case, through appropriate instrumentation of Real-World Device it is possible to create and provision a variety of Real-World Services. However in both cases these are atomic services which are created composite service.

3.1 Concept

The Real-World Service [26] is a feature of the physical world object (Real-World Device), which provides an embedded logic module and a communication module, and is exposed to external systems according to service orientation principles. Such a service allows other components to interact with it dynamically. In contrast to virtual services (such as enterprise services) real-world services provide data about physical things/devices (World of Things) in real time. The Real-World Device is equipped, by means of hardware extensions, with logic and monitoring/management features.

The use of real-world services is associated with the concept of real-world awareness which is defined as follows [29]:

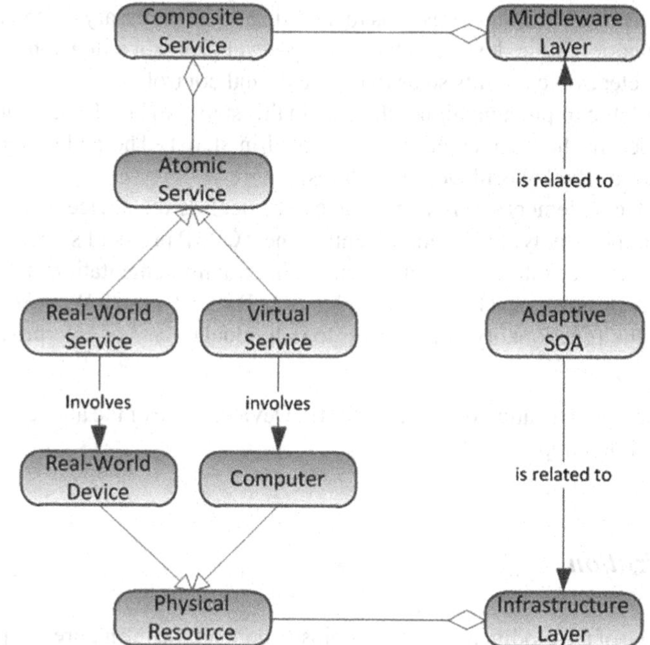

Fig. 5 Concept map introducing Real-World and Virtual Services

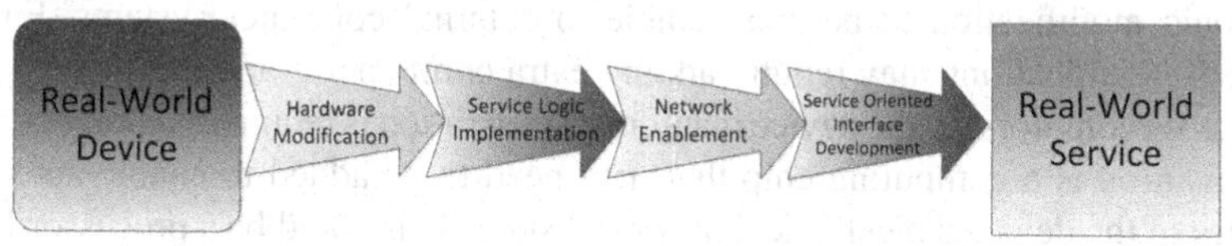

Fig. 6 Process of transforming a Real-World Device into a Real-World Service

> Real World Awareness is the ability to sense information in real-time from people, IT sources, and physical objects—by using technologies like RFID and sensors—and then to respond quickly and effectively.

The Real-World Service is created by adding logic to a physical object (Real-World Device). Network communication and adaptability allow users to obtain information about such an object in real time, and perform suitable management. It seems that the idea of Real-World Services complements the concept of real-world awareness in terms of the ability to sense information in real time, from a variety of physical objects.

Before any Real-World Devices can be used by enterprise systems, they must be modified into Real-World Services. This process adds aspects of service orientation to the Real-World Device and exposes the functionality of the device in the form of a service. In a general case the modification process can be divided into the following stages (Fig. 6):

1. Hardware modification—this calls for addition of the necessary mechanical parts and actuators to the device, as well as low-level protocol extensions by adding various electronic circuits so as to enable digital control.
2. Service logic implementation—the goal in this step is to build a dedicated server embedded in the augmented device created in step 1. The added logic allows exposure of Real-World Device features.
3. Network enablement—this step involves augmenting the device with a communication module, typically implementing the TCP/IP protocol stack.
4. Service oriented interface development—aimed at implementation of the service-oriented protocol stack (for example, SOAP/REST-based Web Services) to expose the functionality of the server to individual clients or enterprise applications.

Depending on the nature of the Real-World Device, some of the above steps might be simplified or skipped.

3.2 Realization

The first stage of the modification process mostly depends on the nature and properties of a Real-World Device. There are many devices in our environment that are intended to be used and controlled manually. These devices usually require mechanical and electronic modification to be connectable to external computer systems. First of all, such modifications may require adding extra components and mechanical parts (such as servomotors) to the device if automatic operation is required. Another typical augmentation is a computing chip that may need to be added to enable automatic changes of the device's mechanical or logical state. This could be a processor-based microcontroller or an FPGA programmable logic matrix. It should also be noted that some devices may already possess inbuilt features which allow a local embedded system to control the device in question. In such cases the device may only require some electronic and electrical modifications. The goal of all these modifications is to expose the full functionality of the Real-World Device on the level of digital logic signals. In some situations there is no need for any modification. In such cases the Real-World Device is already equipped with the proper (simple or advanced) digital interface and is ready to be controlled by external systems. For this type of devices the first stage can be completely skipped.

The second stage of the modification process depends strongly on the selected techniques and hardware used for implementation of the Real-World Device processing logic, such as:

- full hardware implementation (e.g. inside an ASIC or FPGA chip),
- IP core processor (with or without hardware accelerated parts) inside an FPGA chip,
- general-purpose microcontroller (with or without abstraction layers),
- mobile device (e.g. smartphone).

The first approach assumes that service logic and high-level communications required for service exposure will be implemented as "pure hardware". This implies that the service operation algorithm must be converted into a set of properly connected logic gates and flip switches, and then embedded in a programmable logic chip (such as FPGA or ASIC). The use of this technique is very difficult and requires substantial experience (especially in the area of digital circuit design), but in return offers efficiency and performance which is unreachable via any other technique. A description of a sample service implementation exploiting this approach can be found in [59]. The authors present the design of a "system on chip" (SoC) which operates as a Web service (WS). Their proposed system is entirely devoid of software and conceived as a hardware pattern for trouble-free design of network services offered as WS in a service oriented architecture (SOA).

The second approach is very similar to the first one as it also assumes the use of an FPGA device in the implementation of service logic and high-level communication, albeit in a completely different way. The FPGA device delivers a hardware platform for a soft-core (virtual) processor with capabilities adjusted to the requirements of the implemented service. The algorithm of operation is implemented as a program that runs on that processor. In the course of service implementation it is also possible to adjust the configuration of the soft-core processor for specific requirements of the implemented service. One example of such an approach is the hardware-software integrated development platform presented in [65]. The platform is based on the ALTERA Stratix II EP2S60 FPGA chip and dedicated to create SOA-compatible image processing services. The core features of such services are implemented as sequential C++ programs executed on the Nios II soft-core processor, while the most computationally expensive image processing operations are offloaded to a hardware accelerator. An additional advantage of the first two approaches is the ability to use the same flexible hardware equipment to implement completely different services.

The third approach assumes the use of a general-purpose microcontroller. In this case, the hardware architecture is delivered by the microcontroller manufacturer. Service logic and exposure are handled by software that runs on the microcontroller. Many modern consumer electronics (called smart or intelligent devices) are now equipped with capabilities used to share their functionality with PDAs, smartphones or other mobile devices (e.g. an electronic scale equipped with a Bluetooth interface for collecting and exchanging weight data). Unfortunately, the communication protocols used in this scope are usually incompatible with the protocols used for service exposure, and it is often impossible to introduce any modification inside the device itself. Fortunately, modern mobile devices (PDAs, smartphones) usually provide wireless Internet access (through Wi-Fi, GPRS, HSPA or LTE protocols)—hence the fourth scenario for modification and conversion of Real-World Devices into Real-World Services is to use the mobile device as a kind of proxy between the smart device and enterprise systems. In this scenario the service logic and high-level communication protocols are implemented as software running on a mobile device.

To enable the Real-World Device to be controlled over the Internet by enterprise systems (the 3rd stage of the modification process), the device must be equipped with a communication module. There are many such modules on the market—examples include the DigiConnect network module [19] and Tibbo programmable embedded modules [74]. Both solutions come fitted with hardware- or software-based TCP/IP stacks and Wi-Fi communications. While the DigiConnect module only offers one socket connection session at the time, the Tibbo modules are much more robust in that they allow up to 16 concurrent socket communication sessions. Having more than one socket session allows us to add extra features to the Real-World Service, such as service discovery.

An important concept for enabling service orientation on Real-World Devices is contained in the Device Profiles for Web Service specification (DWPS) [52] which is the successor of Universal Plug and Play (UPnP). This technology was developed to enable secure Web Service capabilities on resource-constrained devices. DPWS was mainly developed by Microsoft and some printer device manufacturers. DPWS allows secure messages to be sent to and from Web Services. It also supports dynamic discovery of Web Services, Web Service descriptions, as well as subscribing to and receiving events from a Web Service.

Web Services for Devices (WS4D) is an initiative which brings Service-Oriented Architecture (SOA) and Web Services technologies to the application domains of industrial automation, home entertainment, automotive systems and telecommunication systems. WS4D advances results from the ITEA SIRENA project [84]. The WS4D toolkits available on the project's website complies with DPWS. The toolkit is based on gSOAP and is targeted for small resource-constrained devices and can be used to implement DPWS-compliant devices using the C programming language. Another toolkit, based on J2ME, is available for small and resource-constrained devices, enabling implementation of DPWS-compliant devices in Java. Yet another toolkit, based on Apache Axis2, is targeted for resource-rich implementations to connect DPWS-compliant devices with the Web Services world.

In the SOCRADES (Service-Oriented Cross-layer infRAstructure for Distributed smart Embedded devices) project [34] physical legacy devices are grouped into three categories: non-electronic devices that are not WS-capable, electronic devices which do not support WS due to their limited resources, and WS-capable devices. To expose the features of WS-capable devices DPWS profiles are used. For devices which are not WS-capable, this can be done in two ways: by using the Gateway (dedicated for non-WS-capable electronic devices) or the Service Mediator (originally designed for collecting data from non-electronic devices).

Recently significant effort has been invested in enabling the convergence of sensor networks with the IP world and providing Internet connectivity for "smart objects". The IETF Working Group IPv6 over Low power Wireless Personal Area Networks (6LoWPAN) proposed an RFC [47] to enable IPv6 packets to be carried over IEEE 802.15.4. In addition, the IETF Working Group Routing over Low power and Lossy networks (ROLL) designed a routing protocol named IPv6 Routing Protocol for Low power and Lossy Networks (RPL). RPL was proposed because none of the existing known protocols such as Ad hoc On-Demand Distance Vector (AODV), Optimized Link State Routing (OLSR) or Open Shortest Path First (OSPF) meet the specific requirements of Low power and Lossy Networks (LLN), see [70]. The RPL protocol targets large-scale wireless sensor networks (WSN) and supports a variety of applications e.g. industrial, urban, home and building automation or smart grids.

The Constrained Application Protocol (CoAP) [68] is a specialized web transfer protocol for use with constrained networks and nodes for machine-to-machine applications such as smart energy and building automation. These constrained nodes often have 8-bit microcontrollers with small amounts of ROM and RAM, while networks such as 6LoWPAN often have high packet error rates and a typical throughput of several dozen kbit/s. CoAP provides a method/response interaction model between application endpoints, supports built-in resource discovery, and includes key web concepts such as URIs and content-types. CoAP easily translates to Hypertext Transfer Protocol (HTTP) for integration with the web, while also meeting the specialized requirements such as multicast support, very low overhead and simplicity for constrained environments.

Another approach to combining Real-World Services with Virtual Services is to generalize the Open Services Gateway initiative (OSGi) model into a collection of loosely coupled software modules interacting through service interfaces. While OSGi is a Java based solution, in [63] authors discuss how to turn non-Java-capable devices and platforms into OSGi-like services. They propose an extension to Remote Services for OSGi (R-OSGi) which makes communications to and from services independent of the transport protocol, and implement an OSGi-like interface that does not require standard Java (or even any Java at all). As an example, implementations for Connected Limited Device Configuration (CLDC), embedded Linux, and TinyOS are presented.

The idea behind modern network systems is to use the Internet as a connection space for as many Real-World Devices as possible. Each device, modified to become a Real-World Service, has its own reference address on the Internet and can be contacted by others systems. However, the large number of Real-World Services makes it difficult for users to locate any specific service. To make it easier, a Real-World Service registration and discovery features are required. Each of the available Real-World Services regularly sends information about its reference and description to one or more discovery servers which maintain a database of active Real-World Services on the network (registration process). The information needed to find and use the various Real-World Services is available through the discovery server (discovery process). Among the various technologies that provide service discovery we can mention SLP (Service Location Protocol), Jini, Apple Bonjour and WS-Discovery in DWPS [27].

4 Realization of the Adaptation Loop

This section describes the current realization status of the Adaptation Loop and highlights solutions capable of converting Managed Resources into Virtual Services (as well as Real World Services).

The MAPE-K model introduces five elements of an autonomous loop which, taken together, support development of completely autonomous systems. In order to describe an adaptation loop it is enough to use only four elements: Monitor, Analyse, Plan and Execute (later referred to as MAPE). Figure 7 presents techniques involved in realization of the adaptation loop. They are divided into those related to Monitoring and Execution, and those applicable to Analysis and Planning. The former techniques are used for instrumentation of Managed Resources for the purpose of adding Sensors and Effectors, while the latter are used for interpretation, analysis of data provided by Sensors and planning actions executed by means of Effectors.

The first subsection will present a review of techniques used for Monitoring and Execution, while the second subsection describes techniques of Adaptation and Planning. The final subsection contains a survey of existing work presented in the context of identified techniques.

4.1 Monitoring and Execution

As presented in Fig. 7, the Managed Resource can be either Virtual (related to a Virtual Service) or Real-World (related to a Real-World Service); however in both cases it can be composed of both Software and Hardware elements. Instrumentation techniques are therefore divided into two categories. Software instrumentation is mostly related to the Middleware Layer while hardware instrumentation ties in with the Infrastructure Layer. As presented in Fig. 5 a Real-World Service always involves a Real-World Device, which makes hardware instrumentation especially important for adding adaptability to resources related to the real world. Software instrumentation is commonly used in the case of both Virtual and Real-World Resources.

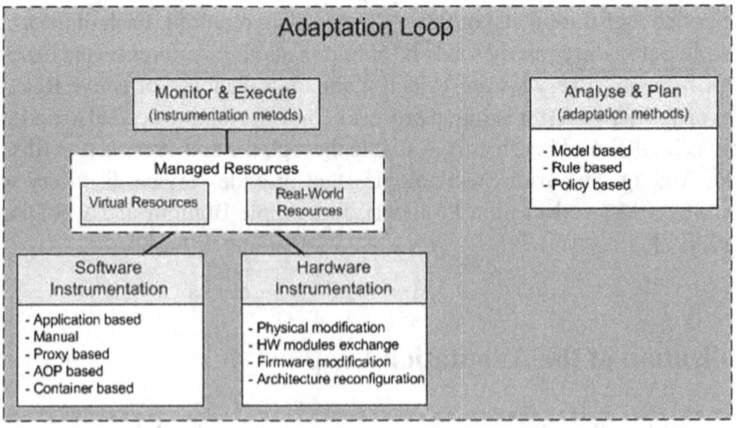

Fig. 7 Adaptation loop realization techniques

4.1.1 Software Instrumentation

One of the key approaches applied to software instrumentation is the Interceptor Design Pattern [4]: a design pattern used when software systems or frameworks need to offer a way to change, or augment, their usual processing cycle. The Interceptor Design Pattern may solve several problems related to software development. For example, it is commonly used for monitoring the internal execution of an application. Another area of usage is the ability to change or extend application behaviour. New features are implemented as interceptors and invoked by the working application. Such extensions do not need to be aware of other parts of the application, nor change existing parts. They also do not affect the design of the system.

The Interceptor Design Pattern may be realized with the use of various approaches, for instance with the Proxy Pattern where the original object is replaced with a proxy with the same interface (or contract) as the original. Usage of this pattern provides several possibilities, such as controlling access to the proxied object or lazily loading such objects. The Proxy Pattern also provides the possibility to monitor original object invocations or even alter them. It is also possible to create a proxy-based solution which enables plugging in of sensors and effectors, facilitating transparent software instrumentation.

Another approach for software instrumentation compliant with the Interceptor Design Pattern is called container-based instrumentation. In contrast to previously described solutions, in this case the subject of instrumentation is not an application per se, but rather a container or execution environment in which the application is being executed. The lowest level of such instrumentation is enrichment of the virtual machine in which the process is executing. Examples of such systems are presented in [24, 82] where Java Virtual Machine (JVM) is instrumented in order to access runtime monitoring data. However, this approach is somewhat dated: nowadays similar techniques are mapped into the domains of services and components. In this case the subject of instrumentation is the container in which services are connected, exposed and able to communicate—such as Service Component Architecture (SCA) or Enterprise Service Bus (ESB). In the case of SCA, instrumentation may be applied to enable exchanging composites from which the service is created. In this way it is possible to manage the Quality of Service (QoS) of services which are executed in the container. In the case of the ESB container, instrumentation may be used to relay service calls to one of several instances of a particular service in order to provide the desired QoS.

One of the main programmatic mechanisms used to realize software instrumentation is the concept of Aspect-Oriented Programming (AOP), first introduced in [38]. In this approach existing software can be interwoven with aspects which change or extend its behaviour in a fully transparent way. This approach is especially useful when the source code does not admit modifications. Initially, aspects were woven into the software during compile or load time and any changes required the software to be stopped. To solve this issue the concept of dynamic aspect weaving was introduced. In [5, 56, 57] the authors presented several approaches to weaving and unveawing aspects at runtime, providing a basis for using aspects in order to realize the adaptation control loop.

It is also possible for the software itself to be written in such a way as to enable instrumentation. However, this solution is strongly limited to mechanisms exposed directly by the software, i.e. retrieving monitoring data via dynamically added sensors. As such, if some features or mechanisms are not enabled, it becomes necessary to use one of the techniques described above.

4.1.2 Hardware Instrumentation

Instrumentation and introduction of new features, including sensors and effectors, to hardware resources may require modifications in the physical design of electronic modules. Modifying and reconfiguring embedded systems in order to add new features might be applied on different levels of design, starting from firmware reconfiguration, through modification of external peripherals and microprocessor design, all the way to selection of hardware modules that the system is composed of. In general, these possibilities can be classified as:

1. Physical modifications—modifications that require additional elements to be installed or modules to be exchanged.
2. Embedded software modifications—modifications performed by software reconfiguration, e.g. changes to firmware or configuration elements.

Some embedded systems can be designed in a modular way, enabling new features (e.g. measurement and debugging elements) to be added by plugging additional modules into the existing platform. In other solutions it might be necessary to solder additional elements manually. Such modifications cannot be performed automatically as they require physical intervention. As programmable logic devices become more and more robust, software modifications might be perceived, to some extent, as physical modifications since they involve modifications in the embedded system's internal architecture. Upon installation effectors may use the same hardware reconfiguration techniques to affect hardware resources during the adaptation process.

In the simplest case, embedded systems might be reconfigured by replacing the firmware of the internal processor causing a change in its functionality and enabling additional features to be used in the adaptation loop. This kind of reconfigurability mirrors typical standalone systems. A more promising approach is to reconfigure the hardware of the embedded system. The concept of reconfigurable computing that combines some of the flexibility of software with the high performance of hardware processing using high-speed reconfigurable computing fabrics, has existed since the 1960s. Gerald Estrin's landmark paper proposed the concept of a computer composed of a standard processor and an array of "reconfigurable" hardware components [21]. The main processor would control the behavior of the reconfigurable hardware which would, in turn, be tailored to perform a specific task, such as image processing or pattern matching, with performance similar to a dedicated hardware platform. Once the given task was completed, the hardware could be adjusted to perform some other task. This results in a hybrid computer architecture, combining the flexibility of software with the speed of hardware. Unfortunately, Estrin's idea was far ahead of its time given the sophistication of electronic devices. In the 1980s and 1990s there was a renaissance in this area of research, with many proposed reconfigurable architectures developed in the industry and academia [11], such as COPACOBANA, Matrix, Garp, Morphosys and PiCoGA [44]. Such designs became feasible due to the relentless progress in silicon-based technologies which finally allowed complex designs to be implemented on a single chip. The world's first commercial reconfigurable computer, Algotronix CHS2X4 [36], was completed in 1991. Algotronix was designed as a low-cost add-on card for the PC. It contained 9 programmable CAL1024 logic chips. The computer found application in a number of areas, including self-reconfiguration, in which the board reconfigured itself from a design stored in RAM as a result of computation. Ultimately the CHS2X4 would not achieve commercial success, but proved promising enough that its core technologies were later bought by Xilinx—inventor of the Field-Programmable Gate Arrays.

FPGAs contain programmable logic components called "logic blocks", and a hierarchy of reconfigurable interconnects that allow the blocks to be "wired together"—akin to changeable logic gates that can be inter-wired in (many) different configurations. Logic blocks can be configured to perform complex combinational functions, or merely simple logic gates like AND and XOR. In most FPGAs, the logic blocks also include memory elements, which may be simple flip-flops or more complete blocks of memory. The main difference compared with custom hardware, i.e. application-specific integrated circuits, is the ability to adapt the hardware at runtime by loading a new circuit on the reconfigurable fabric. This can be achieved through partial reconfiguration, by configuring a portion of a field programmable gate array while another part is still running or operating. Much like software, hardware can be designed modularly, by first creating subcomponents and then higher-level components. In many cases it is useful to swap out one or several subcomponents while the FPGA is operating.

Normally, reconfiguring a FPGA requires it to be held in reset mode. While in that mode an external controller reloads the new design into the FPGA. Partial reconfiguration allows for critical parts of the design to continue operating while a controller (either inbuilt or external) loads a partial design into the reconfigurable module. Partial reconfiguration also can be used to save space for multiple designs by only storing the partial designs that change between sessions. A common situation in which partial reconfiguration might be useful is the case of a communication device. If the device controls multiple connections, some of which require encryption, it would be beneficial to load different encryption cores without bringing the whole controller down. In the scope of design functionality, partial reconfiguration can be divided into two categories:

- dynamic partial reconfiguration, also known as active partial reconfiguration—allows changing parts of the device while the rest of the FPGA is still running;

- static partial reconfiguration—the device is not active during the reconfiguration process. While partial data is sent to the FPGA, the rest of the device is stopped (in shutdown mode) and brought up once the configuration is completed.

There are two styles of partial reconfiguration of an FPGA:

- Module-based partial reconfiguration, which enables reconfiguring distinct modular parts of the design. To ensure communication across reconfigurable module boundaries, special bus macros ought to be prepared. A macro works as a fixed routing bridge that connects the reconfigurable module with the remainder of the platform. Module-based partial reconfiguration requires a set of specific guidelines to be followed at the design stage.
- Difference-based partial reconfiguration, which can be used when a small change is introduced in the design. It is especially useful when exchanging small routines or dedicated memory blocks. The partial bit-stream contains only information about differences between the current design structure (which resides in the FPGA) and the new content of an FPGA.

Maturation of Field Programmable Gate Arrays has led to the concept of open-source hardware, where physical artifacts are designed and offered in the same manner as free and open-source software. Hardware design (i.e. mechanical blueprints, schematics, materials, printed circuit layout data, hardware description language source code and integrated circuit layout data), in addition to the software that drives the hardware, are all released using the open source approach. Since the rise of reconfigurable programmable logic devices, sharing of logic designs has adopted the concept of open-source hardware. Instead of raw schematics, hardware description language (HDL) code is shared. HDL descriptions are commonly used to set up system-on-a-chip platforms using either field-programmable gate arrays (FPGAs) or application-specific integrated circuit (ASIC) designs. HDL modules, when distributed, are called semiconductor intellectual property cores, or IP cores. This allows developers to exploit general purpose electronic modules along with specific hardware descriptions [53] to build custom hardware devices.

4.2 Analysing and Planning

Several approaches to constructing autonomous managers are currently under investigation. One of these is to apply control theory for performance control in complex applications such as real-time scheduling, web servers, multimedia, and power control in CPUs [42]. This methodology is viewed as a promising foundation but its main drawbacks include the need to identify and model managed systems. A different approach is to use policies to guide decisions based on the observed system state and its behaviour. Several languages and specifications have been designed for this purpose. The WS-Policy [75] specification represents a set of specifications that describe the capabilities and constraints of security (and other business) policies on intermediaries and endpoints (this includes e.g. the required security tokens, supported encryption algorithms, and privacy rules) and how to associate policies with services and endpoints. The Ponder [18] language provides common means of specifying security policies that map onto various access control implementation mechanisms for firewalls, operating systems and databases. Understanding of the policies varies between researchers, but usually the term "policy" is used to represent a set of considerations which guide decisions. The policy is provided to the system as a set of rules. A policy information model provides an abstract representation of the information required to define a policy:

- Condition-Action Information Model policies consist of several policy rules that have two elements: conditions and actions. Conditions define when the policy should be applied while Actions define what needs to be done when a particular policy rule is applied. Condition-action policy rules assume the following form: if [list of conditions] then [list of actions]. Such policies are evaluated at regular intervals. Even though this information model is very simple, it has some drawbacks. The frequency of policy evaluation has to be defined, which may influence system reaction time.
- Event-Condition-Action Information Model treat events as conditions. This type of policy model is useful for asynchronous policy evaluation as a response to events generated by the system (e.g. state changes or parameters exceeding threshold values). An event-condition-action policy is denoted as when [list of events and conditions] then [list of actions]. This type of policy becomes particularly useful when determining policy evaluation frequency becomes unfeasible or when event-driven policies are called for.

Policy rules are executed by a rule engine which uses algorithms for efficient pattern matching. Usage of rule engines brings a number of advantages, including separation of business logic from application implementation, the ability to change policies without software recompilation and extending the set of policies at runtime.

Another promising approach for runtime adaptation is to apply mechanisms which leverage software models and the applicability of model-driven engineering techniques to the runtime environment. A runtime model is a causally-connected self-representation of the associated system that emphasizes the structure, behaviour, or goals from a problem space perspective [10]. Models may express several different aspects of the running system. They might express the structure of the underlying system or its behavioural aspects. Structural models show how the system is constructed in terms of objects and invocations. In contrast, behavioral models emphasize how the system executes in terms of flows or traces and events occurring in the working system.

4.3 Survey on Existing Solutions

This section presents a comparison of existing solutions in the context of adaptation loop techniques described previously. Such comparison is performed in order to show the current state of the art concerning instrumentation and adaptation methods in different projects, and to evaluate the prospects of applying them to the SOA domain. The results of this research are presented in Table 2.

As described in the previous sections, several software instrumentation techniques exist. The following paragraphs present selected solutions and their usage in the context of adaptation loops.

In [23] the authors present a framework called Rainbow which can be used for adaptation of software systems. Such adaptation concerns the use of mechanisms located outside of the analyzed system, which enable adaptation strategies to be specified. The Rainbow framework enforces the implementation of its sensors and actuators, thereby enabling adaptation. This solution can be used for systems built in accordance with the SOA paradigm thanks to its modular and flexible architecture. In order to realize an adaptation loop an architectural model of the system has to be prepared. On the basis of this model Rainbow enables adaptation invariants and strategies to be specified. Together, these determine the system's adaptation style and can be evaluated by the adaptation engine. Rainbow is aimed at virtual services, however any set of adaptation strategies created for a hardware adaptation can be used after reorganizing process for the purpose of real-world services.

In [48] the authors present results of their work in the DiVA project which focuses on dynamic variability in complex adaptive systems. DiVA introduces a methodology and tools for runtime QoS management of adaptive systems. It also proposes an approach for specifying and executing dynamically adaptive software systems which combines model-driven and aspect-oriented techniques in order to tame the complexity of such systems. This approach depends on the model of the managed system. In order to apply the proposed approach engineers need to design models independently of the running system and leverage them at runtime to drive the dynamic adaptation process. The DiVA solution does not directly address the SOA domain, but—owing to its flexible architecture—could be easily incorporated into SOA systems. The proposed techniques can be applied to Virtual as well as Real-World Services; however no hardware adaptation metodologies have been published so far.

In [25] the authors describe the Adaptive Server Framework (ASF)—an architectural concept which facilitates the development of adaptive behavior for legacy server applications. ASF provides clear separation between the implementation of adaptive behavior and the business logic of the server application. ASF incurs low CPU overhead and memory usage. It is portable across different J2EE applications servers. The goal of ASF is to provide infrastructure components and services to facilitate the construction of behavioral adaptation. ASF components interact with the application server, monitor the runtime environment, analyse collected data and change the application's behavior by adapting its responses or setting the server's configuration to fulfill business goals. The authors identify two monitoring techniques: adding

Table 2 Instrumentation and adaptation methods used by existing solutions

Solution	Software instrumentation						Hardware instrumentation			Analyse and Plan		
	App. based	Manual	Proxy based	AOP based	Cont. based	Physical mod.	Modules exchange	Firmware mod.	Arch. reconf.	Model based	Rule based	Policy based
[23]			X								X	X
[48]				X						X		
[25]		X	X							X		X
[28]	X									X		
[81]		X								X	X	
[14]		X			X					X		
[43]					X					X		
[79]		X			X					X		
[83]	X	X										
[7]			X									
[6]		X	X		X							
[49]				X	X							
[80]					X							
[60]				X	X							
[55]				X								
[71]				X								
[13]				X								
[41]							X					
[46]									X			
[16]									X			
[12]								X	X			
[50]							X	X	X			
[69]							X	X				
[2]								X				

interceptors by means specific for a given service container (which can be qualified as container-based instrumentation) and wrapping components in JMX MBeans and redirecting client invocations (a combination of proxy and manual instrumentation techniques). In ASF adaptation analysis and planning involve a component model of the system, which can be changed during adaptation, and rely on policies which drive the adaptation process. The proposed approach focuses on components (i.e. the components layer of S3) and can therefore be perceived as partly related to SOA. ASF does not deal with Real-World Services or hardware instrumentation aspects.

In [28] the authors propose a middleware-centric approach to building applications capable of adapting to dynamically changing requirements, pursued in the FAMOUS (Framework for Adaptive Mobile and Ubiquitous Service) project. The authors focus on handheld devices where communication bandwidth or UI preferences change dynamically, depending on the ambient light and noise. The adaptation loop is realized in the following way: when a context change occurs, it is detected by the context monitor which then notifies the adaptation manager. The adaptation manager searches for a configuration which best fits the current context and resource utilization by the application. The search uses a planner component to iterate through plans for all possible application variants. A plan is generated by selecting a specific component for each component role of the application. As some of the selected components can be composites, the planning continues recursively until all leaf nodes are selected. The best configuration is selected by computing a utility value for each plan with respect to the user preferences and properties of the execution environment. This value is returned by the utility function, defined by the developer. The utility funciton is typically a weighted mean of the differences between the offered and required properties. Individual weights in the utility function represent changing user priorities and may be adjusted at runtime. The variant with the highest utility value is chosen. In order to avoid constant changes, the adaptation manager also needs to evaluate whether the perceived improvement is high enough to justify an adaptation. Such evaluation is based on a user-adjustable utility improvement threshold and adaptation delay.

In [81] the authors describe three important steps towards adaptive online systems. First is the assumption that online hardware reconfiguration due to workload changes has the potential to improve performance. Second step assumes that by using uninstrumented middleware and given only raw, low-level system statistics it is possible to predict which of the two configurations will outperform the other at any given time. The last one imposes extending the prediction capability to make precise numerical estimations (i.e. quantitative changes in performance when the system is switched to each of the possible configurations). By fulfilling all three criteria performance gains can be traded off against inevitable reconfiguration costs. The authors start off with a set of experiments using the TPC-W benchmark which shows that given configurations might prove better in different circumstances Subsequently the authors claim that they can infer the optimal configuration on the basis of low-level operating system statistics (with no customized instrumentation). They create a model mapping the current system state (represented by output from the vmstat tool) to the optimal configuration. Using results from previous experiments as training data they apply the WEKA package as an implementation of standard machine learning methods.

In [14] the concept of Adaptable ESB is introduced. It consists of an operations support system that is compliant with NGOSS (Next Generation Operations System and Software) and implements a service-oriented architecture (SOA) that relies on an enhanced enterprise service bus (ESB). This enhanced ESB, referred to as an adaptable service bus (ASB), enables runtime changes to business rules, thus avoiding costly application shutdowns. An implementation of this system has been used by the ChungHwa Telecom Company, Taiwan, since January 2008 and provides complete support for its billing application. As a result the billing process cycle has been reduced from 10–16 days to 3–4 days, paving the way for further business growth.

The paper [43] propose the usage of ESB to build a dependable SOA middleware. Its authors exploit state-of-the-art solutions (i.e. Bayesian networks and fault detection algorithms) to propose a service-based architecture ensuring high dependability of business processes and services. The architecture comprehensively addresses the following challenges: discovering causal relationships, providing high scalability and preventing excessive overhead. The deficiency of the designed middleware is its reliance on the monitoring API of a particular ESB, which is not standardized among different vendors.

Morin et al. have studied the role of runtime models in managing runtime or dynamic variability [22]. Their research focuses on reducing the number of configurations and reconfigurations that need to be considered when planning adaptations of the application. The authors illustrate their approach with a customer relationship management application. Fleurey et al. present preliminary work on modeling and validation of dynamic adaptation [48]. Their proposed approach envisions runtime use of Aspect-Oriented Modelling (AOM). First, the application base and variant architecture models are designed and the adaptation model is built. At runtime the adaptation model is processed to produce the system configuration to be used during execution. Although adaptation to context changes is precisely described, the methods of adapting to changing QoS requirements and capabilities are only mentioned.

There are many different ways to extract monitoring data. Some studies rely on the data provided by the application layer [7, 83], while others focus on the mechanisms of monitored containers [43, 80] or turn to instrumentation of monitored systems [6, 49, 79]. Identifies two main kinds of monitoring data extraction: instrumentation and interception. Interception assumes that the monitored system enables some proprietary way of installing interceptors. Many of the frameworks relies on the AOP instrumentation which has been the subject of numerous studies (cf. [13, 55, 60, 71]).

In order to enforce adaptation in Real-World Services it is necessary to enable hardware instrumentation. Several interesting solutions related to this concept are currently available, as highlighted in the following paragraphs.

The concept of reconfigurable general-purpose hardware can be extended to cover replaceable hardware modules that can be selected for particular usage scenarios. One of the relevant open-source hardware startups is Bug Labs [41]. The company develops a Lego-like hardware platform which tinkerers and engineers can use to create their own digital devices. Development starts with BUGBase, which is a general-purpose Linux computer about the size of a PlayStation Portable, encased in white plastic. It provides four connectors that plug right into the motherboard. The company also manufactures a variety of modules that can plug into the computer—including an LCD screen, a digital camera, a GPS unit, a motion sensor, a keyboard, an EVDO modem, and a 3G GSM modem (There are also extensions for USB, Ethernet, WiFi, and serial ports). Bug Labs intends to produce approximately 80 different modules and hopes that external companies and developers will create their own modules.

A representative general-purpose hardware solution is marketed by the Milkymist [46] project. It is a comprehensive open-source platform for live synthesis of interactive visual effects. The project goes to great lengths to apply the open source principles at every level possible, and is best known for the Milkymist system-on-chip (SoC) platform, which is among the first commercialized system-on-chip designs with free HDL source code. As a result, several Milkymist technologies have been reused in applications unrelated to video synthesis. For example, NASA's Communication Navigation and Networking Reconfigurable Testbed (CoNNeCT) experiment uses the memory controller that was originally developed for the Milkymist system-on-chip in the development of an experimental software-defined radio prototype.

A polar opposite to full reconfiguration of the embedded system architecture is the concept known as Programmable System on Chip (PSoC) [16]. A PSoC integrated circuit is composed of a core, configurable analog and digital blocks and programmable routing and interconnect. Flexible mixed-signal arrays enable signals to be routed to and from I/O pins. This architecture allows designers to create customized peripheral configurations to match the requirements of each individual application, making PSoC substantially different from other microcontroller designs [15].

Reconfigurable hardware platforms are often used as tools for acquisition and processing of data from scientific experiments. Some general (basic) information about the concept and classification of such platforms can be found in [76]. Two specific examples are briefly described below.

The platform described in [12] was designed to acquire and process (in real-time) data from nuclear spectrometry experiments. It uses Xilinx FPGA chips and Texas Instruments Digital Signal Processors. Among the tasks handled by DSP processors is real-time parametrization of the hardware processing algorithms stored in the FPGA. The design also allows the available hardware to be used as a general-purpose acquisition and processing platform.

Another example of a reconfigurable hardware platform for scientific experiments aimed at cellular architectures is called CONFETTI and introduced in [50]. This is a modular, hierarchical platform composed of a number of simple FPGA-based computing units called ECells. ECells can communicate with each other at speeds up to 500 Mbits/s and can be configured independely from many different sources (local FLASH memory, Ethernet, Wi-Fi, etc.) Owing to its modular construction, the user has the ability to replace any of ECell unit with another, compatible unit. Communication between the ECells is also widely configurable.

Another interesting class of devices is represented by Sun SPOTs—Small Programmable Object Technology, developed by Sun Microsystems (now part of Oracle) [69]. SPOTs are designed as an experimental platform for prototyping applications which might be strongly integrated with the environment. Each device comes with a general-purpose sensor board which can—at the developer's discretion—be swapped for a different hardware module such as a flash card reader, a more robust analog input/output extension, or an FPGA board. Another adaptation option is to use the built-in management feature which allows monitoring of working applications, changing device properties and even installing new software remotely.

Address trace analysis is one of the available techniques used to evaluate cache and memory efficiency of computer systems. Address traces are streams of addresses generated during the execution of programs. They can be aggregated using various methods. An interesting way to collect traces—named ATUM—was proposed by Agarwai et al. [2]. Their concept is to modify the microcode of each processor instruction that requires access to memory. In this way the address referenced by that instruction is also stored in a special protected place in memory. This method of collecting address traces can be used in any microprocessor which admits microcode modifications and can be treated as an instrumentation of hardware through firmware modifications (microcode can be treated as processor firmware).

As presented in Table 2 many existing solutions are—in one way or another—related to adaptation loops. While some of them are clearly more mature than others, there is no single technique which tackles all the issues connected with implementing adaptation loops in both the software and the hardware domain. Another significant conclusion is that the hardware domain lacks the requisite Analyze and Plan components. Only a handful of projects address SOA systems, whether directly or indirectly. In order to remedy this issue we have decided to introduce the concept of an Adaptive SOA Solution Stack (AS3) Pattern. It can be applied to heterogeneous environments comprised of both Real-World and Virtual Services in order to enforce the adaptation loop in a seamless and standardized way across all layers of the system which is the subject of adaptation.

5 Adaptive SOA Solution Stack

One of the main models of SOA application development and deployment is the SOA Solution Stack (S3) proposed by IBM [3]. Its core concept is depicted in Fig. 8. The S3 model provides a detailed architectural definition of SOA split into nine layers. Each layer comes with its own logical aspects (including all the architectural building blocks, design decisions, options, key performance indicators, etc.) and physical aspects (which cover the applicability of each logical aspect in reference to specific technologies and products). The S3 model is based on two assumptions:

- The existence of a set of service requirements (functional and nonfunctional) which collectively establish the SOA objective;

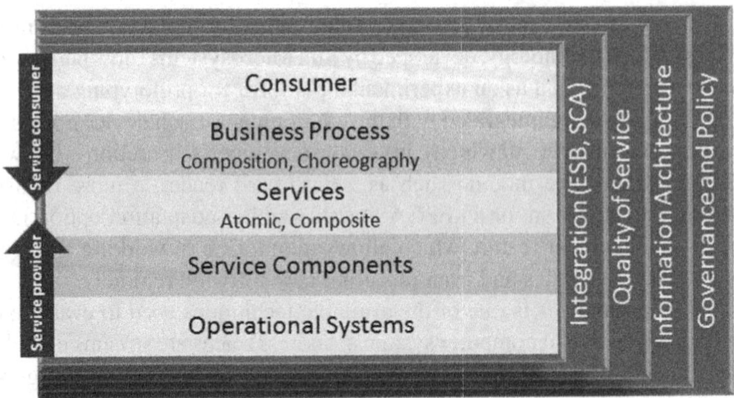

Fig. 8 SOA Solution Stack [3]

- The notion that specific service requirements can be fulfilled by a single layer or some combination of layers. Each layer can satisfy service requirements by way of a layer-specific mechanism.

The nine layers of the S3 stack are as follows: Operational Systems, Service Components, Services, Business Process, Consumer, Integration, QoS, Information Architecture, and Governance and Policy. A broad (non-technical) description of each S3 layer (except the Consumer layer), is provided in the following paragraphs.

- **Operational Systems**—This layer includes all application and hardware assets running in an IT operating environment that supports business activities (whether custom, semicustom or off-the-shelf). As this layer consists of existing application software systems, SOA solutions may leverage existing IT assets. Currently this layer typically includes a virtualized IT infrastructure that results in improved resource manageability and utilization. This property could be effectively exploited in the development of an adaptive virtualized infrastructure, guaranteeing the required level of accessibility of computational or communication resources.
- **Service Components**—This layer contains software components, each of which is an incarnation of a service or service operation. Service components reflect both the functionality and QoS for each service they represent. Each service component:

  - provides an enforcement point for ensuring QoS and service-level agreements;
  - flexibly supports the composition and layering of IT services;
  - conceals low-level implementation details from consumers.

  In effect, the service component layer ensures proper alignment of IT implementations with service descriptions. Service QoS depends on the efficiency of internal components used for service provisioning. It provides a space for adaptability within the Service Component layer. The observed service QoS is not only the result of Service Component activity but also depends on computational resources used during execution. This behaviour illustrates the role of the Operational Systems layer and facilitates multilayer adaptability.

- **Services**—This layer consists of all services defined within SOA. In the broadest sense, services are what providers offer and what consumers or service requestors use. In S3, however, a service is defined as an abstract specification of one or more business-aligned IT functions. This specification provides consumers with sufficient information to invoke the business functions exposed by a service provider. It is necessary to point out that services are implemented by assembling components exposed by the Service Component layer and that this assembly process might be performed dynamically with support from adaptability mechanisms.
- **Business Process**—In this layer the organization assembles the services exposed in the Services layer into composite services that are analogous to key business processes. In the non-SOA world business processes exist as custom applications. In contrast, SOA supports application construction by introducing a composite service which orchestrates information flow among a set of services and human actors. Again, these composite services can be constructed dynamically according to a specific adaptation policy.
- **Integration**—This layer integrates layers 2–4. Its integration capabilities, supported by ESB, enable mediation, routing and transporting service requests from the client to the correct service provider. This layer is particularly well suited for adaptability mechanisms.
- **Quality of Service**—Certain characteristics of SOA may exacerbate well-known IT QoS concerns: increased virtualization, loose coupling, composition of federated services, heterogeneous computing infrastructures, decentralized service-level agreements, the need to aggregate IT QoS metrics to produce business metrics and so on. As a result, SOA clearly requires suitable QoS governance mechanisms.
- **Information Architecture**—This layer covers key data and information-related issues involved in developing business intelligence with the use of data marts and warehouses. It includes stored metadata, which is needed to correctly interpret actual business information.
- **Governance and Policy**—This layer covers all aspects of managing the business operations' lifecycle. It includes all policies, from manual governance to autonomous policy enforcement. It also provides guidance and policies for managing service-level agreements, including capacity, performance, security and monitoring. As such, the Governance and Policy layer can be superimposed onto all other S3 layers. From a QoS and performance standpoint it is tightly connected to the QoS layer. The layer-specific governance framework includes service-level agreements based on QoS and key process indicators, a set of capacity planning and performance management policies to design and fine-tune SOA solutions as well as specific security-enabling guidelines for composite applications.

The decomposition of SOA Systems proposed by S3 can be used for more precise partitioning of the Adaptive Systems Space introduced in Sect. 2. The "where" axis (showing where adaptation mechanisms can be located) may now be split into sections referring to the S3 layers, as presented in Fig. 9. The Operational Systems

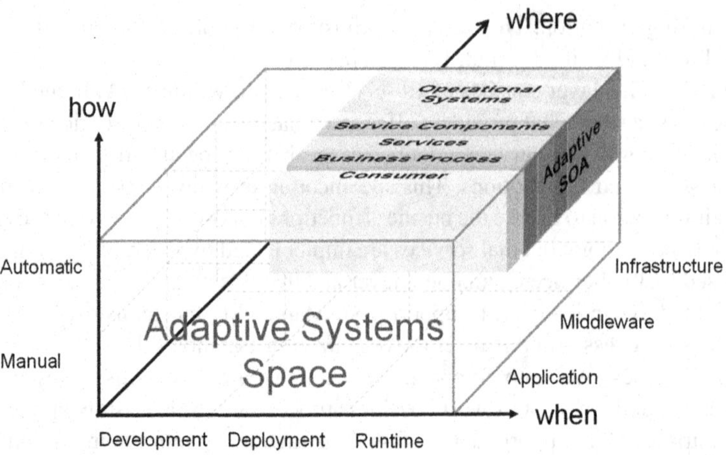

Fig. 9 Mapping of S3 onto the Adaptive Systems Space

layer is assigned to the Infrastructure layer while Service Components, Services, and Business Processes are aggregated by the Middleware Layer. Since the Consumer Layer often contains application-specific mechanisms, it is mapped to the Application Layer. Vertical S3 layers crosscut the entire "where" axis, making them a perfect place for deployment of mechanisms required by the adaptation loop. This is consistent with the fact that the vertical layers (QoS, Information Architecture, Governance and Policy) are directly related to non-functional parameters involved in the adaptation process. The presented mapping clearly highlights the structure of Adaptive SOA in the context of S3.

The S3 Model which coincides with the Adaptive SOA Space is named the Adaptive SOA Solution Stack (AS3) [85]. It could be constructed via uniform introduction of adaptability aspects to each layer of the S3 Model, yielding a multilayer adaptive system which takes advantage of modern software and hardware technologies and offers full control over QoS and QoE parameters. The concept of AS3 has two important constituents:

- AS3 Element Pattern—an architectural pattern used for modelling adaptability aspects in each S3 layer. It contains several components used in the adaptation process.
- AS3 Process—an abstract process defining the transformation of the non-adaptive layer of the S3 stack into its adaptive equivalent.

Each constituent will be described in more detail in the following sections.

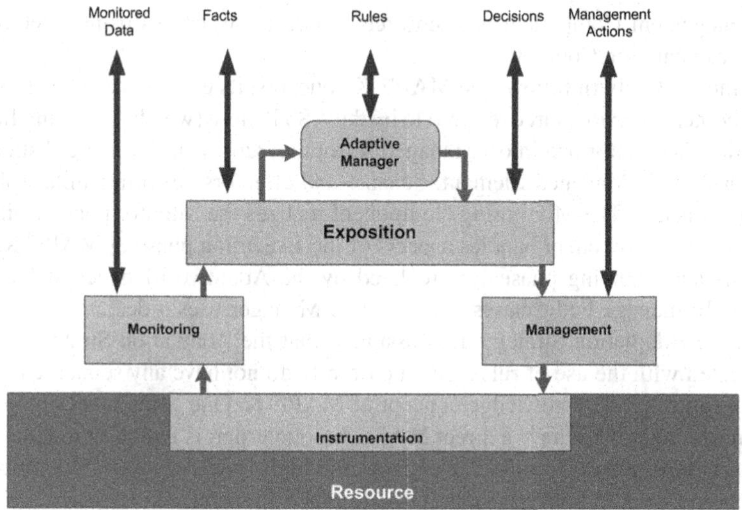

Fig. 10 AS3 Element Pattern

5.1 The AS3 Pattern

In general, the AS3 Pattern depicted in Fig. 10 follows the concept of the MAPE-K control loop and refines it in the context of adaptability-related S3 layers. Throughout the remainder of this chapter we will refer to the AS3 Element Pattern simply as the AS3 Pattern, while the S3 layer to which adaptability is added in accordance with the AS3 Pattern will be referred to as the AS3 Layer. The elements of this pattern could be described as follows.

The Resource is an abstract entity (S3 Layer—e.g. Integration, Service Components or Services). It is transformed into the Managed Resource through instrumentation with sensors and effectors. Sensors expose the state and configuration of the Resource and enable monitoring of its activity. Effectors provide mechanisms for changing Resource parameters or configuration according to actions enforced by the Management Component. Data gathered by sensors is passed to the Monitoring Component which is responsible for calculating selected metrics and processing events. The aggregated data—the output of the Monitoring Component—is forwarded to the Exposition Component.

The Exposition Component cooperates with the Adaptive Manager which is used to select control actions. It transforms monitored data into the format used by the given Manager instance. The Exposition Component therefore acts as a harmonization layer. It is also possible for the Exposition Component to expose some facts to other AS3 Elements and receive high-level decisions which should be enforced in the control loop. The Adaptive Manager Component is used to enact the adaptation loop. Actions selected by the Adaptive Manager are converted by the Exposition Component to a format acceptable by the Management Component. The responsibility of the Management Component is to enforce management actions using effectors of the Instrumentation Component.

As the AS3 Pattern follows the MAPE-K concepts, its elements can be classified in this context. The Resource referred to by the AS3 Pattern (which, following Instrumentation, is transformed into a Managed Resource) represents the same abstraction as the MAPE-K Managed Element. Sensors and effectors perform similar roles in both approaches. The Monitoring Component realizes the Monitor phase while the Management Component handles aspects of the Execution phase of MAPE-K. The Analysis and Planning phases are realized by the Adaptive Manager of the AS3 Pattern. To manage both phases the Adaptive Manager uses a declarative strategy named the Adaptation Strategy. It is assumed that the Adaptation Strategy can be represented with the use of rules which currently do not have any semantic context for realization of the Knowledge concept of MAPE-K. One component of the AS3 Pattern which does not have a direct MAPE-K counterpart is Exposition. It acts as a data harmonization layer and enables facts and high-level decisions to be obtained and exposed by AS3 Patterns located in different S3 Layers.

The full potential of the AS3 Stack is manifested in the cooperation abilities of different AS3 Layers. For instance, it is possible to monitor the whole system by collecting information from Monitoring Components present in each AS3 Layer. Such reasoning may also refer to other types of AS3 components. Thus, the following aspects are inherent in a complete AS3 stack: observability (Monitoring component), manageability (Management component) and policy (Adaptive Manager component) [85]. A key challenge related to leveraging the AS3 Stack is to propose a means of introducing adaptability to layers of the SOA system in accordance with the AS3 Pattern, as well as managing adaptability in an effective way.

The AS3 Pattern can be uniformly applied to systems consisting of both Real-World and Virtual Services. Figure 11 depicts the concept map which reflects the application of the AS3 Pattern in such heterogeneous systems. Blue elements have already been introduced in Sect. 3—they are related to SOA systems. The remaining (green) elements are directly related to the concept of AS3. The AS3 Pattern draws upon Adaptive SOA as one of the possible approaches to modelling adaptation in SOA systems. As presented in the concept map, the AS3 Pattern models enrichment of Middleware and Infrastructure, placed on the "where" axis in Fig. 9, which presents the Adaptive Systems Space. The purpose of enrichment is to enable adaptation of Composite Services and reconfiguration of Physical Resources, with particular focus on Real-World Devices. Enrichment of Middleware concerns all components of the AS3 Pattern, while enrichment of Infrastructure involves only a restricted subset of AS3 Pattern components, i.e. Instrumentation (sensors and effectors), Monitoring and Management. Since hardware is less flexible than software, full implementation of the adaptation loop is rather difficult, as highlighted in Sect. 4. Infrastructure sensors and effectors are enabled by reconfiguration of Physical Resources which could be either Computers or Real-World Devices. Middleware is enriched with all AS3 Pattern and supports the complete adaptation loop. An important point is that instrumented infrastructure enables multi-level adaptation of both Real-World and Virtual Services, performed on the Middleware level.

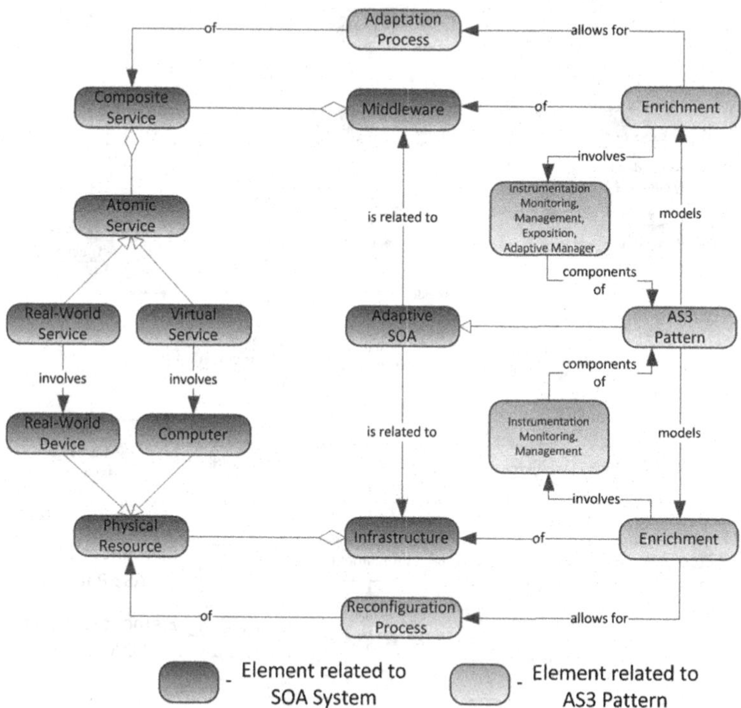

Fig. 11 Application of the AS3 Pattern in heterogeneous systems

The following parts of this section explain some of the concepts introduced in Fig. 11. The first part focuses on the Middleware and explains the enrichment and adaptation enabled by the AS3 Pattern, while the second part is devoted to Infrastructure and its enrichment, highlighting the reconfiguration of Real-World Devices.

Enrichment Supporting the Adaptation Process in the Middleware Layer

Figure 12 presents Middleware enrichment with the use of the AS3 Pattern. First of all, it is assumed that Middleware has a layered structure and that each layer can be divided into a part which provides the Runtime Environment and a part containing the layer's Logic. Logic is delivered by some Artifacts which are specific to the given layer. The Runtime Environment for Artifacts is assumed to be provided in the form of a Container. Layer-specific Artifacts are simply deployed to the Container, which provides them with the Communication feature. As mentioned before, the Middleware enrichment involves all components introduced by the AS3 Pattern. Realization of such enrichment assumes that Monitoring, Management, Exposition and Adaptive Manager components are simply deployed to the Layer's Container in the form of Layer-specific Artifacts. The Instrumentation component is handled

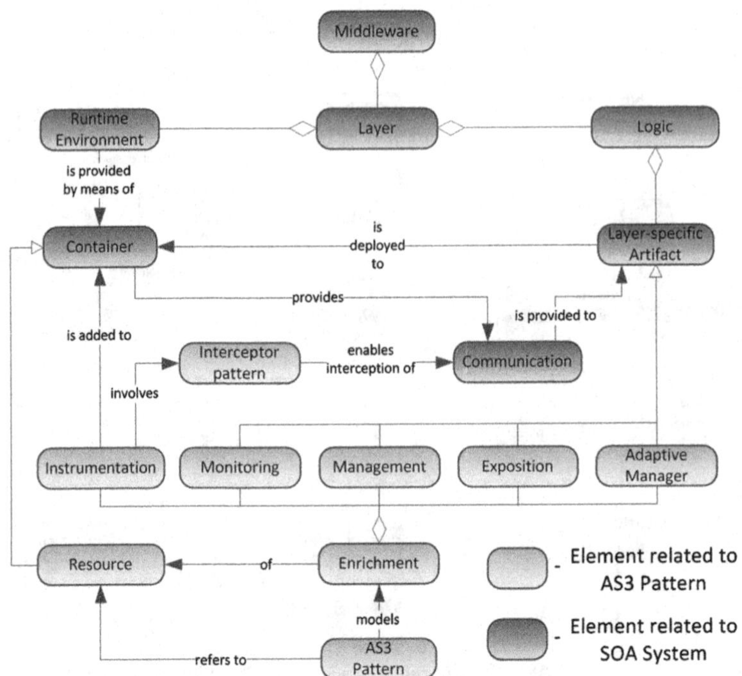

Fig. 12 Enrichment of Middleware using the AS3 Pattern

differently. Since the AS3 Pattern models the enrichment of some Abstract Resource, Middleware Layer Containers are treated as Resources to which Instrumentation is added. It is therefore assumed that Instrumentation is introduced in the Container (by whatever means) and that it applies the interceptor design pattern. The use of this pattern allows for interception of communication performed by Artifacts and leverages this for implementation of Sensors and Effectors which provide monitoring and management features for respective components of the AS3 Pattern. The presented design of Middleware enrichment carries several important advantages:

- Communication between Layer-specific Artifacts can be monitored and managed in an non-intrusive way, which is transparent for applications.
- Monitoring, Management, Exposition and Adaptive Manager components can be easily deployed and managed owing to management features provided by the Runtime Environment of a given Layer.
- Communication between AS3 Pattern components can be easily performed with the use of the Communication feature provided by the Container.

The final purpose of Middleware enrichment is providing mechanisms required by adaptation of Composite Services and used to implement the application's logic. In order to realize this goal, AS3 assumes a certain structure of Composite Services. Specifically, it is assumed, that on a higher level of abstraction each Composite

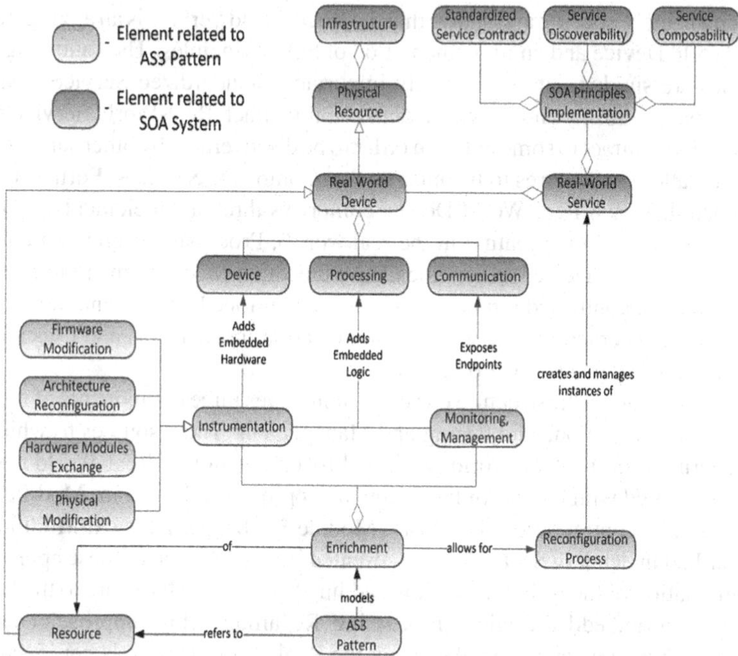

Fig. 13 Enrichment and Reconfiguration of the Infrastructure using the AS3 Pattern

Service can be described as an abstract composition in which abstract services are, in turn, described by their features without referring to a particular instance. For each abstract service several different instances can be deployed and used during execution of an application. Middleware enrichment allows the decision subsystem to dynamically recompose the application by selecting a set of service instances for each service that belongs to the Composite Service. When decisions are made locally, i.e. without regard to how a particular service instance may influence the overall application, the results often lead to unsatisfactory solutions. To improve the outcome of adaptation, approaches such as usage of stochastic models may be used to estimate global system behaviour.

Enrichment Supporting the Reconfiguration Process of the Infrastructure Layer

Figure 13 presents the concept of the Infrastructure enrichment using the AS3 Pattern. The figure focuses on the aspect of Real-World Services and Real-World Devices. Mirroring the relations depicted in Fig. 11, the Infrastructure aggregates Physical Resources, some of which may be Real World Devices. Figure 13 shows that the Real-World Service always involves some Real World Device. Figure 13 makes this involvement more specific and shows that the Real-World Service is an aggregation of a Real World Device and an implementation of SOA Principles. The following SOA Principles are singled out as especially important: Standardized Service Contract, Service Discoverability and Service Composability. Each Real-World Service needs to publish its Contract to some entity in order to be discoverable by other services, and as a result, allow its features to be integrated in Composite Services. Furthermore, it is assumed that each Real-World Device comprises three main elements: a Device (which implements some feature in the real world), Processing Logic (which deals with controlling the Device) and some Communication Mechanisms (which enable the Device to be controlled remotely). The aforementioned implementation of SOA Principles is concerned mostly with Communication mechanisms and uses them to expose Processing in a service-oriented way.

As stated before, Infrastructure enrichment involves three components of the AS3 Pattern: Instrumentation, Monitoring and Management. The resources to which the AS3 Pattern refers are Real-World Devices. Instrumentation of Real-World Devices can be performed with the use of the following approaches: Firmware Modification, Architecture Reconfiguration, Hardware Module Exchange, Physical Modification (all described in detail in Sect. 4.1). As presented in Fig. 13, each of these approaches is a combination of the following actions: adding Embedded Hardware to the Device and/or adding Embedded Logic to Processing. Regardless of the approach used, the result of such actions is always the same: the Real-World Device is equipped with Sensors and Effectors. Monitoring and Management components are handled in a somewhat different manner than Instrumentation. While they can also involve some additional Embedded Logic, they mostly focus on exposure of endpoints by means of Communication mechanisms. The presented realization of Infrastructure enrichment allows for Reconfiguration, which can be used to spawn new Real-World Services and manage them in the context of a specific Real-World Device.

5.2 *The AS3 Process*

The purpose of the AS3 Process is transforming systems built in accordance with the S3 Model into adaptive environments and managing the adaptation process across different S3 Layers. The assumption of the AS3 Process is that the AS3 Pattern may be applied only to selected layers. Leveraging the potential of specific layers in the context of the adaptation loop may enhance the SOA environment with (among others) the following features:

- Dynamic service sizing [64]: scaling the service to adapt to changing load conditions, either by (i) scaling up (i.e. resizing a running service component, for example by increasing or decreasing its memory allocation), or (ii) scaling out (adding or removing instances of service components).

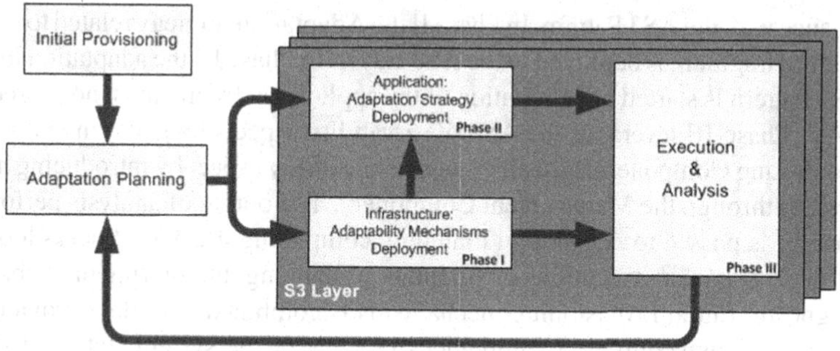

Fig. 14 Abstract view of the AS3 Process

- Policy-driven operation optimization [85]—flexibility can be controlled by rules, following an "event-condition-action" approach in which certain conditions trigger automatic actions to alter the service's capacity.
- Cross-business process monitoring and management [61] e.g. to enforce a close feedback loop control paradigm.

The AS3 Process involves the selection of layers that need to be enhanced with adaptability. Such an approach alleviates the overhead incurred by any adaptation-related activities that do not contribute to improving the performance of the application. An abstract view of the AS3 Process in presented in Fig. 14. The Process relies on the following two concepts related to adaptability deployment supported by the AS3 Pattern:

- Adaptability Mechanisms (AM)—these affect all components of the AS3 Pattern, i.e. Instrumentation, Monitoring, Management, Exposition and Adaptive Manager. Deployment of those components into the S3 Layer transforms it into the AS3 Layer, capable of enforcing the Adaptation Strategy.
- Adaptation Strategy (AS)—configuration of Adaptability Mechanisms which drives the adaptation loop. The strategy always refers to sensors and effectors needed to monitor and manage a fragment of the application whose adaptation is defined in the strategy. This strategy is enforced by the Adaptive Manager.

The central activity of the AS3 Process is Adaptation Planning which selects S3 Layers in which adaptation has to be introduced and prescribes an Adaptation Strategy for each layer. A prerequisite for starting the AS3 Process is Initial Provisioning. The purpose of Initial Provisioning is installation of agnostic monitoring and management mechanisms which cut across all S3 Layers. These mechanisms enable discovery of resources belonging to S3 Layers and monitoring of selected QoS parameters. With their help it becomes possible to identify S3 Layers which require Adaptability Mechanisms and design an initial Adaptation Strategy for selected application fragments.

Parts of the AS3 Process related to selected S3 Layers can be divided into three phases. In Phase I, Adaptability Mechanisms are deployed to the infrastructure in accordance with the AS3 Pattern. In Phase II the Adaptation Strategy related to a given application fragment is deployed to the AS3 Layer. In Phase III the adaptation loop of the AS3 Pattern is started and execution of the application fragment is monitored and analyzed. Phase III leverages the AS3 observability aspects by gathering data from the Monitoring Component, as well as its manageability aspect by introducing minor corrections through the Management Component. The output of analysis performed in Phase III is passed to Adaptation Planning, completing the AS3 Process loop.

During subsequent executions of Adaptation Planning, monitoring data obtained from agnostic Initial Provisioning mechanisms is combined with data provided by AS3 Layers. This results in a comprehensive view of the system state and shows how the adaptation process influences the fulfillment of consumer requirements. As a result of Adaptation Planning, some (or all) of the following actions may be executed:

- deploying Adaptability Mechanisms to an S3 Layer which had not been instrumented before;
- modifying the previously-deployed Adaptation Strategy or Adaptability Mechanisms (for instance by adding more mechanisms);
- removing the previously-deployed Adaptation Strategy or Adaptability Mechanisms which are no longer needed.

In a given S3 Layer, execution of actions enforced by the Adaptation Planning may involve Phase I, Phase II or both phases in such a way that Phase I occurs before Phase II. If Adaptation Planning does not enforce any changes in the infrastructure and on the application level then the Process progresses directly to Phase III where the execution of the adaptation loop is monitored. Phases of the AS3 Process can be performed at different stages of the system's lifecycle. e.g. during provisioning and execution. This distinction is important as some extensions can be introduced either during deployment or at runtime. The AS3 process reflects this fact by introducing three different models (Static, Hybrid and Dynamic) for each phase of the AS3 Process.

The Static Model assumes that a given phase cannot be performed at runtime. In Phase I this means that the infrastructure (or part thereof) has to be shut down and then, once appropriate changes are applied, the infrastructure is again provisioned and returns to the execution phase. In Phase II the same applies to the application. The application has to be stopped and redeployed to support adaptability required by the Adaptation Strategy. The static model can be imposed e.g. by execution containers and applications which do not support runtime modifications. The Dynamic Model assumes that a given phase of the AS3 Process can be performed at runtime. In Phase I the Adaptability Mechanisms are deployed to the infrastructure without the need for a restart. In Phase II deployment of the Adaptation Stategy does not involve halting the application. The Hybrid Model assumes that some modifications of the adaptation process might be performed using the Dynamic Model while others may need to follow the Static procedure. Both Phases (I and II) can be handled in Static, Dynamic or Hybrid Models. Phase III is performed exclusively in the Dynamic Model since it is closely related to execution of the system and oversight of the adaptation process.

Additionally, Phases II and III are executed depending on the implementation of a given S3 Layer as well as the capabilities of Adaptability Mechanisms designed for that layer.

Having presented an abstract view of the AS3 Process and discussed its related aspects, we can summarize the exact steps of the Process in the following list:

1. Performing Initial Provisioning of the whole system;
2. Discovering resources present in all layers;
3. Performing adaption planning which influences steps 5 and 6;
4. Deploying/modifying/undeploying Adaptability Mechanisms in selected S3 Layers according to a suitable execution model;
5. Deploying/modifying/undeploying Adaptation Strategy Agents in selected S3 Layers according to a suitable execution model;
6. Deploying/modifying/undeploying Adaptation Strategy in selected AS3 Layers according to a suitable execution model;
7. Executing the adaptation process and analysing monitoring data provided by AS3 Layers and Initial Provisioning.

The process then continues by jumping to step 2.

The complexity of real-life SOA systems calls for tools which support effective realization of the AS3 Process. The most important core features required from such tools are as follows:

- Selective non-intrusive monitoring installable on demand across different layers of the S3 Model;
- Discovery of services and their interconnection topology during system operation;
- Flexible mechanisms for presenting and managing monitoring data in order to support system response evaluation and adaptation strategy planning;
- Dynamically defined and pluggable adaptability policies for execution of operations;
- On-demand installation of effectors to enforce adaptation decisions.

A toolkit which supports the AS3 Process and meets all the listed requirements—namely, the AS3 Studio—is presented in the next section.

6 AS3 Studio

The AS3 Process is a high-level concept which is platform-independent. However, tools that automate it have to be platform-specific and support a selected set of technologies. Recently, many vendors of SOA-related solutions have begun to focus on supporting dynamic and manageable software environments executed within OSGi [54] containers. More importantly, some of those solutions available as open-source software can be used for implementation of different layers of the S3 Model: examples include Fuse ESB (Services and Integration Layers), Apache Tuscany (Service

Components Layer) and Business Process engines: Apache ODE, JBoss jBPM (Business Process Layer). In light of this, OSGi emerges as the natural choice for the base implementation technology of AS3 Studio.

6.1 OSGi Monitoring and Management Platform

At its core, OSGi [54] is a dynamic component-oriented Java platform for applications developed in accordance with service-oriented design principles [20]. The OSGi framework provides an execution environment for applications, which are called bundles. Bundles expose their features as services according to the "publish, bind, and find" model [30]. Each bundle can be deployed and activated at runtime. Bundles can dynamically select services to be used. Furthermore, the OSGi Framework enforces strict modularization of bundles, which entails that there is no need to shut down the entire JVM when a particular bundle is modified.

The AS3 Studio is a suite of several components deployed over the OSGi Monitoring and Management (OSGiMM) platform:

- AS3 Tools for the Middleware Layer,
- AS3 Tools for Real-World Services,
- AS3 Console.

OSGiMM [62] provides mechanisms for monitoring and managing OSGi containers federated by means of Message Oriented Middleware consisting of a network of message brokers. It enables cooperation of services deployed in OSGi containers distributed across a Federated OSGi system.

OSGiMM consists of core instrumentation and a set of bundles. The core instrumentation is required for dynamic management and monitoring. During installation, the instrumentation has to be added to each OSGi container with the use of provided scripts. Accordingly, OSGiMM bundles have to be deployed to each container of federation. The fundamental feature of OSGiMM is discovering information about all services, bundles and containers of the Federated OSGi as well as their structural relations.[1] Such information is later referred to as a topology. The implementation also provides efficient invocations of service groups, which are used as a foundation for typical management and monitoring patterns [86] in the OSGiMM.

In summary, OSGiMM provides generic features which affect the implementation of the following adaptability aspects:

- Declarativity—the user specifies a monitoring scenario, indicating which parameters need to be monitored and which topology elements are provided. A scenario is specified declaratively and can be exported to a distributed repository for future use.

[1] For example, a container may comprise bundles while a bundle may consist of specific services.

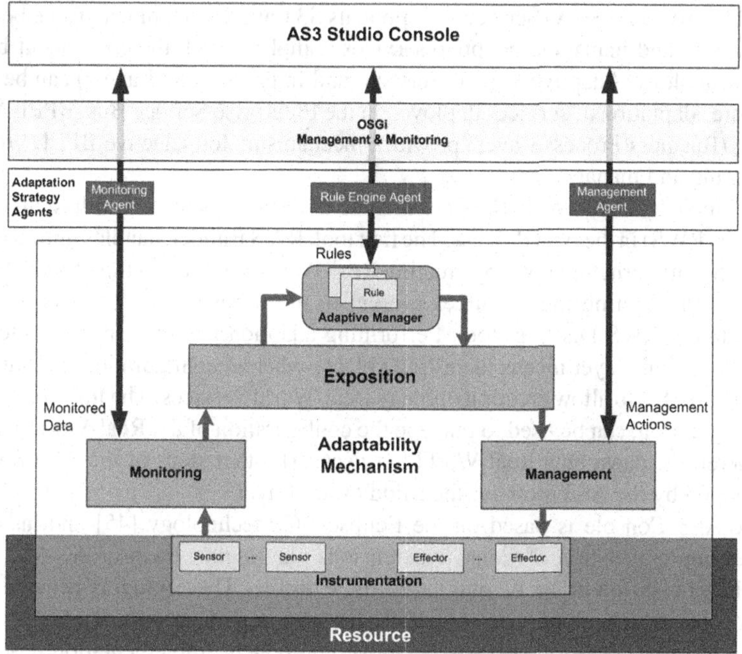

Fig. 15 AS3 toolkit architecture

- Dynamism—monitoring scenarios can be activated at any time and activation does not involve halting the application that is to be monitored. Activation triggers realization of on-demand instrumentation.
- Selectivity—instrumentation is only triggered in locations which are important for a given monitoring scenario; therefore the overhead incurred by monitoring is restricted to a minimum.
- Self-configuration—the federation may change, e.g. when a new container is added or when some services are undeployed. Regardless, the realization of the monitoring scenario is ensured.
- Flow aggregation—when there are multiple users who wish to perform the same monitoring or management activities, the data flow related to the task is aggregated in order to reduce the cost of transmission.

All those features make the OSGiMM a good solution for managing and monitoring OSGi services in a unified manner, which is important in the context of adaptability aspects.

Each of the AS3 tools for the Middleware Layer is a set of bundles that implement Adaptability Mechanisms for a particular S3 Layer, and a set of agents for controlling these mechanisms. The specified architecture of an AS3 Tool is depicted in Fig. 15. There are three different agents: Monitoring, Rule Engine, and Management, all referred to as Adaptation Strategy Agents (ASA). Currently, three AS3 Tools are available. Adaptive SCA (Service Components S3 Layer) is a tool which can be used for building and managing adaptive services compliant with the component-based SCA technology. Adaptive VESB (Services and Integration S3 Layers) can be used to ensure adaptation of services deployed in the Enterprise Service Bus. BPEL Monitoring (Business Process Layer) provides mechanisms for selective BPEL process monitoring and management.

AS3 tools for Real-World Services include three utilities for use with Real-World Services (RWS) in the AS3 Process. The first tool, RWS Builder, handles initialization within the infrastructure layer by creating Real-World Services on top of Real-World Devices. Discovering the resources present in the infrastructure layer is possible thanks to the RWS Discover tool. Performing adaption planning in the context of the infrastructure layer means identifying places where reconfiguration mechanisms should be added to allow creation of many Real-World Services. The final tool, RWS Reconfiguration, can be used to manage the configuration of the Real-World Device and therefore spawn new Real-World Services. All other steps of the AS3 Process are covered by the AS3 tools for the Middleware Layer.

The AS3 Console is based on the Eclipse RCP technology [45] and, as such, enables implementation of Console extensions via plugins. Each of the AS3 tools contributes a different set of plugins to the Console. These plugins provide GUI components which support layer-specific features, e.g. definition of SCA or ESB adaptation rules, creation of facts, adaptive component implementation, complex service modelling, BPEL engine instrumentation and discovery, etc. The AS3 Console is also extended with OSGiMM plugins, which allow it to discover, monitor and manage whole Federated OSGi. In order to connect to the Federated OSGi it is necessary to provide the address of the federation container which will function as the entry point for the console. Many AS3 Consoles can be connected to federation containers simultaneously and it is also possible to connect more than one Console to a single container. In this way it is possible to use the AS3 Studio from many points of the federation at the same time.

As listed in the previous section, the AS3 Process involves execution of several steps. These steps are supported by the AS3 Studio according to the workflow presented in the previous section. Step 1 (Initial Provisioning) has to be performed manually. During this step OSGiMM bundles, along with their core instrumentation, are installed in each of the OSGi containers. Subsequently the configuration of message brokers has to be provided. During this process a logical topology of connections between federation nodes is created which combines the containers into a Federated OSGi. Steps 2 (Discovery) and 3 (Adaptation Planning) are described later on in the section, as they involve the continuous adaptation process. The AS3 Studio automates steps 4 and 5 (Deploying/modifying/undeploying Adaptability Mechanisms and Adaptation Strategy Agents). The user may choose (using the AS3 Console) which mechanisms or agents should be installed in each container. Since OSGiMM discovers the topology of the federation it is possible to transparently transfer specially prepared bundles to remote containers and install them automatically. Once this is done, the user may check whether all operations finished successfully and whether AM/ASA are present in the discovered federation topology. Furthermore, AS3 Studio automates Step 6 (Deploying/modifying/undeploying Adaptation Strategy) by employing the preinstalled Adaptation Strategy Agents. The Adaptation Strategy is specific to a particular layer of the AS3 Model and will therefore be presented along with overall description of the tools.

Steps 2 (Discovery) and 3 (Adaptation Planning) are both performed in a continuous manner during runtime, with full support of the AS3 Studio. This support exploits the concept of the Dynamic Monitoring Framework [86] proposed in our earlier work. The Monitoring Agent in each AS3 Layer, as well as OSGiMM itself, implement an interface which provides topology discovery of resources available in this layer. Additionally it is possible to declaratively specify a Monitoring Scenario which can contain two types of Monitoring Subscriptions:

- Topology Subscription—related to monitoring changes in a topology fragment specified in the subscription,
- Metric Subscription—related to monitoring of topology element metrics (performance, availability, reliability).

The Monitoring Scenario is created with the use of the AS3 Console. When a Scenario is activated, OSGiMM sends subscriptions to appropriate containers where they are passed to the Monitoring Agents. Each agents starts a monitoring process on behalf of a given subscription. Results are communicated to the AS3 Console. The Console provides configurable monitoring panels which can be tailored to a given Scenario, enabling visualization of monitoring data collected from different parts of the federation. The described mechanisms ensure continuous discovery and monitoring, thus allowing the operator to identify elements of the system where adaptation should be applied.[2] When the system changes, the monitoring processes adapt automatically and monitoring panels continue to relay information relevant to Adaptation Planning.

Steps 6 and 7 of the AS3 Process are specific to each AS3 tool and will be described separately for each of the AS3 tools in the following paragraphs. .

6.2 Adaptive VESB

Adaptive VESB is the AS3 tool which introduces adaptability mechanisms within the Integration Layer of the S3 Model. The tool exploits model-driven adaptation [72] for SOA. The system analyzes composite services deployed in the execution environment and adapts to QoS and QoE changes. The user composes the application in a selected technology and provides an adaptation policy along with a service execution model. The architecture-specific service composition layer continuously modifies the deployed service in order to enforce the adaptation policy. The abstract plan, providing input for architecture-specific service composition, can be hidden and used only by IT specialists during application development. System behaviour is represented by the service execution model. The composite service execution model

[2] Further details can be found on the AS3 Studio website and in our previous paper [86].

is an abstraction of the execution environment. It covers low-level events and relations between services, exposing them as facts in the model domain. Decisions taken in the model domain are translated to the execution environment domain and then executed. An adaptation strategy is specified, according to a user-provided service execution model, taking into account the quality of execution metrics. It relies on configuring ESB [73] elements that are responsible for dynamic selection of service instances for particular use cases in order to provide the desired quality.

Adaptability Mechanisms for ESB are constructed around the AS3 Pattern and consist of the following elements: instrumentation component, monitoring component, management component and adaptive manager. The instrumentation layer enriches ESB with additional elements providing the adaptability transformations necessary to achieve adaptive ESB. These elements are responsible for managing sensors and effectors installed in ESB. Sensors gather information about running applications by intercepting messages, while effectors are used to influence message flow in ESB by modifying the routing table of NMR (which is a core element of ESB involved in message processing). The monitoring layer supplies notifications of events occurring in the execution environment. As the volume of monitoring information gathered from ESB might overwhelm the Exposition layer, events are correlated with one another using Complex Event Processing and notifications pertain only to such complex events. The Exposition Layer is responsible for maintaining and updating the composite service execution model. Facts representing the state of the system or events occurring within are supplied to the Adaptive Manager which analyses them and infers decisions to be implemented in the execution environment by the Management Layer.

Adaptation Strategy Agents for ESB provide interfaces for managing adaptability mechanisms. The Monitoring Agent provides high-level features for management and configuration of sensors deployed in ESB. The Management Agent can be used to influence message routing in ESB, while the Rule Engine Agent manages the composite service execution model and deploys adaptation strategies. The Adaptation Strategy for Adaptive ESB consists of the following elements:

- Definition of a composite service execution model,
- Definition of new facts that represents events occurring in the integration layer,
- Definition of an adaptation policy that uses the previously defined and deployed facts.

All these elements can be configured using the ASA provided by VESB tools. To simplify strategy definition, VESB Tools also provide a GUI for the AS3 Studio Console.

6.3 Adaptive SCA

Adaptive SCA is an AS3 tool designed to enhance the Service Components Layer of the S3 Model with adaptability features. It enables service designers to assemble services from components according to the SCA specification. SCA is a technology-independent solution and therefore supports components created using many different technologies as well as various communication protocols. A set of components connected with one another within a service is also called a composition.

SCA uses the Dependency Injection Pattern [58] to model a composition: adaptation mechanisms need to be introduced on the level of references between components. Adaptive SCA instruments these references and uses OSGiMM to monitor communication between components, providing suitable mechanisms to choose which component should handle a particular reference.

The Adaptation Strategy for Adaptive SCA defines different sets of components which are to be used in case of specific situations discovered by the monitoring system. These sets are also referred to as composition instances. For instance, if a service composed of particular components becomes too slow, it may temporarily switch over to other components which provide better QoS (however at a higher cost).

The Adaptation Strategy definition for Adaptive SCA consists of the following elements:

- Service model definition, i.e. a composition and a set of its instances,
- Definition of adaptation policies that use predefined policy templates and metrics gathered by OSGiMM.

If required, it is possible to create policy templates tailored to a particular service. All these features are supported by the ASA Adaptive SCA tool. To simplify their management, Adaptive SCA also provides a GUI for the AS3 Studio Console.

Deployment of a service with adaptability mechanisms requires additional actions performed automatically by the AS3 Studio. Upon defining the composition and its instances, the Adaptive SCA Tool modifies the SCA deployment descriptor by injecting monitoring and management agents into components' references. Afterwards, service provisioning can be performed and the service executed in accordance with the AS3 Process.

6.4 BPEL Monitoring

The Business Process Layer of the S3 Model supports composition of business processes from the available services describing steps that need to be executed to complete a given process. This allows non-technical users to declaratively describe business process flows either in terms of a document in a dedicated language (such as BPEL) or by using graphical tools. Formally defined business processes can be executed by business process execution engines which interact with services according to the specified flow. Modern business processes are often highly dynamic, which means that proper aggregation and analysis of performance indicators representing their execution is important from the managerial viewpoint.

The AS3 Studio offers a highly configurable monitoring system called the Business Process Monitoring Platform, which support efficient capture, propagation and visualization of the data needed for the business process execution analysis. This tool is implemented within an OSGiMM container and utilizes important system services in order to satisfy the following functional requirements:

- discovery and presentation of the existing business process engines, deployed processes and running process instances,
- on-demand monitoring of the selected discovered elements (mainly business processes),
- presentation and up-to-date view of business process definitions and the execution state of running process instances.

The Business Process Monitoring Platform was developed with the following nonfunctional requirements in mind:

- only the necessary data is transmitted between system elements,
- the monitored business processes are not affected by the monitoring process,
- the architecture remains scalable and extensible for multi-vendor systems,
- a standard data model is in place for all the monitored components.

The system relies heavily on mechanisms provided by the OSGi standard which naturally create a SOA environment in a single Java Virtual Machine. System elements are exposed as OSGi services capable of dynamic discovery and runtime reconfiguration. The core layers in the presented architecture are:

- ESB with OSGiMM—a communications backbone that provides seamless integration of system components, their discovery as well as transport of monitoring data.
- BPEL Monitoring Domain—consists mainly of business process engines along with their respective sensors: BPM Engine Monitors, business processes, independent event processing components and infrastructure nodes where such elements are deployed.
- Monitoring Console—a GUI system element that supports configuration of the monitoring system and exposes key system features to the user.

Each monitored business process engine is associated with a dedicated monitor which generates standardized notifications of changes in the executing process.

The communications layer is responsible for propagation of events, as well as filtering them when no subscriber is interested in a particular kind of event. The presented monitoring platform uses OSGiMM and supports model-based declarative definitions of the monitoring process by means of a monitoring scenario. In light of the above, business process monitoring can be treated as an example of a well-defined scenario.

There are two types of loosely coupled clients connected by the OSGiMM backbone: event sources (mainly business process engine-specific monitors—BPM Engine Monitors) and event consumers such as Monitoring Consoles. The architecture also covers event interceptors, e.g. rule-based event processors that are hybrids of these two client types: they intercept the flow of events from other event sources to the Monitoring Console for the purpose of processing.

In the context of the presented work installing a monitoring scenario is equivalent to creating a monitoring subscription for the events whose flows is enabled by the activation of the monitoring scenario. As the main monitoring goals are twofold (monitoring specific topology elements and the entire topology), topology subscription is also supported.

In comparison with the layered architecture of OSGiMM, we can observe relocation of the business process analysis logic. Whereas in standard OSGiMM metric events are processed by building a CEP processor, in the presented platform analytical logic (rule-based event processors) is moved upstream in the layered architecture, becoming an optional and reconfigurable topology element, to enable better management by business users.

6.5 Infrastructure Enrichment Tools

In this subsection two software tools for creating and managing Real-World Services are presented. The first tool (RWS Builder) supports creating RWS by adding service logic, transforming a Real-World Device into a Real-World Service. The second tool (RWS Reconfiguration) enriches the infrastructure with a reconfiguration feature which can be used in the service adaptation process or for incarnation of new Real-World Services. Both tools operate in a specific hardware environment containing FPGA (Field Programmable Gate Array) and CPLD (Complex Programmable Logic Devices) chips from Altera Corp., microcontrollers with the ARM-Cortex core from ST-Microelectronics, Tibbo Ethernet and Wi-Fi modules on custom circuit boards.

6.5.1 RWS Builder

The process of creating a Real-World Service on the basis of a Real-World Device consists of several steps described in Chap. 3. One of those steps involves generating the service logic. Here, the programmer's effort can be greatly reduced by using the tool described in the following section.

Real World Service logic is typically implemented as a hardware description of an FPGA chip set up as a software module for a microcontroller-based device. Developing this logic manually to produce a hardware description is a tedious job which requires knowledge of the underlying hardware architecture and proficiency in using hardware description languages (i.e. VHDL, Verilog).

RWS Builder is an example of a software tool for high-level code synthesis. Its functionality is tailored to two specific types of RWS: motion detection and object classifier services. The generated project files depend on the selected functionality and can be filled with service-specific code by the developer. RWS Builder integrates multiple heterogeneous design tools from different vendors into a single applica-

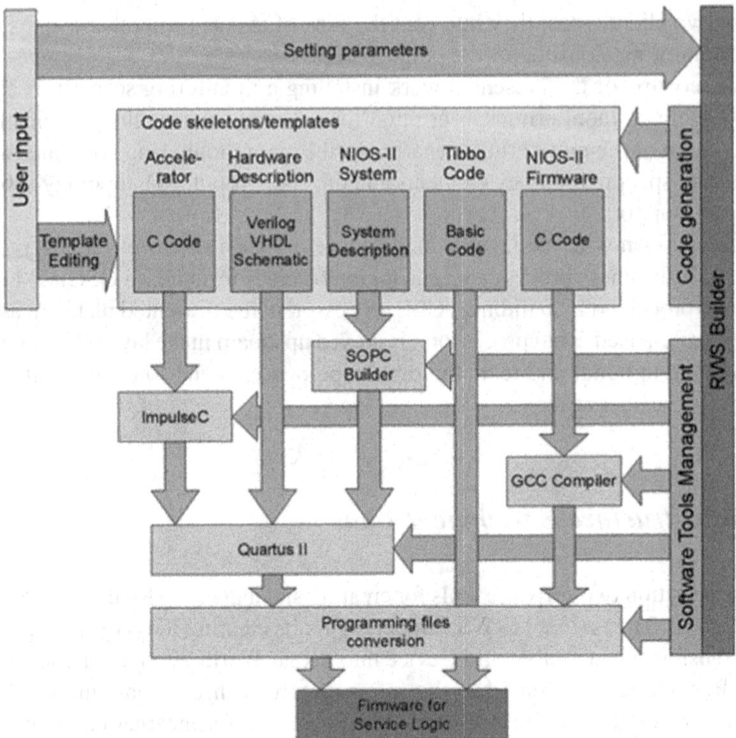

Fig. 16 Service logic generation using RWS Builder

tion that manages all required design, implementation and installation steps. It is responsible for creating an optimized code skeleton to be compiled by the ImpulseC C-to-HDL compiler. It also generates an adequate standard hardware design for the Altera Quartus-II environment and an optimized software template for embedded FPGA microcontrollers and network communication modules. The service logic generation procedure is visually depicted in Fig. 16.

6.5.2 RWS Reconfiguration

An important aspect of providing RWS is introducing solutions which enable discovery of such services along with their reconfiguration. This is the key added value of AS3. Discovery tools are necessary to locate RWS in a network and to recognize their potential capabilities and what kind of logic they can handle.

In order to provide a list of all Real-World Services to users and adaptation tools a live repository has been implemented. The selected approach follows the same paradigm as in Service Location Protocol (SLP)-and Simple Service Discovery Pro-

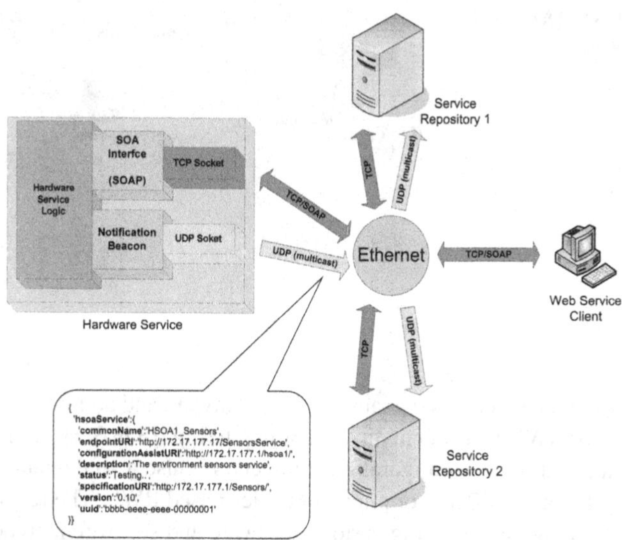

Fig. 17 RWS Discovery method

tocol (SSDP)-based discovery schemes. The general architecture of the proposed discovery solution is presented in Fig. 17.

Each of the services available in the network announces its presence using User Datagram Protocol (UDP) advertisements sent out to a well-known multicast address. Such an advertisement, called a notification beacon, carries service description, including metadata (endpoint information, Universal Unique Identifier (UUID), timestamp, lifetime, human-readable description etc.) and a link to a service specification document. Notification beacons are received by dedicated multicast listeners which in turn update the Lightweight Directory Access Protocol (LDAP) service repository. Data stored in different repositories can be synchronized using LDAP servers tools. In this way one repository can collect service announcements from more than one IP network. Service announcements may employ addresses accessible from outside the local network if this type of client is expected. Access to the repository is usually provided using a Transmission Control Protocol (TCP) connection. In the absence of network configuration restrictions computers from all over the Internet can browse the repository and utilize hardware services. From the repository point of view, any external entity which needs to update the repository contents needs to provide security credentials and also pass authorization checks. The repository itself was implemented as an LDAP database, leveraging existing security mechanisms to provide Authentication, Authorization, and Accounting (AAA) services. It can be used by automatic tools as well as end users interested in browsing through the repository contents and able to leverage LDAP query and search mechanisms.

Once the hardware device is detected, it might be necessary for the user to choose and upload firmware and configuration files which provide service-specific logic. The

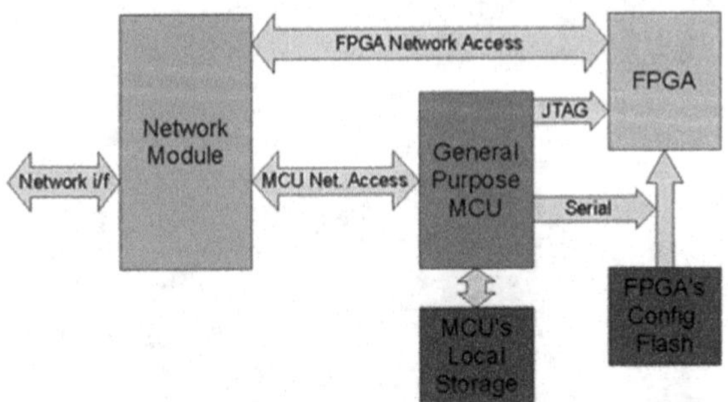

Fig. 18 The RWS Reconfiguration subsystem

reconfiguration process covers two approaches: firmware and architecture modification, both part of RWD instrumentation according to Fig. 13 in Sect. 5. The ability to remotely change hardware configuration parameters is also a vital feature for implementing adaptability in FPGA- or microcontroller-based RWS. The reconfiguration process can be implemented using various hardware and software methods.

In FPGA-based RWS, configuration is stored on add-on Flash memory configuration chips which upload their contents to the FPGA upon power-up or on request. The reconfiguration process itself can proceed in several ways. The remote reconfiguration feature used by the the presented tool (RWS Reconfiguration) relies on a general-purpose microcontroller unit (MCU) locally connected to the FPGA chip and responsible for uploading new configurations (Fig. 18).

In this case the microcontroller is equipped with a wired or wireless network interface and locally accesses FPGA using the Joint Test Action Group (JTAG) interface, the Flash chip using JTAG or dedicated serial interface, or both of these interfaces.

The RWS reconfiguration subsystem can use multiple blocks of local configuration memory in order to store temporary configuration data. As the local FPGA's Flash configuration memory reduces the service reconfiguration downtime, additional local memory in the MCU (Fig. 18) can decrease total reconfiguration time by storing many configuration files. When a new configuration is required, the current one does not need to be overwritten, but can instead be preserved for future reuse. This enables implementation of simple configuration caching, eliminating the need to transfer a new configuration file for each feature. In some implementations the MCU's local storage might be shared between the MCU and an FPGA, and therefore used locally to store data required by the service running on the FPGA.

7 Case Study

The goal of this section is to illustrate the issues discussed earlier on in this chapter. We will show how the AS3 Studio provides support for development of sophisticated applications (see also [17]) and how its adaptability mechanisms can be exploited to transparently maintain the specified quality level of the application's operation.

As already mentioned, advanced service-oriented systems need to provide support for seamless integration of both virtual and real-world services into a single application. Thus, the case study scenario will utilize both types of services and address selected aspects of safety management in the real world (e.g. in an enterprise). Although reliant on specific components, the application presented here can be perceived as a representative of a broader class of solutions related to safety management. We will demonstrate how software services can enhance operation of real-world services, taking over their tasks when necessary—to increase application performance. We will also show that the adaptive infrastructure allows easily introducing various procedures to satisfy application's QoS requirements.

The real-world devices used in this case study have been implemented and instrumented by the authors themselves—their short description is given below.

7.1 Characteristics of the Entrance Protection Application

The general idea behind the application called Entrance Protection is as follows: the face of a person wishing to enter a protected area is captured by a camera. If it is recognized as belonging to an authorized entrant, access is granted and the door lock disengages for several seconds. To ensure sufficient light for the camera a lighting system is installed and connected to a power switch. The switch is activated if a light level detector—enabled just prior to powering up the camera—detects unsatisfactory illumination. The whole process is triggered by a pyrometer which detects rapid, significant temperature changes within the observed area. Figure 19 presents a logical view of this layout with particular elements accessible as services connected to Enterprise Service Bus.

The most interesting part of the system is its pattern recognition feature—so this aspect will be discussed in more detail. A common practice when constructing video surveillance systems is to send the video stream to a central point for processing—in our scenario this involves pattern recognition. In the presented case, however, a much more reasonable approach is to process the data locally at its source avoiding network congestion during data transfer. A digital camera with an embedded pattern recognition module and a library of stored patterns may instead dispatch a simple event, triggering the downstream parts of the business process. Figure 20 presents the Entrance Protection business process in the BPMN notation. The process is triggered by an event generated by the Pyrometer Service when the measured temperature is approximately equal to the normal temperature of a human body. In the "Measure Illuminance" task the Light Detector Service is called to measure illumination. If necessary, the Power Switch service may be used to provide additional lighting. In the next step, the Camera Service is invoked to begin recording and run the recognition algorithm. The service returns its results within a predefined period of time. Upon successful recognition the identifier of the matching pattern is returned and the process progresses to the next step; otherwise it is aborted. In the "Check Access" task an external virtual service is called to check if permission can be granted. If so, the Door Lock service is called to release the door lock and, after 3 more seconds, called again to lock the door.

7.2 Properties of the Real World Devices and Services Used

This section briefly characterizes the Real World Services utilized in our case study. They have been instrumented and are controlled using our universal RWS module providing them with IP communication feature. Consequently, they expose their functionality as SOAP endpoints, which is a common solution applied by third-party devices. Such an approach is fully satisfactory for controlling each device separately; however to be able to easily build sophisticated applications more advanced mechanisms are necessary. Using RWS Builder tool described in the previous section, all Real World Services used in the discussed scenario have been equipped with a dedicated proxy exposing their functionality in the OSGi environment and thus enabling their utilization in the OSGi Monitoring and Management layer of the AS3 Studio. In the presented case study several real-world services are used such as camera, door lock, light detector, power switch and temperature detector.

Camera

The smart camera can be used to implement video surveillance and environment monitoring services. In the presented scenario the smart camera is equipped with an

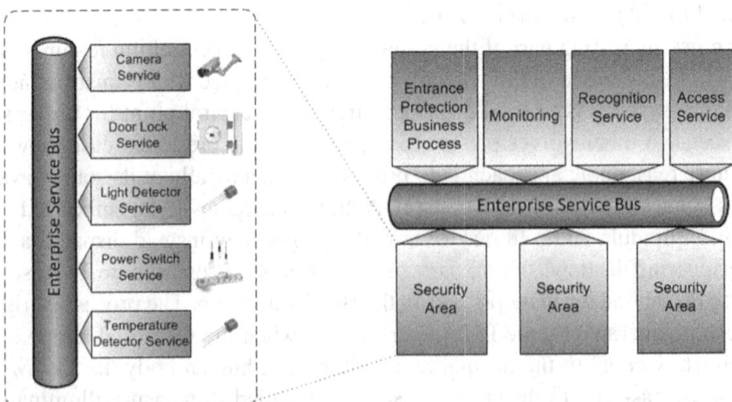

Fig. 19 A single instance of the entrance protection application (*left-hand figure*) and integration of its multiple instances (*right-hand figure*)

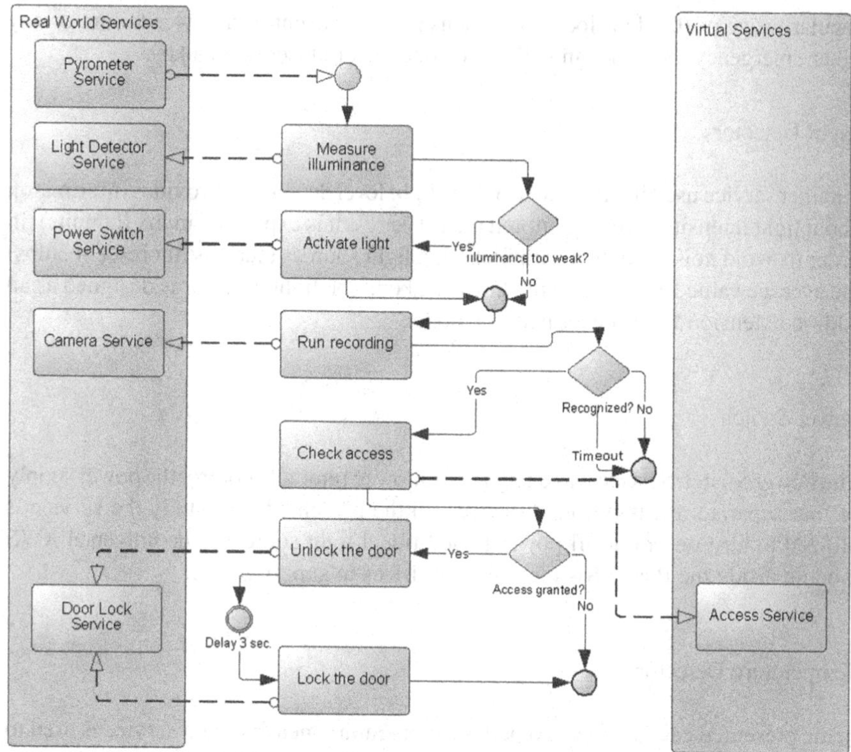

Fig. 20 Entrance Protection business process in the BPMN notation

object classifier functionality and the internal classification algorithm is trained for facial recognition.

The object classifier service is designed to perform classification of similar objects, e.g. faces, car models or road signs. The service applies an algorithm based on Kernel Regression Trees, introduced in [66] and implemented in a fast and massively parallel manner in the FPGA. The classifier's interface consist of methods to begin recognition of an object in the captured scene and retrieve the results of recent recognition runs.

Door Lock

The Door Lock Service is the main actuator unit in the security system described in this section. It is able to remotely lock or unlock the door leading to a restricted area in response to client requests. It assumes the form of an anti-burglary lock service, designed as an add-on module for ordinary door locks available on the market. The system consists of an anti-burglary door lock connected to a servomotor with the help of a custom aluminum hitch. It should be noted that this design does not prevent regular usage of the door lock's mechanism. In the event of a power failure or any other emergency the lock can still can be opened or closed with a key.

Light Detector

Another service used in this scenario is the light level detector. It provides information about light intensity at the monitored area. The result is expressed in lux (lx units). In order to avoid noise and deal with flickering light sources (such as fluorescent bulbs) the average value from several readings is taken. The light detector is designed as an add-on extension for a sensor network node.

Power Switch

The Power Switch Service was designed to allow to remotely control the power supply of four separate mains-powered devices. In the presented case study the service is utilized to turn on or cut off power to additional light sources. The universal RWS module inside the Power Switch uses opto-triacs to supply power.

Temperature Detector

In the presented case study a contactless temperature measurement service is used to measure the temperature of an object in the monitored area. Results reported by this service are used to trigger the whole system's logic. The pyrometric detector used to build the service can measure temperatures between -50 and 350° C. The detector is connected to a Sun SPOT device which is one of the nodes of a sensor network.

7.3 Application Quality Considerations

A system implementing the presented concept may appear quite simple, but to operate correctly in a large enterprise with many entry points and thousands of employers it must properly take into account such aspects as scalability, extendibility and ease of integration with other systems running in the enterprise. Business processes are characterized, among others, by their QoS parameters which are specified in the Service Level Agreement. In the adaptive systems introducing such parameters may trigger automatic adaptation mechanisms whose goal is to meet quality criteria regardless of the complexity of the task or any other external factors. In the discussed case the most critical element of the business process is pattern recognition and thus possibility to introduce adaptation in its operation will be here discussed in more detail. An obvious limitation of the presented approach is camera memory capacity—in large systems storing all relevant patterns in its memory may prove prohibitively expensive, or even impossible. To increase the solution's cost-effectiveness without giving up the advantage of local processing some patterns (for example those with the lowest frequency of matches) may be stored outside of the hardware and processed in software. (In light of this concession, transmission of visual data appears unavoidable, although in order to limit data volume only selected images should be sent—rather than the entire video stream.) Fortunately, the pattern recognition task can easily be distributed among many processing (worker) nodes and performed concurrently. The camera pattern recognition module can either return an identifier associated with the matching pattern or fail to do so if it is unable to find an association. The latter case may occur in one of three situations:

- the person whose face was scanned is not known to the pattern recognition module,
- the quality of the captured image was too low (it was too noisy, etc.),
- the recognition process was not able to complete within the required time limit - what is dependent not only on the number of patterns to be analyzed but also on complexity of the transformations applied to each image.

It is evident that it is possible to influence the percentage of recognition failures mainly in the last case. For a set amount of processing power the relation between processing time and recognition failures is inversely proportional. Two QoS parameters can be introduced to describe this behavior:

- maximum processing time (MPT),
- maximum level of recognition failures (MRF).

To be able to control both parameters' values independently it is necessary to influence accessible processing power. In our case additional virtual services can take some processing from the camera. Our goal is to show how to effectively use adaptability mechanisms provided by AS3 Studio to control the amount of additional processing power automatically and dynamically—satisfying imposed QoS parameters.

7.4 Business Process and Infrastructure Enhancements to Satisfy the Application's Quality Requirements

Taking into consideration the discussed QoS requirements the business process should be enhanced to utilize virtual services which implement pattern recognition algorithms similar to the one provided by the Camera Service - the result of its enhancement is depicted in Fig. 21. If the recognition process performed by the Camera Service fails due to a timeout (see the previous subsection), the Entrance Protection business process may download the image currently being analyzed and distribute it to a number of virtual services along with subsets of pattern identifiers to be examined. Successful recognition by any of the nodes triggers an event which immediately aborts the entire processing task on other nodes. The remainder of the business process is realized without any changes. Operation of this business process

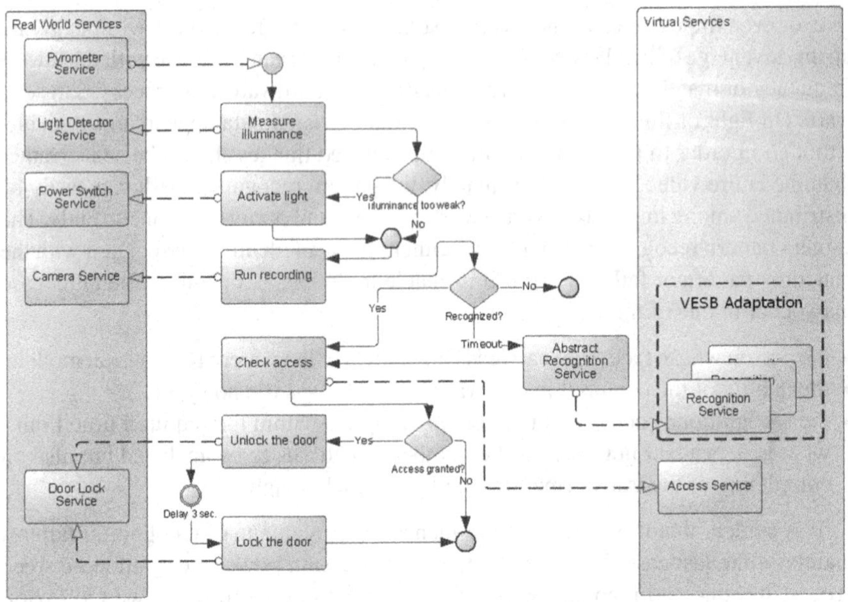

Fig. 21 Enhanced Entrance Protection business process in the BPMN notation

is partially controlled by Adaptation Manager being the part of the Virtual ESB (VESB).

The adaptability mechanisms introduced by VESB enable the business process to be described in an abstract way, without referring to particular service instances. When an abstract composition is instantiated both the number of instances of virtual services and their locations are specified at deployment time. The number of virtual services that perform image recognition may change according to the number of images being processed at the same time [72]. Accordingly, the number of service instances is controlled by the AM deployed in VESB. The Camera Service may contain a database of photos of registered employees, while the photos of trainees might be stored in a database that is accessed by virtual services. When the Camera Service fails to recognize an entrant, that person's picture can instead be processed by a dynamically changing pool of virtual service instances (Recognition Service) implementing the image recognition algorithm. During morning hours, when employees typically arrive at the workplace, the volume of images to be processed is higher than during the rest of the day and the number of virtual service instances should dynamically adapt in order to achieve the desired quality of service. An important issue is how to estimate the number of instances engaged in a particular business process. In this case, a fully sufficient approach is to perform historical analysis of previous executions of the application in the specified timeframe. This data can be gathered transparently by monitoring business process execution and conducting analysis of communication patterns in the Virtual ESB instance dedicated to this application. The presented adaptation strategy may be used to decrease the cost of service maintenance [73]. Additional virtual service instances might be brought online when necessary and then turned off, reducing resource and power consumption while other working instances are capable of providing satisfactory Quality of Experience (QoE).

Monitoring the activity of a VESB instance can lead to enhancements of the recognition process. Many different heuristics can be applied (e.g. noting that people usually enter the building in groups, etc.) The gathered statistics can trigger reconfiguration of the camera service, providing it with a different set of patterns.

7.5 Another Adaptation Procedures

The presented adaptation procedure in not the only one possible; the Entrance Protection application can be extended in many other ways. Table 3 presents some of them, the first one (1) has already been discussed in the previous section and the following ones will be shortly characterized below.

An unusually high number of unsuccessful entry attempts may trigger an alarm or activate additional, more in-depth surveillance mechanisms. Such functionality can be achieved by further enhancing the business process (2a); however a more interesting option is to exploit the monitoring and management features of AS3 Studio (2b). Communication between federated ESB services can be intercepted and analyzed (by a dedicated business interceptor) and sent to the Complex Event Processor which then performs continuous statistical analysis and triggers appropriate actions.

The same approach can be used to determine whether the lighting system works correctly (3a, 3b). If, despite its use, there is still an excessive number of recognition failures, further corrections can be introduced, whether manually or automatically (e.g. turning on additional lamps).

The adaptation process may also affect lower levels of the infrastructure. If software processing is frequently able to cope with cases not correctly handled by hardware, this could be interpreted as an incentive to replace the FPGA-based pattern recognition algorithm or set of stored patterns (4).

8 Conclusions

Adaptive SOA Systems construction is fully justified by complexity of the enterprise class applications deployed nowadays and their changing business and execution requirements. This approach is very much in line with one of an emerging paradigm aiming at simplifying and reducing efforts required to deploy and maintain of complex computer systems such as Autonomic Computing (AC). The detailed analysis shown that exists direct relation between AC properties and the service orientation principles. It is also evident that the MAPE-K pattern could be used as a foundation

Table 3 Various adaptation procedures applied to the Entrance Protection business process

	Observed behavior	Sensor	Effector	Adaptation strategy	Behavior implemented by
1	MRT/MRF exceeded	Monitoring activity in VESB instance	ESB routing mechanism	To keep optimal number of recognition service instances	VESB
2a	Excessive number of unsuccessful entry attempts	Security service (being the part of the business process)	Alarm service	To ensure desired security level	Security service
2b		Business interceptor connected to ESB, OSGiMM			CEP
3a	Excessive number of recognition failures due to noise	Environment service (being the part of the business process)	Light service/manual setting of light	To ensure desired level of MRF	Environment service
3b		Business interceptor connected to ESB, OSGiMM			CEP
4	Considerable number of cases not handled by hardware	Monitoring of the business process, OSGiMM	Trigger replacement of FPGA-based pattern recognition memory	To ensure defined percentage of cases handled in hardware	RWS Builder

of the adaptive SOA system construction. These considerations lead to the adaptive systems space definition and location of Adaptive SOA in this space. This approach is further refinement in the context of S3 layer model of SOA systems.

Pragmatic usage of the MAPE-K Pattern across S3 layer results in AS3 Element definition. This element exposes implementation aspects of the MAPE-K pattern deployment in context of SOA systems. It is used by the AS3 Process which transforms systems built in accordance with the S3 Model into their adaptive versions and managing the adaptation process across different S3 Layers.

The AS3 Pattern may be applied only to selected layers. Leveraging the potential of particular layers adaptation loop execution can enhance the SOA environment with such features as: dynamic service flexibility, policy-driven operation optimization, and cross-business process monitoring and management.

The proposed solution considers adaptability aspects of SOA systems in uniform way referring to SOA applications composed with software services, named also Virtual Services, and hardware components being specified as Real World Services. Such approach is justified by increasing importance of pervasive systems bringing interaction from enterprise systems back to the real world. In this context, adaptive behavior of Real Word Services plays a critical role combining adaptive interaction, adaptive composition and task automation, by involving knowledge regarding user's profile, intentions, and previous use of the system. To clarify the proposed approach the reference model of the Adaptive SOA referring to Real-World and Virtual Services has been proposed and presented in the form of the concept maps.

Taking into account rather complex process of enrichment of the SOA System with adaptability functionality, dedicated software tools which support this transformation are important. The presented approach fully exploits the separation of concerns paradigm which isolates adaptability aspects from application business logic in the development and deployment phases and at the run-time. Such approach is strongly supported by the dynamic and flexible software execution environment offered by OSGi which allows the software modification at the run-time. The similar role plays the FPGA as far as Real-World Services are considered. The properly designed FPGA boards enable possibility for remote firmware modification on demand. Combining these two technologies lead to successful deployment of SOA adaptive systems in practice.

Acknowledgments This work has been performed by many contributors of IT-SOA Projects. The authors wants to thanks especially to: Paweł Bachara, Marcin Jarząb, Robert Szymacha, Dominik Radziszowski, Jacek Kosiński, Kornel Skałkowski, Przemysław Wyszkowski, Sławomir Zieliński. The research presented in this paper was partially supported by the European Union in the scope of the European Regional Development Fund program no. POIG.01.03.01-00-008/08.

References

1. An architectural blueprint for autonomic computing (IBM Corp.), IBM Corp., (2006)
2. Agarwal, A., Sites, R.L., Horowitz, M.: ATUM: a new technique for capturing address traces using microcode. In: ISCA, pp. 119–127 (1986)
3. Arsanjani, A., Zhang, L.J., Ellis, M., Allam, A., Channabasavaiah, K.: S3: A service-oriented reference architecture. IT Prof. **9**, 10–17 (2007)
4. Avgeriou, P., Zdun, U.: Architectural patterns revisited a pattern language. In: 10th European Conference on Pattern Languages of Programs (EuroPlop 2005), Irsee, pp. 1–39 (2005)
5. Baker, J., Hsieh, W.: Runtime aspect weaving through metaprogramming. In: Proceedings of the 1st International Conference on Aspect-riented Software Development, AOSD '02, pp. 86–95. ACM, New York (2002)
6. Barbon, F., Traverso, P., Pistore, M., Trainotti, M.: Run-time monitoring of instances and classes of web service compositions. In: Proceedings of the IEEE International Conference on Web Services, ICWS '06, pp. 63–71. IEEE Computer Society, Washington (2006)
7. Baresi, L., Guinea, S.: Towards dynamic monitoring of ws-bpel processes. In: Proceedings of the Third International Conference on Service-Oriented Computing, ICSOC'05, pp. 269–282. Springer, Berlin (2005)
8. Bieberstein, N., Bose, S., Fiammante, M., Jones, K., Shah, R.: Service-Oriented Architecture (SOA) Compass: Business Value, Planning, and Enterprise Roadmap. IBM Press, Upper Saddle River (2005)
9. Bigus, J.P., Schlosnagle, D.A., Pilgrim, J.R., Mills, Diao, Y.: Able: a toolkit for building multiagent autonomic systems. IBM Sys. J. **41**(3) (2002). doi:10.1147/sj.413.0350
10. Blair, G.S., Bencomo, N., France, R.B.: Models@run.time. IEEE Comput. **42**(10), 22–27 (2009)
11. Bobda, C.: Introduction to Reconfigurable Computing: Architectures, Algorithms, and Applications, 1st edn. Springer Publishing Company, New York (2007)
12. Cardoso, J.M., Simoes, J.B., Correia, C.M.B.A., Combo, A., Pereira, R., Sousa, J., Cruz, N., Carvalho, P,. Varandas, C.A.F.: A high performance reconfigurable hardware platform for digital pulse processing (2004)
13. Chen, C., Li, L., Wei, J.: Aop based trustable sla compliance monitoring for web services. In: Proceedings of the Seventh International Conference on Quality Software, QSIC '07, pp. 225–230. IEEE Computer Society, Washington (2007)
14. Chen, I.Y., Ni, G.K., Lin, C.Y.: A runtime-adaptable service bus design for telecom operations support systems. IBM Syst. J. **47**(3), 445–456 (2008)
15. Jayapandian, J.: Embedded control and virtual instrument simplifies laboratory automation. Curr. Sci. **90**(6), 765–770 (2006)
16. Corporation, C.S.: PSoC—Technical Reference Manual (TRM) (2006)
17. Czekierda, L., Masternak, T., Zielinski, K.: Evolutionary approach to development of collaborative teleconsultation system for imaging medicine. IEEE Trans. Inf. Technol. Biomed. **16**(4), 550–560 (2012). doi:10.1109/TITB.2012.2194506
18. Damianou, N., Dulay, N., Lupu, E., Sloman, M.: The ponder policy specification language. In: M. Sloman, J. Lobo, E. Lupu (eds.) POLICY Lecture Notes in Computer Science, vol. 1995, pp. 18–38. Springer, Berlin (2001)
19. DIGI: http://www.digi.com/ (2012)
20. Erl, T.: Service-Oriented Architecture: Concepts, Technology, and Design. Prentice Hall PTR, Upper Saddle River (2005)
21. Estrin, G.: Reconfigurable computer origins: the ucla fixed-plus-variable (f+v) structure computer. IEEE Ann. Hist. Comput. **24**(4), 3–9 (2002). doi:10.1109/MAHC.2002.1114865
22. Fleurey, F., Dehlen, V., Bencomo, N., Morin, B., Jzquel, J.M.: Modeling and validating dynamic adaptation. In: M.R.V. Chaudron (ed.) MoDELS Workshops Lecture Notes in Computer Science, vol. 5421, pp. 97–108. Springer, Berlin (2008)
23. Garlan, D., Cheng, S.W., Huang, A.C., Schmerl, B., Steenkiste, P.: Rainbow: architecture-based self-adaptation with reusable infrastructure. Computer **37**, 46–54 (2004)

24. Goldberg, A., Havelund, K.: Instrumentation of java bytecode for runtime analysis. In: Proc. Formal Techniques for Java-like Programs. Technical Reports from ETH vol. 408(2003).
25. Gorton, I., Liu, Y., Trivedi, N.: An extensible and lightweight architecture for adaptive server applications. Softw. Pract. Exper. **38**(8), 853–883 (2008)
26. Guinard, D., Trifa, V., Karnouskos, S., Spiess, P., Savio, D.: Interacting with the soa-based internet of things: Discovery, query, selection, and on-demand provisioning of web services. IEEE Trans. Serv. Comput. **3**(3), 223–235 (2010)
27. Guinard, D., Trifa, V., Spiess, P., Dober, B., Karnouskos, S.: Discovery and on-demand provisioning of real-world web services. In: IEEE International Conference on Web Services, ICWS 2009, Los Angeles, CA (2009)
28. Hallsteinsen, S., Floch, J., Stav, E.: A middleware centric approach to building self-adapting systems. In: Proceedings of the 4th International Conference on Software Engineering and Middleware, SEM'04, pp. 107–122. Springer, Berlin (2005)
29. Heinrich, C.: RFID and beyond—growing your business through real world awareness. Wiley, Heinrich (2005)
30. Hayman, C.: The benefits of an open service oriented architecture in the enterprise, (IBM Corp.), IBM Corp., (2005)
31. Jacob, B., Lanyon-Hogg, R., Nadgir, D.K., Yassin, A.F.: A Pratical Guide to IBM Autonomic Computing Toolkit. IBM Corp. (2004).
32. Janiesch, C., Niemann, M., Steinmetz, R. (eds.): The TEXO Governance Framework, SAP Research Brisbane, White Paper, Version 1.1, Working Draft (2011)
33. Kaiser, G.E., Parekh, J.J., Gross, P., Valetto, G.: Kinesthetics extreme: An external infrastructure for monitoring distributed legacy systems. In: Active Middleware Services, pp. 22–31. IEEE Computer Society, USA (2003)
34. Karnouskos, S., Bangemann, T., Diedrich, C.: Integration of legacy devices in the future soa-based factory. In: 13th IFAC Symposium on Information Control Problems in Manufacturing (INCOM), Moscow, Russia (2009)
35. Karnouskos, S., Savio, D., Spiess, P., Guinard, D., Trifa, V., Baecker, O.: Real world service interaction with enterprise systems in dynamic manufacturing environments. In: L. Benyoucef, B. Grabot (eds.) Artificial Intelligence Techniques for Networked Manufacturing Enterprises Management. Springer (2010) ISBN: 978-1-84996-118-9
36. Kean, T., Buchanan, I.: The use of fpga's in a novel computing subsystem. In: First International ACM Workshop on Field Programmable Gate Arrays (1992)
37. Kephart, J.O., Chess, D.M.: The vision of autonomic computing. Computer **36**(1), 41–50 (2003). doi:10.1109/MC.2003.1160055
38. Kiczales, G., Lamping, J., Mendhekar, A., Maeda, C., Lopes, C., Loingtier, J.M., Irwin, J.: Aspect-oriented programming. In: ECOOP. Springer, Berlin (1997)
39. Klein, C., Schmid, R., Leuxner, C., Sitou, W., Spanfelner, B.: A survey of context adaptation in autonomic computing. In: Proceedings of the Fourth International Conference on Autonomic and Autonomous Systems, ICAS '08, pp. 106–111. IEEE Computer Society, Washington (2008) doi:10.1109/ICAS.2008.23
40. Kon, F., Campbell, R.H., Mickunas, M.D., Nahrstedt, K., Dennis, M., Nahrstedt, M.K., Ballesteros, F.J.: 2k: A distributed operating system for dynamic heterogeneous environments. In: 9th IEEE International Symposium on High Performance, Distributed Computing, pp. 201–210 (1999)
41. Bug Labs: Bug Labs: modular, open source hardware (2009)
42. Lee, K., Sakellariou, R., Paton, N.W., Fernandes, A.A.A.: Workflow adaptation as an autonomic computing problem. In: Proceedings of the 2nd workshop on Workflows in Support of Large-Scale Science, WORKS '07, pp. 29–34. ACM, New York (2007)
43. Lin, K.J., Panahi, M., Zhang, Y., Zhang, J., Chang, S.H.: Building accountability middleware to support dependable soa. IEEE Internet Comput. **13**(2), 16–25 (2009)
44. Lodi, A., Toma, M., Campi, F., Cappelli, A., Canegallo, R., Guerrieri, R.: A vliw processor with reconfigurable instruction set for embedded applications. IEEE J. Solid-State Circuits **38**(11), 1876–1886 (2003)

45. McAffer, J., Lemieux, J.M.: Eclipse Rich Client Platform: Designing, Coding, and Packaging Java Applications. Addison-Wesley, Upper Saddle River (2005)
46. milkymist.org: milkymist.org (2009)
47. Montenegro, G., Kushalnagar, N., Hui, J., Culler, D.: Transmission of IPv6 Packets over IEEE 802.15.4 Networks. RFC 4944 (Proposed Standard) (2007)
48. Morin, B., Barais, O., Jzquel, J.M., Fleurey, F., Solberg, A.: Models@run.time to support dynamic adaptation. IEEE Comput. **42**(10), 44–51 (2009)
49. Moser, O., Rosenberg, F., Dustdar, S.: Non-intrusive monitoring and service adaptation for ws-bpel. In: Proceedings of the 17th International Conference on World Wide Web, WWW '08, pp. 815–824. ACM, New York (2008).
50. Mudry, P.A., Vannel, F., Tempesti, G., Mange, D.: Confetti : a reconfigurable hardware platform for prototyping cellular architectures. In: IPDPS, pp. 1–8. IEEE (2007)
51. Niemann, M., Eckert, J., Repp, N., Steinmetz, R.: Towards a generic governance model for service oriented architectures. In: AMCIS'08, pp. 361–361 (2008)
52. OASIS Web Services Discovery and Web Services Devices Profile (2005)
53. Opencores.org. http://opencores.org/
54. OSGi Alliance: OSGi Service Platform Release 4. [Online]. http://www.osgi.org/Main/HomePage. (2007)
55. Patel, S.V., Pandey, K.: Soa using aop for sensor web architecture. In: Proceedings of the 2009 International Conference on Computer Engineering and Technology—Volume 02, ICCET '09, pp. 503–507. IEEE Computer Society, Washington (2009)
56. Popovici, A., Alonso, G., Gross, T.: Just-in-time aspects: efficient dynamic weaving for Java. In: Proceedings of the 2nd International Conference on Aspect-Oriented Software Development, AOSD '03, pp. 100–109. ACM, New York (2003). doi:10.1145/643603.643614
57. Popovici, A., Gross, T., Alonso, G.: Dynamic weaving for aspect-oriented programming. In: Proceedings of the 1st International Conference on Aspect-Oriented Software Development, AOSD '02, pp. 141–147. ACM, New York (2002)
58. Prasanna, D.R.: Dependency Injection, 1st edn. Manning Publications Co., Greenwich (2009)
59. Prez, F.M., Abarca, J.A.G.M., Morillo, H.R., Gimeno, F.J.M., Jorquera, D.M., Iglesias, V.G.: Wake on lan over internet as webservice system on chip. IEEE Trans. Industr. Electron. **16**(1), 45–69 (2012)
60. Psiuk, M.: AOP-based monitoring instrumentation of JBI-compliant ESB. In: Proceedings of the 2009 Congress on Services—I, SERVICES '09, pp. 570–577. IEEE Computer Society, Washington (2009)
61. Psiuk, M., Bujok, T., Zielinski, K.: Enterprise service bus monitoring framework for soa systems. IEEE Trans. Serv. Comput. **5**(3), 450–466 (2012). doi:10.1109/TSC.2011.32.
62. Psiuk, M., Zmuda, D., Zieliski, K.: Distributed OSGI built over message-oriented middleware (2011). doi:10.1002/spe.1148. http://dx.doi.org/10.1002/spe.1148
63. Rellermeyer, J.S., Duller, M., Gilmer, K., Maragkos, D., Papageorgiou, D., Alonso, G.: The software fabric for the internet of things. In: C. Floerkemeier, M. Langheinrich, E. Fleisch, F. Mattern, S.E. Sarma (eds.) IOT. Lecture Notes in Computer Science, vol. 4952, pp. 87–104. Springer, Berlin (2008)
64. Rochwerger, B., Breitgand, D., Epstein, A., Hadas, D., Loy, I., Nagin, K., Tordsson, J., Ragusa, C., Villari, M., Clayman, S., Levy, E., Maraschini, A., Massonet, P., Mun andoz, H., Tofetti, G.: Reservoir—when one cloud is not enough. Computer **44**(3), 44–51 (2011). doi:10.1109/MC.2011.64
65. Ruta, A., Brzoza-Woch, R., Zielinski, K.: On fast development of fpga-based soa services—machine vision case study. Design Autom. Emb. Syst. **16**(1), 45–69 (2012)
66. Ruta, A., Li, Y., Liu, X.: Robust class similarity measure for traffic sign recognition. IEEE Trans. Intell. Transp. Syst. 846–855 (2010)
67. Satyanarayanan, M.: Pervasive computing: vision and challenges. IEEE Pers. Commun. **8**, 10–17 (2001)
68. Shelby, Z., Bormann, C., Frank, B.: Constrained application protocol (coap). IETF Internet draft, 1–81 (2011)

69. Smith, R.B.: Spotworld and the sun spot. In: Proceedings of the 6th International Conference on Information Processing in Sensor Networks, IPSN '07, pp. 565–566. ACM, New York (2007). doi:10.1145/1236360.1236442.
70. Song, H., Lee, S.H., Lee, S., Lee, H.S.: 6lowpan-based tactical wireless sensor network architecture for remote large-scale random deployment scenarios. In: Proceedings of the 28th IEEE Conference on Military Communications, MILCOM'09, pp. 1044–1050. IEEE Press, Piscataway (2009)
71. Sun, M., Li, B., Zhang, P.: Monitoring BPEL-based web service composition using AOP. In: Proceedings of the 2009 Eigth IEEE/ACIS International Conference on Computer and Information Science, ICIS '09, pp. 1172–1177. IEEE Computer Society, Washington (2009)
72. Szydlo, T., Zielinski, K.: Method of adaptive quality control in service oriented architectures. In: Proceedings of the 8th International Conference on Computational Science, Part I, ICCS '08, pp. 307–316. Springer, Berlin (2008)
73. Szydlo, T., Zielinski, K.: Adaptive enterprise service bus. New Generation Comput. **30**(2–3), 189–214 (2012)
74. Tibbo: http://www.tibbo.com/(2012)
75. Vedamuthu, A.S., Orchard, D., Hirsch, F., Hondo, M., Yendluri, P., Boubez, T., Yalinalp, M.: Web services policy framework (wspolicy) http://www.w3.org/TR/ws-policy (2007)
76. Voros, N.S., Masselos, K. (eds.): System Level Design of Reconfigurable Systems-on-Chip. Springer-Verlag New York, Inc., Secaucus, NJ, USA (2005)
77. Vuković, M.: Context aware service composition, PhD dissertation, Univ. of Cambridge (2007)
78. Walsh, W.E., Tesauro, G., Kephart, J.O., Das, R.: Utility functions in autonomic systems. pp. 70–77 (2004). doi:10.1109/ICAC.2004.1301349.
79. Wang, Q., Shao, J., Deng, F., Liu, Y., Li, M., Han, J., Mei, H.: An online monitoring approach for web service requirements. IEEE Trans. Serv. Comput. **2**(4), 338–351 (2009)
80. Wetzstein, B., Strauch, S., Leymann, F.: Measuring performance metrics of ws-bpel service compositions. In: J.L. Mauri, V.C. Giner, R. Tomas, T. Serra, O. Dini (eds.) Proceedings of the Fifth International Conference on Networking and Services, ICNS 2009, 20–25 April 2009, Valencia, Spain, pp. 49–56. IEEE Computer Society, Washington (2009)
81. Wildstrom, J., Witchel, E.J., Mooney, R.: Towards self-configuring hardware for distributed computer systems. In: Proceedings of the Second International Conference on Automatic Computing, ICAC '05, pp. 241–249. IEEE Computer Society, Washington (2005)
82. Yeung, K., Kelly, P.H.J., Bennett, S.: Performance Analysis and Grid computing. Dynamic Instrumentation for Java Using a Virtual JVM, pp. 175–187. Kluwer Academic Publishers, Norwell (2004)
83. Yuan, H., Choi, S.W., Kim, S.D.: A practical monitoring framework for esb-based services. In: Proceedings of the 2008 IEEE Congress on Services Part II, SERVICES-2 '08, pp. 49–56. IEEE Computer Society, Washington (2008)
84. Zeeb, E., Bobek, A., Bohn, H., Prter, S., Pohl, A., Krumm, H., Lck, I., Golatowski, F., Timmermann, D.: Ws4d: Soa-toolkits making embedded systems ready for web services, In: Proceedings of the Open Source Software and Product Lines Workshop (OSSPL07) (2007)
85. Zielinski, K., Szydlo, T., Szymacha, R., Kosinski, J., Kosinska, J., Jarzab, M.: Adaptive SOA solution stack. IEEE Trans. Serv. Comput. **5**, 149–163 (2012). http://doi.ieeecomputersociety.org/10.1109/TSC.2011.8
86. Zmuda, D., Psiuk, M., Zielinski, K.: Dynamic monitoring framework for the SOA execution environment. Procedia CS **1**(1), 125–133 (2010)

Printed in the United States
By Bookmasters